The Incompleteness Phenomenon

The Incompleteness Phenomenon

A New Course in Mathematical Logic

Martin Goldstern
Haim Judah

Bar Ilan University
Ramat Gan, Israel

CRC Press
Taylor & Francis Group
Boca Raton London New York

CRC Press is an imprint of the
Taylor & Francis Group, an **informa** business
AN AUERBACH BOOK

CRC Press
Taylor & Francis Group
6000 Broken Sound Parkway NW, Suite 300
Boca Raton, FL 33487-2742

First issued in paperback 2019

ISBN-13: 978-1-56881-029-4 (hbk)
ISBN-13: 978-0-367-40169-6 (pbk)

Library of Congress Cataloging-in-Publication Data

Goldstern, M. (Martin)
The incompleteness phenomenon : a new course in mathematical logic / Martin Goldstern, Haim Judah.
p. cm.
Includes index.
ISBN 1-56881-029-6
1. Incompleteness theorems. I. Judah, Haim. II. Title.
QA9.54.J83 1995
511.3--dc20 95-39361
CIP

Cover illustration: Gödel's Uber formal unentscheidbare Sätze der Principia Mathematica und verwandter Systeme, page one. Courtesy of the Archives, Institute for Advanced Study.

Visit the Taylor & Francis Web site at
http://www.taylorandfrancis.com

and the CRC Press Web site at
http://www.crcpress.com

This book is dedicated to
Rafael Guendelman

Contents

Foreword

You may have wondered: does not the shoemaker go barefoot? Mathematics boasts of being the epitome of exactness, but what is the exact meaning of proof? Construction? Computation?
Or you may be very ambitious and wonder whether we can prove theorems concerning the collection of all possible mathematical theories.
Or you may have resigned yourself to having no exact answer, as you cannot "pull yourself out of the mud," at most you can philosophize about it.
Or you may wonder: is mathematics one body or is it fragmented into many branches; i.e. can we put it all in one framework?
Or you may be philosophically inclined and wonder whether having a proof and being true are the same.

However there is a branch of mathematics dealing exactly with those problems: LOGIC. It is one of the oldest intellectual disciplines (see Aristotle) yet also one which has developed enormously in this century.

Yes! Mathematics can deal with these problems and give exact answers with proof; i.e. we can define relevant notions and give answers.
Yes! We can define what a proof is, and show in a sense that being true and having a proof are the same (Gödel's completeness theorem).
Yes! We cannot raise ourselves out of the mud: we cannot prove in our system that it does not have a contradiction (Gödel's incompleteness theorem).
Yes! We can have a general theory of mathematical theories (model theory).
Yes! We can define what it means to be computable i.e. having an algorithm (for this purpose mathematical machines were invented, and you probably have met their offspring, the computers).

Saharon Shelah

Introduction

This book is a course in Mathematical Logic. It is divided into four chapters which can be taught in two semesters. The first two chapters provide a basic background in mathematical logic. All details are explained for students not so familiar with the abstract method used in mathematical logic. The last two chapters are more sophisticated, and here we assume that the reader will be able to fill in more details; in fact, this ability is an essential step for this sphere of mathematical thinking.

Mathematical logic is the most abstract branch of mathematical thought, the most abstract human discipline. The main objective in this area is to understand the logic implicit in all mathematical thought. The difference between logic (considered as a branch of philosophy) and mathematical logic is that in mathematical logic we use and develop mathematical methods. That is, we use mathematical theorems to investigate and explain the logic implicit in mathematics. It should be clear that some of the results can also shed light on more general questions in epistemology and philosophy of science, but this is not the subject of this book. The main result in basic mathematical logic is that every "reasonable" mathematical system is intrinsically incomplete. This means that axiom systems cannot capture all semantical truths. This can also be expressed in the following way: If we assume that the human mind works in a way similar to an ideal computer, then there are mathematical problems which can never be solved by mathematicians. This is one aspect of Gödel's famous incompleteness theorem, and the study of this phenomenon of incompleteness will be the main focus of this book. We think that the material of this book should be part of the basic background of every student in any discipline which employs deductive and formal reasoning as a part of its methodology. This definitely includes a large part of the social sciences.

Chapter 1 contains the basic material a student has to know about mathematical languages and logical systems. The most fundamental tool in this book is the concept of mathematical induction. Section 1.1 introduces this concept in a very general way through the notion of an "inductive structure." This notion is essential for everything in the rest of the book. In Section 1.2 we study propositional logic and tautologies. In Section 1.3 we present "first order logic" as a typical example of a mathematical language, and the notion of a "model," i.e., a mathematical structure which allows us to interpret symbols of the language. For example, a model for the language of groups (in which we can speak about "multiplication" and "inverses") will be a group. In this section we also study the concept of validity, i.e., semantical truth. In Section 1.4 we will study the concept of a "formal proof." We will define an axiom system and a deductive tool called modus ponens. Together they try to capture the notion of logical truth.

The main objective of Chapter 2 is to show that the syntactical concept of "provability" and the semantical concept of "validity" coincide. This theorem is called the completeness theorem (since it shows that the logical system presented in 1.4 is "complete"). An equivalent version of this theorem says that every axiom system which does not contain a contradiction will be realized in some model. This theorem was discovered by Gödel; the proof we present is due to Henkin. The main idea is as follows: If we want a sentence such as "Reagan is Batman" to be true in some model, we can consider a model with an element x, which has two names, "Reagan" and "Batman." More generally, we will build a model from names (i.e., syntactical objects), and two names will denote the same object only if our axioms tell us that they have to. (A philosophical question for the reader: Is Reality no more than a Henkin model? We doubt it.) We start this chapter with a study of the concept of "enumerability," which plays an important role in the proof of the completeness theorem. In Section 2.2 we present Henkin's proof. We close the chapter by applying the completeness theorem to show that there are nonstandard models of number theory.

In Chapter 3 we deal with model theory. Model theory investigates the relation a logical theory has to its models. We exhibit several tools used in model theory, and we show how to apply them to classical problems of model theory such as finding the number of nonisomorphic countable models of a first order theory.

The last chapter is mainly devoted to Gödel's famous incompleteness theorem, which says that any reasonable axiom system for the natural numbers is intrinsically incomplete. We start by proving that the Peano axioms for number theory are incomplete and then give a more general version of this theorem. We prove the incompleteness theorem in a simplified form to make the proof more accessible to beginners who are not especially interested in mathematical logic. The proof of the incompleteness theorem is conceptually based on the "liar's paradox," which was already known to the ancient Greeks. Gödel's novel idea was to encode the language into the formal system itself. We present the details of this main tool of Gödel's proof. The main change from the traditional proof is that we prove everything semantically, working in the model of natural numbers rather than talking about derivations from the Peano axioms. In this way we can simplify the proofs, while keeping near to the main idea underlying the incompleteness phenomenon. We close Chapter 4 with an introduction to recursion theory, a branch of mathematical logic which is closely related to the methods used in the proof of the incompleteness theorem.

We have included exercises at the end of each section. They are an intrinsic part of this book. We believe that it is impossible to *understand* mathematics without actually *doing* mathematics.

This book is based on Judah's logic lectures, given in Berkeley and Bar Ilan. We want to thank all our friends who have read the manuscript at various stages and have made important contributions. We especially thank Boaz Tsaban for preparing the index and Tzvi Scarr and Andy Lewis, who wrote early parts of the first two chapters, for their dedication to the project.

Martin Goldstern
Haim Judah

Chapter 1

The Framework of Logic

1.1. Induction

Induction is the main tool used to prove theorems in mathematical logic. The best way to develop an intuitive feel for *inductive proofs* is to look at some examples. We start this section with examples that are close to ordinary experience, followed by mathematical examples. In the middle of the section we establish our *induction principle* by defining *inductive structures*. We conclude this section with a proof that the usual language for sentential logic is an inductive structure.

1.1.1. Example. Everyone has a name

We begin with the fact that everyone has parents.
Let's assume that parents with names always give names to their children.
Adam and Eve had names.
So we conclude that every person who ever lived had a name.
For if not, let Person be the first person with no name.
Person was not Adam or Eve, who had names.
So Person had parents.
Person's parents had names, since Person is the first person without a name.
But by assumption, the parents must have given Person a name.
So it cannot be that Person had no name.
Thus there cannot have been a first person without a name.
So everyone had a name.

In the previous example we were using a strong assumption about reality, namely that the initial conditions determine the future of the system forever. Clearly, systems like this do not exist in reality, but in mathematics we deal with ideal objects that are not subject to any external influence or the influence of time.

The first mathematical objects were the numbers:

$$1, 2, 3, 4, 5, 6, 7, 8, 9, 10$$

What are the natural numbers? This is a good question. The first thing we can say is that it is **not a mathematical question**, but rather, a philosophical question about mathematics. There is controversy, as always in philosophy, about the nature of the natural numbers, and the various opinions are strongly influenced by the positions the philosophers have on the existence of objects in reality. We mention a few of these positions without further remarks:

(1) The numbers are abstract objects in a world of ideas.
(2) The number 5, say, is what all objects with 5 components (or all sets with 5 members) have in common
(3) The number 5 is a human category used to communicate.

Are there infinitely many numbers? Again this is a philosophical question. The following argument is usually given to "prove" that there are infinitely many numbers.

Assume that there are not infinitely many numbers. Then there must be the biggest one, call it n. Then $n + 1$ is bigger than n, a contradiction.

There are several problems with this argument. One objection is: If n is a number, why should it follow that $n + 1$ is a number? Let us replace the concept of "number" by "conceivable number," where we call a number n "conceivable" if we can imagine somebody owning n dollars.

Thus, 5 is a conceivable number, 10000 is a conceivable number, and even 10^9, a thousand millions (also called billion or milliard), is a conceivable number. What about 10^{12}? What about $n + 1$, where n is the total value of all property anybody on earth owns? And if this is still conceivable, is 10^{100} conceivable? $10^{10^{100}}$?

We can also consider the following (related) argument: The universe, as perceived by us, is finite. So how can there be infinitely many numbers?

A second problem is the following: Even if we agree that for every number n there is a number $n+1$, does that mean that the *infinite* totality of all numbers exists? Assume we want to make a list of all numbers. Even if we know that whenever we write down the number n, there will be room for the number $n+1$ (and time to write it down, too), does that mean we will ever have a complete list?

It is possible to avoid all this discussion by using the axiomatic method and stipulating that the natural numbers are any universe of objects and operations that satisfies our list of axioms.

For our purpose we will assume that the reader has a good feel for the natural numbers, the operations of addition, multiplication, and exponentiation, and the $<$-relation.

We will write $\mathbb{N}$ for the set of natural numbers (including 0). Thus, $\mathbb{N} = \{0, 1, 2, \ldots\}$.

1.1.2. Example. Show that for all $n > 0$ the following holds:

$$(1 + 2 + \cdots + n) = \frac{n \cdot (n+1)}{2} \qquad (*)$$

Proof: by induction on n.
First Stage: $n = 1$. The sum on the left side consists of the single term 1, and the expression on the right side is $= 1 \cdot (1+1)/2 = 1$.
Second Stage:

$n = k+1$. Assume that we already know that $1 + \cdots + k = \frac{k(k+1)}{2}$. We need to show that:

$$(1 + \cdots + k + (k+1)) = \frac{(k+1)((k+1)+1)}{2}.$$

So we start with:

$$(1 + \cdots + k + (k+1)) = (1 + \cdots + k) + (k+1).$$

We already know that the first sum is equal to $\frac{k(k+1)}{2}$, so we get

$$= \frac{k(k+1)}{2} + (k+1) = \frac{k(k+1)}{2} + \frac{(k+1)2}{2} = \frac{(k+1)(k+2)}{2}$$

which concludes the proof.

Why is the proof complete? If $(*)$ does not hold for all n, then there must be a first number n for which $(*)$ does not hold. But we just showed that there can be no such "first number": n cannot be 1, by the "first stage," and if $n > 1$, then $(*)$ must hold for $n-1$. In the second stage we showed that $(*)$ must then hold for n also.

1.1.3. Example. Every convex polygon with $n \geq 3$ vertices has exactly $\frac{n(n-3)}{2}$ diagonals.

Proof: Here the first case is for $n = 3$. The polygon in this case is a triangle and it has no diagonals. This is also what the formula says.

Now, assume that the formula is correct for a polygon with n vertices. Given a convex polygon with $n+1$ vertices, we use our "induction hypothesis" in the following way: take a subset of n vertices and connect it (using one diagonal of the original polygon) to get a closed polygon. This polygon has $\frac{n(n-3)}{2}$ diagonal all of which are also diagonals of the original polygon. In addition, one edge on this polygon is a diagonal of the original polygon. Finally we need to add $n-2$ diagonals going from the extra vertex to each one of the other vertices except its two neighbors. The total number is therefore:

$$\frac{n(n-3)}{2} + 1 + (n-2) = \frac{(n+1)(n-2)}{2}$$

The above examples are typical inductive proofs in mathematics. This argument can in general be used in the following way to show that all the natural numbers have a certain property P:

(a) Show that the number 0 has the property P.

(b) Use the assumption that k has the property P in order to show that $k+1$ has the property P.

(c) Conclude from (a) and (b) that all natural numbers have the property P.

Why do we accept (c)? We do so because mathematical intuition says that the numbers can be generated as follows:

First we have the number 0 (by (a), 0 has property P).

From 0 we can go to the number $0+1(=1)$ (by (b) and the above, 1 also has the property P).

From 1 we can go to the number $1+1(=2)$ (by (b) and the above, 2 also has the property P).

From 2 we can go to the number $2+1(=3)$ (by (b) and the above, 3 also has the property P).

$\vdots$

From k we can go to the number $k+1$ (using (b) and the above).

$\vdots$

Looking at this process of "generating numbers," we can see that an inductive proof is a **schematization** of an infinite proof starting from 0 and continuing through all the numbers.

The point here is that the above description of the natural numbers may be schematized as follows: We start by assuming the existence of 0 and the existence of the operator +1 (this means to add 1). Then we say that n is a number if:

(a) n is 0, or

(b) There is a number m such that $n = m + 1$.

We can see that in this process two different concepts are involved, namely: "0" is an object given *a priori*, and it is the basis from which we build the rest of the numbers, which are formed from "0" by repeatedly applying the process of "adding 1." This process of "adding 1" is the second concept implicit in this presentation of the natural numbers. This process is also given *a priori*, and we can think of it as the "method" of getting a new object from a previously given object. Such "methods" are usually described by functions.

An abstract view of this situation gives us the following important concepts that will be the main ingredient of all our mathematical constructions:

(1) **Blocks:** are the objects, given *a priori* (like the object "0" in the above example).

(2) **Operators:** are the methods, given *a priori* and used to create new objects from the previously created objects (like "adding 1" in the above example).

Now we may assume that we have a set B of blocks and a set K of operators.

What can we do with B and K? We want to form a collection of objects using the elements of B and the operations in K. The first objects will be the objects in B. Then we can apply the operations of K to the elements of B to get new objects. Then we can again apply the operations of K to these new objects to get more objects, and we can continue this process for ever, to get a collection of objects which is denoted by $C(B, K)$.

To exemplify this construction, define the following operation:

$$\begin{aligned} s : \mathbb{N} &\to \mathbb{N} \\ n &\mapsto s(n) = n + 1. \end{aligned}$$

This function s is called the successor function. Now if $B = \{0\}$ and $K = \{s\}$, then $\mathbb{N} = C(B, K)$.

We will give two more examples of inductive structures:

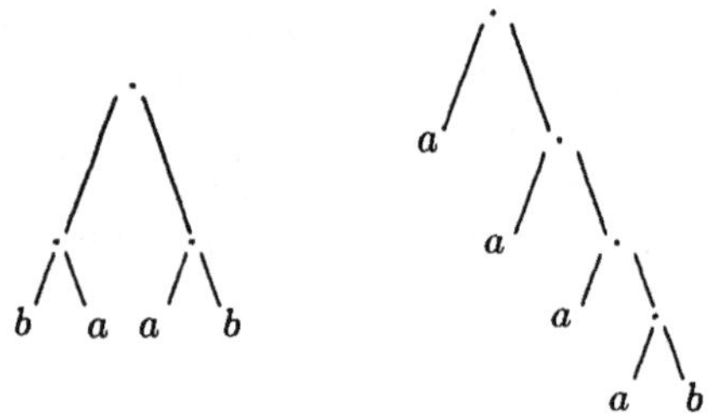

Figure 1.

1.1.4. Example. Let us consider a two element set $B = \{a, b\}$, and let D be the collection of all finite sequences of members of B. Let $K = \{f, g\}$, where

$$\begin{aligned} f : D &\to D \\ x &\mapsto f(x) = axa \end{aligned}$$

and

$$\begin{aligned} g : D &\to D \\ x &\mapsto g(x) = bxb. \end{aligned}$$

Then the $C(B, K)$ is the set of all sequences of odd length which are their own mirror image.

1.1.5. Example. Our set of blocks will again be the two-element set $\{a, b\}$. We will consider finite sequences that contain a, b, ., and the square brackets [,]. Our only operator will be the 2-place function f defined by

$$f(x, y) = [x.y] .$$

The following are examples of elements of $C(\{a, b\}, \{f\})$:

$$a$$
$$b$$
$$[a.b]$$
$$[a.a]$$
$$[[b.a].[a.b]]$$
$$[a.[a.[a.[a.b]]]] .$$

When we deal with an inductive structure, it is sometimes convenient to associate to each element of $C(B, K)$ its "syntax tree." We will not formalize this concept, but only give a few examples. The syntax trees for the last two objects from example 1.1.5 are given in figure 1.

Now we will give a formal definition of $C(B,K)$:

1.1.6. Definition.

(a) Every block is in $C(B,K)$ (that is, every element in B is also an element of $C(B,K)$).

(b) If F is an n-place operator in K, and $c_1, \ldots, c_n$ are elements of $C(B,K)$ then $F(c_1, \ldots, c_n)$ is an element of $C(B,K)$.

(c) Every element of $C(B,K)$ is obtained by (a) or (b).

We call $C(B,K)$ the set **generated** from B by K. If $C = C(B,K)$, then (B,K) is called an **inductive structure** on C.

Note that a given set C can be generated by different inductive structures (see exercise 7).

For example, every set can be viewed as an inductive structure by simply taking $B = C$ (i.e. taking each element of the set as a block). We also do not exclude the possibility that the result of applying an operator is again a "block." For example, we could view the natural numbers as an inductive system with the set of blocks $= \{0,2\}$ and one operator (the successor operation). However, this is very unnatural: Why do we need 2 as a block, if it can already be obtained from the other block, 0, by applying the successor operation twice?

In general we want an inductive structure to be a **simple description of a set** — the simpler the description, the easier it is to prove properties by induction. Often there is a unique "natural" way to define an inductive structure on a set C.

1.1.7. Example. The natural inductive structure on the set of natural numbers is given by choosing $\{0\}$ as the set of blocks, and the successor operation as the only operator.

1.1.8. Definition. Let $C = C(B,K)$ be an inductive structure, and let P be a property that elements of C may or may not have. Let F be an n-place operator in K.

We say that "F preserves P," if:

Whenever $a_1, \ldots, a_n$ satisfy the property P, then $F(a_1, \ldots, a_n)$ also satisfies the property P.

1.1.9. Induction law. Let $C = C(B,K)$ be an inductive structure with B as the set of blocks and K as the set of operators such that the following is true:

(a) Every block satisfies the property P, and
(b) Every operator preserves the property P.

Then:

(∗) Every element of C satisfies the property P.

1.1.10. Notation. When we prove that a property P holds for all elements x of an inductive structure $C(B, K)$, we usually start by saying:

We will prove P by induction on $C(B, K)$

or

We will prove $P(x)$ by induction on x.

Such a proof consists of two parts: In the first part we deal with the blocks, i.e., we show that all blocks have the property P. (This is called the *induction basis.*) In the second part we deal with the operators, i.e., we show that all operators preserve P. This is called the *inductive step.*

Usually the essence of these proofs is in the second part. Sometimes we can give a general argument that works for all operators, sometimes we have to deal with each operator separately.

To show that an n-ary operator F preserves the property P, we assume that $a_1, \ldots, a_n$ are arbitrary elements in our inductive structure satisfying P, and we have to show that $F(a_1, \ldots, a_n)$ satisfies P.

The assumption "$a_1, \ldots, a_n$ satisfy P" is often called the "induction hypothesis" or "inductive assumption."

Induction is one of the most natural ways to deal with infinite objects. There are many examples of mathematical objects which can be seen as inductive structures. One example that we have in mind is the language for "sentential logic." We will study the mathematical properties of this language below. We will then generalize these properties to other inductive structures. The language of sentential logic will be defined starting from a set of blocks and a set of operators. The blocks will be basic sentences like:

New York is a city.

2 is odd.

The operators will be the logical connectives used to build more complex sentences from the blocks and other sentences. For instance, "and" and "if and only if" are logical connectives.

Before we define the language of sentential logic, we will make a few remarks about formal languages. We consider an "alphabet" or "set of symbols" S. The set of all "words" in our formal language will be all elements of S^+, the set of all finite sequences of elements from S. (We allow only sequences of length ≥ 1, i.e., we will not consider the empty sequence as a word.) It does not matter what the true nature of these "symbols" is, we only demand that

No symbol is also a finite sequence of symbols. $(*)$

We do not strictly distinguish between a symbol x and the one element sequence containing only the symbol x. This causes no ambiguities because of $(*)$.

If x and y denote symbols, we write xy to denote the sequence with first element x and second element y, i.e., (x, y). (Here, x may be equal to y.) We call the sequence (x, y) a "pair" and say that the sequence (x, y) has "length" 2. The set of all pairs of elements of S is called S^2.

Similarly, we call a sequence (x, y, z) of length 3 a "triple" or "3-tuple." A sequence $(x_1, \ldots, x_n)$ of length n will be called "an n-tuple" or "word of length n." The set of all n-tuples is written as S^n. So S^1 is the set of all words of length 1, which is essentially the same as S.

For two sequences x and y we write xy for the concatenation of these two sequences. Similarly for xyz, etc.

We will not explain what a "finite sequence" is. We assume that the reader is familiar with basic facts such as $(xy)z = x(yz)$ whenever x, y, z are finite sequences, and

$$\text{If } xr = ys, \text{ then } x = y \text{ and } r = s$$

whenever x and y are symbols and r and s are sequences.

We will often use letters such as "x" to denote symbols of our language. It is important to keep in mind that such a letter "x" is not the symbol itself, but only a name for the actual symbol. Thus it is possible that the letter "x" on one page denotes a different symbol than the letter "x" on some other page. Also, different letters may denote the same symbol (or sequence of symbols).

Before giving the explicit definitions of the language for sentential logic, let us introduce some sets.

Let $B = \{\mathtt{A}_1, \mathtt{A}_2, \mathtt{A}_3, \ldots\}$ be a set of distinct symbols (i.e., whenever i and j are distinct natural numbers, then "$\mathtt{A}_i$" and "$\mathtt{A}_j$" stand for distinct symbols) and let F be the following set with 6 elements: $F = \{\neg, \wedge, \vee, \rightarrow, \leftrightarrow, |\}$.

The elements of B are called **sentential symbols**. They will serve as blocks when we build our language as an inductive structure. The elements of F are called **connectives**. $\neg$ is called "not", $\wedge$ is called "and", $\vee$ is called "or", $\rightarrow$ is called "implies", $\leftrightarrow$ is called "iff" (= if and only if), and $|$ is called "nand".

1.1.11. Definition. The alphabet for the sentential language will be the set $S = B \cup F \cup \{(,)\}$. S^+ is the collection of finite sequences from S.

The following are examples of members of S^+:

$$\mathtt{A}_1$$
$$\mathtt{A}_3$$
$$(\leftrightarrow \wedge \mathtt{A}_1(|))$$
$$()\mathtt{A}_1 \rightarrow \mathtt{A}_2.$$

1.1.12. Definition. We define some operations over S^+. For any two elements α, β of S^+ we define $F_\neg(\alpha), F_\wedge(\alpha,\beta), F_\vee(\alpha,\beta), F_\rightarrow(\alpha,\beta), F_\leftrightarrow(\alpha,\beta)$ and $F_|(\alpha,\beta)$ by

$$\begin{aligned}
F_\neg(\alpha) &= (\neg\alpha)\\
F_\wedge(\alpha,\beta) &= (\alpha \wedge \beta)\\
F_\rightarrow(\alpha,\beta) &= (\alpha \rightarrow \beta)\\
F_\vee(\alpha,\beta) &= (\alpha \vee \beta)\\
F_\leftrightarrow(\alpha,\beta) &= (\alpha \leftrightarrow \beta)\\
F_|(\alpha,\beta) &= (\alpha|\beta)
\end{aligned}$$

Each of these operators has domain S^+ and range in S^+.

1.1.13. Definition. Let $K = \{F_\neg, F_\wedge, F_\vee, F_\rightarrow, F_\leftrightarrow, F_|\}$. Then we define $\mathcal{L}$ to be $C(B,K)$.

We will call $\mathcal{L}$ the *sentential language*, or the language of sentential logic. The elements of the sentential language will be called *sentential formulas*.

The following are examples of sentential formulas:

$$(\mathtt{A}_1 \vee (\mathtt{A}_2 \vee \mathtt{A}_3))$$
$$(((\neg\mathtt{A}_1) \wedge \mathtt{A}_2) \vee (\mathtt{A}_1 \wedge (\neg\mathtt{A}_2)))$$
$$((\mathtt{A}_1 \vee \mathtt{A}_2) \leftrightarrow (\mathtt{A}_2 \vee \mathtt{A}_1))$$
$$(\mathtt{A}_1|\mathtt{A}_1).$$

For example $(\mathtt{A}_1 \vee (\mathtt{A}_2 \vee \mathtt{A}_3))$ is a sentential formula, because

$$(\mathtt{A}_1 \vee (\mathtt{A}_2 \vee \mathtt{A}_3)) = F_\vee(\mathtt{A}_1, F_\vee(\mathtt{A}_2, \mathtt{A}_3)).$$

Again we will give examples of "syntax trees" of elements of $C(B,K)$.

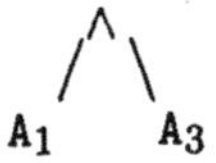

$A_1 \wedge A_3$

Figure 2.

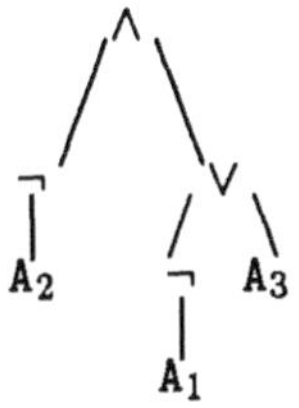

$((\neg A_2) \wedge ((\neg A_1) \vee A_3))$

Figure 3.

For example, the syntax tree of $(A_1 \wedge A_3)$ is given in figure 2, and the syntax tree of $((\neg A_2) \wedge ((\neg A_1) \vee A_3))$ is given in figure 3.

It is not immediately obvious which elements of S^+ are in $C(B, K)$. The next proposition will help us in deciding which elements of S^+ are sentential formulas.

1.1.14. Lemma. Let α be in $\mathcal{L}$. Then the number of right parentheses in α is equal to the number of left parentheses.

Proof: by induction. Let P be the property of having an equal number of right and left parentheses.

(a) If α is a block then α has the property P (since the number of left parentheses = the number of right parentheses = 0).

(b) Assume that α and β have the property P, then

$$\begin{aligned} F_\neg(\alpha) &= (\neg\alpha) \text{ has the property } P \\ F_\wedge(\alpha,\beta) &= (\alpha \wedge \beta) \text{ has the property } P \\ &\cdots \\ F_|(\alpha,\beta) &= (\alpha|\beta) \text{ has the property } P. \end{aligned}$$

So by the induction law every element of $\mathcal{L}$ has the property P (i.e. has an equal number of right and left parentheses). This ends the proof of 1.1.14.

If α and β are two strings of symbols from $\mathcal{L}$, say $\alpha = \alpha_1 \cdots \alpha_n$ and $\beta = \beta_1 \cdots \beta_m$, then by $\alpha\beta$ we mean the concatenation of the two strings, so in this case, $\alpha\beta = \alpha_1 \cdots \alpha_n \beta_1 \cdots \beta_m$.

1.1.15. Definition. We say that α in S^+ is an *initial segment* of β in S^+ if there exists γ in S^+ such that $\alpha\gamma = \beta$.

1.1.16. Example. $\wedge A_1$ is an initial segment of $\wedge A_1)A_2$.
$(A_1\wedge$ is a initial segment of $(A_1 \wedge A_2)$.
No string is an initial segment of itself.

1.1.17. Lemma. If α is in $\mathcal{L}$ and α' is an initial segment of α, then α' has more left parentheses than right parentheses.

Proof: by induction.
Induction Basis:

If α is in B, then there are no initial segments, so α has the property.
Induction Step, Case 1: $\alpha = (\neg\beta)$. α' can be:

(1.a) "("
(1.b) "($\neg$"
(1.c) "($\neg\beta'$" (where β' is an initial segment of β)
(1.d) "($\neg\beta$".

In cases (1.a) and (1.b), α' has one left parenthesis and no right parentheses, so it has more left parentheses than right parentheses.

Case (1.c): If α' is $(\neg\beta'$ then by the induction hypothesis β' has more left parentheses than right parentheses. α' has the same number of right parentheses as β', but one more left parenthesis. Hence also α' has more left than right parentheses.

Case (1.d): If $\alpha' = (\neg\beta$, then as β has the same number of left and right parentheses (by lemma 1.1.14), so α' has more left than right parentheses, since it has one left parenthesis more than β.

Induction Step, Case 2: $\alpha = (\beta \wedge \gamma)$. So α' is an initial segment of $(\beta \wedge \gamma)$. α' must be one of the following:

(2.a) "("
(2.b) "(β'", where β' is an initial segment of β
(2.c) "(β"
(2.d) "($\beta\wedge$"
(2.e) "($\beta \wedge \gamma'$", where γ' is an initial segment of γ
(2.f) "($\beta \wedge \gamma$"

In cases (2.a), (2.c), (2.d), (2.f), the number of left parentheses of α' is exactly one more than the number of right parentheses, because β (and γ) have the same number of right and left parentheses.

Case (2.b): β' has more left parentheses than right parentheses, so α', having an additional left parentheses (but the same number of right parentheses as β'), has more left parentheses than right parentheses.

Case (2.e): γ' has more left parentheses than right parentheses, and β has the same number of left and right parentheses, so $\beta \wedge \gamma'$ has more left than right parentheses. So α', having an additional left parentheses (but the same number of right parentheses as $\beta \wedge \gamma'$), has more left parentheses than right parentheses.
Induction Step, Cases 3, 4, 5 and 6: α' is an initial segment of $(\beta \vee \gamma)$, $(\beta \rightarrow \gamma)$, $(\beta \leftrightarrow \gamma)$ and $(\beta|\gamma)$ respectively. These cases are similar to case 2.

1.1.18. Corollary. No initial segment of a member of $\mathcal{L}$ is a member of $\mathcal{L}$.

Proof: If α' is an initial segment of a member of $\mathcal{L}$, then α' has more left than right parentheses, by 1.1.17, so α' is not in $\mathcal{L}$, by 1.1.14.

1.1.19. Lemma. If α is in $\mathcal{L}$, then α is a block, or the first symbol of α is the left parenthesis.

Proof: If $\alpha \in \mathcal{L}$ is not a block, then α is of the form $F_{\neg}(\gamma)$, $F_{\wedge}(\gamma, \delta)$, $F_{\vee}(\gamma, \delta)$, $F_{\rightarrow}(\gamma, \delta)$, $F_{\leftrightarrow}(\gamma, \delta)$ or $F_{|}(\gamma, \delta)$. In all these cases α starts with a left parenthesis.

1.1.20. Remark. From now on, we will write @ to denote a member of $\{\wedge, \vee, \rightarrow, \leftrightarrow, |\}$, i.e., @ will stand for any binary connective.

1.1.21. Theorem. Let α be in $\mathcal{L}$. Then α falls into exactly one of the following cases:

(a) α is in B, or
(b) there exists a unique γ in $\mathcal{L}$ such that $\alpha = (\neg\gamma)$, or
(c) there exist unique β, γ in $\mathcal{L}$ and a unique @ in $\{\wedge, \vee, \rightarrow, \leftrightarrow, |\}$ such that $\alpha = (\beta @ \gamma)$.

Proof: If α is in B we are done. Otherwise, to prove the theorem, we must show that representations such as (b) and (c) are unique. If uniqueness does not hold, then we have the following cases:
Case 1: $\alpha = (\neg\delta)$ and also $\alpha = (\beta @ \gamma)$.
Therefore $(\neg\delta) = (\beta @ \gamma)$, hence $\neg\delta) = \beta @ \gamma)$ and this implies that the first element of β is $\neg$, therefore β cannot be in $\mathcal{L}$, by 1.1.19.
Case 2: $\alpha = (\neg\delta)$ and $\alpha = (\neg\beta)$. Therefore $\neg\delta) = \neg\beta)$, so $\beta) = \delta)$, so $\beta = \delta$.
Case 3: $\alpha = (\varepsilon @ \delta)$ and $\alpha = (\beta \Delta \gamma)$, where Δ is in $\{\wedge, \vee, \rightarrow, \leftrightarrow, |\}$
We have to show that in this case $\varepsilon = \beta$, @ $= \Delta$ and $\delta = \gamma$.
From the assumption: $\varepsilon @ \delta = \beta \Delta \gamma$. Therefore, ε is an initial segment of β or β is an initial segment of ε or $\varepsilon = \beta$. But the first two cases contradict

1.1.18, so $\varepsilon = \beta$. Therefore $@\delta = \Delta\gamma$, so $@ = \Delta$ and $\delta = \gamma$, so uniqueness does hold.

What we just proved about $\mathcal{L}$ is useful enough to deserve its own name:

1.1.22. Definition. Let C be an inductive structure and let B, K be such that $C = C(B, K)$. We say that C satisfies *unique readability* if for every b in C exactly one of the following alternatives holds:

(a) b is in B, or

(b) there are unique $a_1, \ldots, a_n$ in C and a unique F in K such that $b = F(a_1, \ldots, a_n)$, and no element of B is of the form $F(a_1, \ldots, a_n)$.

1.1.23. Corollary. $\mathcal{L}$ satisfies unique readability.

It is easy to see that if we do not use parentheses, the resulting language does not satisfy unique readability:

1.1.24. Example. We define an inductive structure as follows: The blocks will be elements of the set

$$B = \{\mathtt{A}_1, \mathtt{A}_2, \mathtt{A}_3, \ldots\}$$

(an infinite set), and the set of operations is

$$K' = \{G_\neg, G_\wedge, G_\vee, G_\rightarrow, G_\leftrightarrow, G_|\}$$

defined as follows:

$$\begin{aligned} G_\neg(\alpha) &= \neg\alpha \\ G_\wedge(\alpha, \beta) &= \alpha \wedge \beta \\ G_\rightarrow(\alpha, \beta) &= \alpha \rightarrow \beta \\ G_\vee(\alpha, \beta) &= \alpha \vee \beta \\ G_\leftrightarrow(\alpha, \beta) &= \alpha \leftrightarrow \beta \\ G_|(\alpha, \beta) &= \alpha|\beta. \end{aligned}$$

The inductive structure $C(B, K')$ does not satisfy unique readability. For example, we have

$$G_\wedge(\mathtt{A}_1, G_\vee(\mathtt{A}_2, \mathtt{A}_3)) = \mathtt{A}_1 \wedge \mathtt{A}_2 \vee \mathtt{A}_3 = G_\vee(G_\wedge(\mathtt{A}_1, \mathtt{A}_2), \mathtt{A}_3).$$

Thus, $\mathtt{A}_1 \wedge \mathtt{A}_2 \vee \mathtt{A}_3$ also has two syntax trees, shown in figure 4.

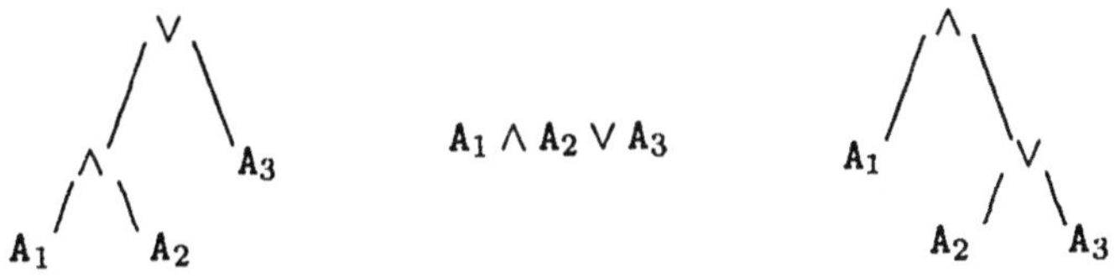

Figure 4.

However, it is possible to leave out the parentheses and still get a language satisfying unique readability, if we switch from "infix notation" (where an operator symbol, such as $\vee$, is written **between** its two operands, as in $(A_1 \vee A_2))$ to "prefix notation." In prefix notation, operators are written before the operands.

1.1.25. Definition. We let our set of blocks be $B := \{A_1, A_2, \dots\}$ as before, and we let our set of operators be $K^{\mathrm{prc}} := \{P_\vee, P_\wedge, P_\rightarrow, P_\neg, P_\leftrightarrow, P_|\}$, where

$$\begin{aligned} P_\neg(\alpha) &= \neg\alpha \\ P_\wedge(\alpha,\beta) &= \wedge\alpha\beta \\ P_\rightarrow(\alpha,\beta) &= \rightarrow\alpha\beta \\ P_\vee(\alpha,\beta) &= \vee\alpha\beta \\ P_\leftrightarrow(\alpha,\beta) &= \leftrightarrow\alpha\beta \\ P_|(\alpha,\beta) &= |\alpha\beta. \end{aligned}$$

We let $\mathcal{L}^{\mathrm{prc}}$ be the inductive structure $C(B, K^{\mathrm{prc}})$.

1.1.26. Example. The following are elements of $\mathcal{L}^{\mathrm{prc}}$:

$$\begin{gathered} A_1 \\ \neg A_1 \\ A_2 \\ \vee\neg A_1 A_2 \\ A_3 \\ \rightarrow A_2 A_3 \\ \wedge\vee\neg A_1 A_2 \rightarrow A_2 A_3. \end{gathered}$$

The last expression corresponds to the formula $(((\neg A_1) \vee A_2) \wedge (A_2 \rightarrow A_3))$, as can be seen from the syntax tree in figure 5.

We now show that the language $\mathcal{L}^{\mathrm{prc}}$ satisfies unique readability. We begin by observing the following:

$(((\neg A_1) \vee A_2) \wedge (A_2 \rightarrow A_3))$

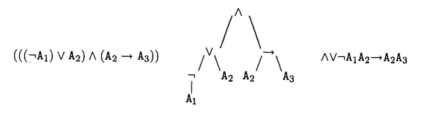

$\wedge\vee\neg A_1 A_2 \rightarrow A_2 A_3$

Figure 5.

1.1.27. Fact. Every element α of $\mathcal{L}^{\text{pre}}$ is either of the form A_n for some n, or it starts with an element of $\{\neg, \wedge, \vee, \rightarrow, \leftrightarrow, |\}$ (but not both). Moreover, if α starts with $\wedge$, it must be of the form $\wedge\beta\gamma$, where β and γ are in $\mathcal{L}^{\text{pre}}$, and similar for the other connectives.

Proof: By induction.

Now we need a lemma similar to lemma 1.1.17.

1.1.28. Lemma. If α, α' are in $\mathcal{L}^{\text{pre}}$, then α' is not an initial segment of α.

Proof: We will prove this "by induction on the length of α" (see also exercise 8). In other words, we will assume that the lemma is false, and choose a shortest possible counterexample. (I.e., we find the smallest n such that there is a formula α consisting of n symbols which does have an initial segment in $\mathcal{L}^{\text{pre}}$. If there is more than one such formula with the same length n, we choose just any one. We write "$|\alpha| = n$" to abbreviate "α consists of n symbols.") Then we will show that such a shortest possible counterexample cannot exist.

So let α and α' be formulas, α' an initial segment of α, and assume that:

Whenever β is in $\mathcal{L}^{\text{pre}}$, $|\beta| < |\alpha|$, then no initial segment of β is in $\mathcal{L}^{\text{pre}}$

<u>Case 1</u> $\alpha = A_n$. This is impossible, because A_n does not have any initial segments.
<u>Case 2</u> $\alpha = \neg\beta$. Now $\alpha' \in \mathcal{L}^{\text{pre}}$ is an initial segment of α, and it also starts with $\neg$, so we must have $\alpha' = \neg\beta'$, where β' is in $\mathcal{L}^{\text{pre}}$. But then $\neg\beta'$ is an initial segment of $\neg\beta$, so β' is an initial segment of β, which is impossible since β is shorter than α.
<u>Case 3</u> $\alpha = \wedge\beta\gamma$.
α' is an initial segment of α, $\alpha' \in \mathcal{L}^{\text{pre}}$, so α' must be of the form $\wedge\beta'\gamma'$, for some β', γ' in $\mathcal{L}^{\text{pre}}$. $\beta'\gamma'$ is an initial segment of $\beta\gamma$. Now both β' and β

are strictly shorter than α, so by our assumption it is impossible that one of them is an initial segment of the other. So we must have $\beta' = \beta$. Hence $\beta\gamma = \beta\gamma'$, so γ is an initial segment of γ', which is impossible.
Cases 4,5,6 $\alpha = \rightarrow \beta\gamma$, $\vee\beta\gamma$, $\leftrightarrow \beta\gamma$, or $|\beta\gamma$. These cases are similar to case 3.

So we have proved unique readability for the inductive structure $\mathcal{L}^{\text{pre}}$.

1.1.29. Lemma. The natural numbers carry an inductive structure that satisfies unique readability.

Proof: The inductive structure is given in 1.1.7. We leave the proof that it satisfies unique readability to the reader. (Exercise 9.)

The following theorems are true in general for inductive structures satisfying unique readability. We will prove them only for the special case of the natural numbers and leave the general case to the reader.

Suppose we want to find out if a natural number is even or odd, using only the information that the inductive structure of the natural numbers gives us (i.e., we are not allowed to use multiplication or division, only the successor function). Instead of considering the property of "being even," we will consider the function f that assigns to even numbers the value 0, and to odd numbers the value 1, or in other words: $f(n)$ is the remainder after dividing n by 2.

How can we compute $f(n)$ for n in the inductive structure of the natural numbers?

If $n = 0$, then $f(n) = 0$.

Otherwise, n is the successor of some natural number, $n = s(k) = k+1$. If k was even, then n is odd, and if k was odd, then n is even. Thus,

$$f(k+1) = \begin{cases} 0 & \text{if } f(k) = 1 \\ 1 & \text{if } f(k) = 0. \end{cases}$$

Writing G for the function that maps 0 to 1 and 1 to 0, we have the following two conditions on f:

$$f(0) = 0 \tag{1}$$

$$\text{for all } k, \quad f(k+1) = G(f(k)). \tag{2}$$

Thus, we have described the function f by the much simpler function G.

Our intuition about the natural numbers tells us that there is a unique function f satisfying (1) and (2). In fact, this is true for any function G:

We have to have

$$\begin{aligned} f(0) &= 0 \\ f(1) &= G(f(0)) = G(0) \\ f(2) &= G(f(1)) = G(G(0)) \qquad\qquad (**) \\ f(3) &= G(f(2)) = G(G(G(0))) \\ &\text{etc.} \end{aligned}$$

But it is exactly the precise meaning of this "etc." that we wanted to get a grip on when we defined inductive systems. So the following lemma — although intuitively clear — needs a proof.

1.1.30. Lemma (Definition by induction on $\mathbb{N}$). Assume that G_s is a function with domain A and range $\subseteq A$, and let a_0 be an element of A. Then there is a unique function f with domain $\mathbb{N}$ that satisfies

$$f(0) = a_0 \tag{1}$$

$$\text{for all } k, \quad f(k+1) = G_s(f(k)) \tag{2}$$

Proof: We will imitate the iterative construction in $(**)$ using the following concept: We say that a function f^* is "good," if

(i) 0 is in the domain of f^*, $f^*(0) = a_0$, and
(ii) Whenever $k+1$ is in the domain of f^*, then also k is in the domain of f^*, and $f^*(k+1) = G_s(f^*(k))$.

Note that the domain of a "good" function may be quite small — for example, there is a (unique) "good" function whose domain is the singleton $\{0\}$.

Now consider the property (of a natural number n): "there is a 'good' function f^* with $n \in \text{dom}(f^*)$."

Subclaim: For all natural numbers n there is a "good" function f^* with $n \in \text{dom}(f^*)$.

Proof of the subclaim: 0 has the property, as we can choose the function f^* with domain $\{0\}$, and with $f(0) = a_0$.

Assuming that k has the property, we have to show that also $k+1$ has the property. So let f^* be a "good" function with $k \in \text{dom}(f^*)$. If it happens that also $k+1 \in \text{dom}(f^*)$, then we are done. Otherwise we can define a good function f' by demanding $f'(n) = f^*(n)$ for all $n \in \text{dom}(f^*)$, and $f'(k+1) = G(f^*(k)))$.

Next we claim that whenever f_1^* and f_2^* are "good" functions, then for all $n \in \text{dom}(f_1^*) \cap \text{dom}(f_2^*)$, f_1^* and f_2^* agree on n, i.e., $f_1^*(n) = f_2^*(n)$. To prove this fact, we consider the following property of a natural number n:

$$n \notin \text{dom}(f_1^*) \cap \text{dom}(f_2^*), \text{ or } f_1^*(n) = f_2^*(n).$$

We leave it as an exercise to the reader to show that **all** natural numbers have this property.

Finally, we can define f: For every number n, we let $f(n)$ be the (common) value $f^*(n)$ of all "good" functions f^* that have n in their domain.

We have to check that f satisfies (1) and (2).

(1) is clear, because all "good" functions f^* satisfy $f^*(0) = a_0$.

To prove (2), consider $f(k+1)$. There must be some "good" function f^* with $f(k+1) = f^*(k+1)$. By the definition of "good," $f^*(k+1) = G(f^*(k))$, and since f^* is a good function we also have $f(k) = f^*(k)$. Thus, $f(k+1) = G(f(k))$.

A similar proof establishes the following lemma:

1.1.31. Lemma (Definition by induction on $\mathbb{N}$, generalized). Assume that G is a binary function with domain $\mathbb{N} \times A$ and range $\subseteq A$, and let a_0 be an element of A.
Then there is a unique function f with domain $\mathbb{N}$ that satisfies

$$f(0) = a_0 \qquad (1)$$

$$\text{for all } k, \quad f(k+1) = G(k, f(k)) \qquad (2)$$

We leave the proof as an exercise to the reader.

The last two lemmas are in general true for inductive structures satisfying unique readability. We will only formulate an analogue of the first lemma:

1.1.32. Lemma (Inductive Definitions). Assume that $C = C(B, K)$ satisfies unique readability. Let A be a set, and let f_0 be a function with domain B and range included in A. Assume that for each F in K we also have a function G_F satisfying:

(1) If F is an n-place function, then $\mathrm{dom}(G_F) = A^n$.

(2) $\mathrm{range}(G_F) \subseteq A$.

Then there is a unique function f satisfying

(1) $f(b) = f_0(b)$ for all blocks $b \in B$.

(2) For all n, all n-ary operators $F \in K$, all $a_1, \ldots, a_n$ in C:

$$f(F(a_1, \ldots, a_n)) = G_F(f(a_1), \ldots, f(a_n)).$$

Inductive definitions play an important role in the following sections. We will use this tool in order to extend truth functions defined for a block to all possible formulas.

1.1.33. Example. We want to define the function "the number of parentheses in the formula α" by induction on α. Since the number of parentheses of $(\beta @ \gamma)$ is 2 plus the number of parentheses in β plus the number of parentheses in γ, and the number of parentheses in $(\neg\beta)$ is the number of parentheses in β plus 2, we let f be the unique function satisfying

$$f(\,(\beta @ \gamma)\,) = f(\beta) + f(\gamma) + 2 \qquad \text{for @ in } \{\wedge, \vee, \rightarrow, \leftrightarrow, |\}$$
$$f(\,(\neg\beta)\,) = f(\beta) + 2$$

Formally, we invoke the previous lemma using the functions $G_{F_@}$ (where @ ranges over $\{\wedge, \vee, \rightarrow, \leftrightarrow, |\}$) and the function $G_{F_\neg}$, where $G_{F_\vee}$ is the function with domain the set of all pairs of natural numbers, defined by

$$G_{F_\vee}(n, m) = n + m + 2.$$

$G_{F_\wedge}$, $G_{F_\rightarrow}$, $G_{F_\vee}$, $G_{F_\leftrightarrow}$, $G_|$ are the same function, and $G_{F_\neg}$ is defined by

$$G_{F_\neg}(n) = n + 2.$$

1.1.34. Example. What does the function f, which is specified by the requirements

$$f(\beta @ \gamma) = f(\beta) + f(\gamma) \qquad \text{for } @ \in \{\wedge, \vee, \rightarrow, \leftrightarrow, |\}$$
$$f(\neg\beta) = f(\beta)$$
$$f(\mathtt{A}_n) = 1 \qquad \text{for } n \in \omega$$

describe?

Answer: The first two requirements show that f counts something in the formula α. This count never increases when two formulas are joined by $\vee$, $\wedge$, etc., and by the third requirement it is $= 1$ for all sentential symbols. So $f(\alpha)$ is the number of blocks (sentential symbols) in α.

Exercises

Induction on $\mathbb{N}$

1. Give two examples of "inductive" processes in reality.

2. Prove by induction:
(a) For all $n \geq 1$:

$$\sum_{i=0}^{n} i^2 = \frac{n(n+1)(2n+1)}{6}.$$

(b) For all $n \geq 1$:

$$\frac{1}{2} + \frac{1}{4} + \frac{1}{8} + \cdots + \frac{1}{2^n} < 1.$$

3. Prove by induction:
(a) The number of regions in the plane formed by n straight lines (such that each two lines have a common intersection but no three lines have common intersection) is:

$$\frac{n(n+1)}{2} + 1.$$

(b) It is possible to color the regions formed by any number of lines in the plane with only two colors such that every two neighboring regions have different colors. (Here, two regions are called "neighbors" if they have a common border line segment, not just a single common point.)

4. What is wrong with the following inductive proof?
Theorem: All the elements of any set are identical.
Proof: By induction on the size of the set. Clearly it is true for sets of size one. Suppose the theorem is true for all sets of size n. Fix a set α with $n+1$ elements, say $\alpha = \{a_1, a_2, \ldots, a_{n-1}, a_n, a_{n+1}\}$. Then the sets $\{a_1, \ldots, a_{n-1}, a_n\}$ and $\{a_1, \ldots, a_{n-1}, a_{n+1}\}$ both have n elements, so (by induction assumption) every a_i is equal to a_1.

5. What is wrong with the following inductive proof?
Theorem: For any finite set of points in the plane there is a straight line passing through all of them.
Proof: By induction on the number of points. Clearly it is true if there are one or two points. Suppose the theorem is true for n points. Fix a $n+1$ distinct points, say $P_1, P_2, \ldots, P_n, P_{n+1}$. By induction hypothesis the points $P_1, P_2, \ldots, P_{n-1}, P_n$ all lie on one line l_1, and also by induction hypothesis the points $P_1, P_2, \ldots, P_{n-1}, P_{n+1}$ all lie on one line l_2. Since

both l_1 and l_2 pass through P_1 and P_2, l_1 and l_2 must be the same line. So $P_1, \ldots, P_{n+1}$ all lie on the same straight line.

General Inductive Structures

6. What is the natural inductive structure for the set $\{2, 5, 8, 11, 14, \ldots\}$?

7. Let $B = \{00, 01, 10, 11\}$ be a set of blocks.
Let $K_1 = \{F_1, F_2, F_3, F_4\}$ where the F_i are operators over sequences of 0's and 1's defined as follows:

$$\begin{aligned} F_1(x) &= 0x0 \\ F_2(x) &= 0x1 \\ F_3(x) &= 1x0 \\ F_4(x) &= 1x1 \end{aligned}$$

Let $K_2 = \{G_1, G_2, G_3, G_4\}$ where the G_i are operators over sequences of 0's and 1's defined as follows:

$$\begin{aligned} G_1(x) &= 00x \\ G_2(x) &= 01x \\ G_3(x) &= 10x \\ G_4(x) &= 11x \end{aligned}$$

Show that $C(B, K_1) = C(B, K_2)$ (i.e., the two sets are the same, although they carry a different structure).

8. "Induction on the length of α": Assume that P is a property such that for all α in the sentential language $\mathcal{L}$:

If every $\beta \in \mathcal{L}$ whose length is smaller than the length of α has the property P, then α has the property P.

Show

that every α has the property P.

Unique Readability and the Induction Theorem

9. Prove that the inductive structure on the natural numbers satisfies unique readability.

10. Let $S = \{a, b, c\}$ be a set of three symbols.
(a) We define an inductive structure with the three blocks a, b, c and the three operators F_a, F_b, F_c ($F_a(x) = xa$, etc.). This structure yields S^+, the set of all finite sequences from S. Show that this structure satisfies unique readability.

(b) Define the concatenation function by induction on S^+. That is, give an inductive definition (using the inductive structure from (a)) of the function f that maps two sequences to their concatenation, $f(x, y) = xy$.
(c) Define "x is an initial segment of y" by induction.
(d) Show that if x, y, r, s are in S^+ and $xy = rs$, then either $x = r$ or one of x and r is an initial segment of the other.

11. Complete the proof of lemma 1.1.30.

12. Prove lemma 1.1.31. Where did we use the fact that we have unique readability?

13. Prove lemma 1.1.32 ,

<u>The Language of Sentential Logic</u>

14. Show that no sentential formula has two consecutive logical connectives.

15. Show that the number of blocks in every sentential formula is one greater than the number of binary logical connectives.

16. What is the meaning of each one of the following functions?
($\mathtt{A}_n$ can be any block; α, β range over $\mathcal{L}$; and @ is in $\{\wedge, \vee, \rightarrow, \leftrightarrow, |\}$).

(a)

$$\begin{aligned} F(\mathtt{A}_n) &= 0 \\ F((\alpha @ \beta)) &= F(\alpha) + F(\beta) + 1 \\ F((\neg\alpha)) &= F(\alpha) + 1 \end{aligned}$$

(b)

$$\begin{aligned} G(\mathtt{A}_n) &= 1 \\ G((\neg\alpha)) &= G(\alpha) + 3 \\ G((\alpha @ \beta)) &= G(\alpha) + G(\beta) + 3 \end{aligned}$$

17. Define the following functions by induction:
(a) The number of blocks in a sentential formula.
(b) The maximal n such that $\mathtt{A}_n$ is a block in α.

18. Give an explicit description of the set:

$$A = \{n \mid \text{There is a sentential formula of length } n\}$$

19. Suppose we omitted all the right parentheses from all the sentential formulas.
(a) Give an inductive definition of the new language.
(b) Show that it satisfies unique readability.

1.2. Sentential Logic

The objective of this section is to study the assertions which are true because of the syntactical structure of the assertion. There are many examples of such assertions in our common language. For instance:

(1) "Batman is President of the U.S.A.,
or Batman is not President of the U.S.A."

is always true. This kind of expression is called a tautology. To discover when an expression is a tautology is not an easy task. The difficulty may increase significantly with the length of the expression. In this section, we will not discuss the truth of such assertions as:

"The white horse of Napoleon is blue"

since truth depends in these cases on semantic considerations. Instead, we will regard the truth or falsity of blocks as being given. So a block like "Batman is the President of the U.S.A." will be either true or false.

If we denote the last expression about Batman by "A", then (1) becomes

"A or not A."

To formalize mathematical assertions, we use combinations of the following logical expressions:

not and or if – then – if and only if.

The meaning of each of these logical expressions is clear once we point out that in mathematics, "or" is always inclusive – that is, when we say "A or B", we mean either A or B or both.

We will use the following customary symbols for each of these English language expressions. We also include a symbol for the expression "not both – and –":

$\neg A$	means	not A (the "negation" of A)
$A \wedge B$	means	A and B (the "conjunction" of A and B)
$A \vee B$	means	A or B, possibly both (the "disjunction" of A and B)
$A \rightarrow B$	means	If A then B (or: A implies B)
$A \leftrightarrow B$	means	A if and only if B (or: A iff B)
$A \vert B$	means	not both A and B, possibly neither (or: A nand B)

With this in mind, the language that has been developed for the study of tautologies is the language $\mathcal{L}$ introduced in Section 1. $\mathcal{L} = C(B, K)$, where $B = \{\mathtt{A}_1, \mathtt{A}_2, \ldots\}$ is a collection of blocks, and the set of operators is given by $K = \{F_\neg, F_\wedge, F_\vee, F_\rightarrow, F_\leftrightarrow, F_|\}$.

(The reader may wonder whether it is necessary to use all these connectives for building the language of sentential logic. For example, is not

$(A \leftrightarrow B)$ the "same" as $(A \to B) \wedge (B \to A)$? And what is this strange symbol $|$? Can we not use $\neg(A \wedge B)$ instead of $A|B$? The answer is that (as we shall see later) it would indeed be enough to use, for example, only the two connectives $\neg$ and $\wedge$. But we include other connectives such as $\to$ because of their practical importance. $|$ is included mainly because by itself it generates a "complete" language, see 1.2.41 below.)

According to the rules given in section 1, the above example (1) is $(\mathtt{A}_1 \vee (\neg\mathtt{A}_1))$. Therefore, the blocks $\mathtt{A}_1, \mathtt{A}_2, \ldots$ will be expressions that are true or false. We don't plan in this section to study the truths of these blocks, but rather the truths of the expressions built up from these blocks by using the operators $F_\neg, F_\wedge, F_\vee, F_\to$, $F_|$ and $F_\leftrightarrow$, when the truth value of each of $\mathtt{A}_1, \mathtt{A}_2, \ldots$ is given.

1.2.1. Definition. A **complete truth assignment** S is a function from the set of all blocks $\mathtt{A}_1, \mathtt{A}_2, \ldots$ to the two element set $\{T, F\}$. The intended meaning of $S(\mathtt{A}_1) = T$ is that the truth assignment S makes $\mathtt{A}_1$ true and $S(\mathtt{A}_2) = F$ means that S makes $\mathtt{A}_2$ false.
A **truth assignment** S is a function from a subset of the set of all blocks $\mathtt{A}_1, \mathtt{A}_2, \ldots$ to $\{T, F\}$.

If the elements of B are interpreted as the set of all possible statements, then a truth assignment S selects certain statements and calls them "true," while all other statements are "false." Notice that when we construct a truth assignment S, the values of $S(\mathtt{A}_i)$ and $S(\mathtt{A}_j)$ are completely independent, if $i \neq j$.

1.2.2. Definition. Let S be a complete truth assignment. We define the **truth function** $\bar{S}$,

$$\bar{S} : \mathcal{L} \to \{T, F\}$$

by induction on $\mathcal{L}$ as follows:
Let α be a member of $\mathcal{L}$.

(1) if α is a sentential symbol (that is, a block) then $\bar{S}(\alpha) = S(\alpha)$;

(2) if $\alpha = (\neg\beta)$ for some β in $\mathcal{L}$, then

$$\bar{S}(\alpha) = \begin{cases} T & \text{if } \bar{S}(\beta) = F, \\ F & \text{if } \bar{S}(\beta) = T. \end{cases}$$

(3) if $\alpha = (\beta \wedge \gamma)$ for some β, γ in $\mathcal{L}$,

$$\bar{S}(\alpha) = \begin{cases} T & \text{if } \bar{S}(\beta) = T = \bar{S}(\gamma), \\ F & \text{otherwise.} \end{cases}$$

(4) if $\alpha = (\beta \vee \gamma)$ for some β, γ in $\mathcal{L}$,

$$\bar{S}(\alpha) = \begin{cases} F & \text{if } \bar{S}(\beta) = F = \bar{S}(\gamma), \\ T & \text{otherwise.} \end{cases}$$

(5) if $\alpha = (\beta \rightarrow \gamma)$ for some β, γ in $\mathcal{L}$, then

$$\bar{S}(\alpha) = \begin{cases} F & \text{if } \bar{S}(\beta) = T \text{ and } \bar{S}(\gamma) = F, \\ T & \text{otherwise.} \end{cases}$$

(6) if $\alpha = (\beta \leftrightarrow \gamma)$ for some β, γ in $\mathcal{L}$, then

$$\bar{S}(\alpha) = \begin{cases} T & \text{if } \bar{S}(\beta) = \bar{S}(\gamma), \\ F & \text{otherwise.} \end{cases}$$

(7) if $\alpha = (\beta|\gamma)$ for some β, γ in $\mathcal{L}$, then

$$\bar{S}(\alpha) = \begin{cases} F & \text{if } \bar{S}(\beta) = T = \bar{S}(\gamma), \\ T & \text{otherwise.} \end{cases}$$

Notice that for example $\bar{S}((\beta \rightarrow \gamma)) = T$ whenever $\bar{S}(\beta) = F$. Thus, a statement "If 0=1 then ..." (where the truth value of "0=1" is F) will always be assigned the truth value T.

1.2.3. Remark.

(1) In definition 1.2.2, we are using the inductive structure of $\mathcal{L}$ to ensure that $\bar{S}$ is defined for all γ in $\mathcal{L}$. That is, if we define the functions $G_{\neg}$, $G_{\wedge}$, ... as in the tables below, then we can phrase the conditions (1)–(7) above as

$$\bar{S}(\neg\beta) = G_{\neg}(\bar{S}(\beta))$$
$$\bar{S}(\beta@\gamma) = G_{@}(\bar{S}(\beta), \bar{S}(\gamma)) \quad \text{for } @ \in \{\wedge, \vee, \rightarrow, \leftrightarrow, |\}.$$

(2) A proof by induction makes it clear that, for *any* @ in $\{\wedge, \vee, \rightarrow, \leftrightarrow, |\}$, and for *any* complete truth assignment S, the value of $\bar{S}(\beta@\gamma)$ depends only on the values of $\bar{S}(\beta)$ and $\bar{S}(\gamma)$.

x	$G_{\neg}(x)$
T	F
F	T

x	y	$G_{\wedge}(x,y)$
T	T	T
T	F	F
F	T	F
F	F	F

x	y	$G_{\vee}(x,y)$
T	T	T
T	F	T
F	T	T
F	F	F

x	y	$G_{\rightarrow}(x,y)$
T	T	T
T	F	F
F	T	T
F	F	T

x	y	$G_{\leftrightarrow}(x,y)$
T	T	T
T	F	F
F	T	F
F	F	T

x	y	$G_{\mid}(x,y)$
T	T	F
T	F	T
F	T	T
F	F	T

The next definition will tell us what it means that a sentential formula α is true for structural reasons.

1.2.4. Definition. An expression α in $\mathcal{L}$ is a **tautology** if and only if $\bar{S}(\alpha) = T$ for every complete truth assignment S.

1.2.5. Examples.

(i) $\mathtt{A}_1$ is not a tautology.
(ii) $(\mathtt{A}_1 \vee (\neg \mathtt{A}_1))$ is a tautology.
(iii) $(\mathtt{A}_1 \rightarrow \mathtt{A}_2)$ is not a tautology.
(iv) $(\mathtt{A}_1 \rightarrow \mathtt{A}_1)$ is a tautology.
(v) $(((\neg \mathtt{A}_2) \rightarrow \mathtt{A}_2) \rightarrow \mathtt{A}_2)$ is a tautology.
(vi) $((\mathtt{A}_1 \rightarrow \mathtt{A}_2) \vee (\mathtt{A}_2 \rightarrow \mathtt{A}_1))$ is a tautology.
(vii) $((\mathtt{A}_1 \rightarrow \mathtt{A}_2) \rightarrow (\mathtt{A}_2 \rightarrow \mathtt{A}_1))$ is not a tautology.
(viii) $(\neg(\mathtt{A}_1 \wedge (\neg \mathtt{A}_1)))$ is a tautology.

Proof: Exercise 2.

1.2.6. Definition. If α, $\beta \in \mathcal{L}$, then the statement

$$\alpha \Rightarrow \beta$$

means:

for every complete truth assignment S: If $\bar{S}(\alpha) = T$, then $\bar{S}(\beta) = T$

or, in other words:

There is no complete truth assignment S with $\bar{S}(\alpha) = T$ and $\bar{S}(\beta) = F$.

We write $\alpha \not\Rightarrow \beta$ iff it is not the case that $\alpha \Rightarrow \beta$. In other words, $\alpha \not\Rightarrow \beta$ means that there is a complete truth assignment S with $\bar{S}(\alpha) = T$ and $\bar{S}(\beta) = F$.
We write $\alpha \Leftrightarrow \beta$ if $\alpha \Rightarrow \beta$ and $\beta \Rightarrow \alpha$, or in other words, if for all truth assignments S, $\bar{S}(\alpha) = \bar{S}(\beta)$. In this case we say that α and β are equivalent.
We write $\Gamma \Rightarrow \alpha$ (where Γ is a set of formulas) if there is no complete truth assignment S such that for all $\gamma \in \Gamma$, $\bar{S}(\gamma) = T$ and $\bar{S}(\alpha) = F$.

1.2.7. Remark. $\alpha \Rightarrow \beta$ holds iff the formula $\alpha \rightarrow \beta$ is a tautology. Similarly, $\alpha \Leftrightarrow \beta$ is true iff $\alpha \leftrightarrow \beta$ is a tautology.
Please note that there is a subtle but fundamental difference between $\alpha \rightarrow \beta$ and $\alpha \Rightarrow \beta$:
$\alpha \rightarrow \beta$ is ONE formula,
whereas $\alpha \Rightarrow \beta$ is a STATEMENT ABOUT TWO formulas.

1.2.8. Lemma.
(1) If $\alpha \Rightarrow \beta$ and $\beta \Rightarrow \gamma$, then $\alpha \Rightarrow \gamma$.
(2) If α is a tautology, and $\alpha \Rightarrow \beta$, then also β is a tautology.

Proof: Exercise.

1.2.9. Definition. Let α be in $\mathcal{L}$. We define $BKS(\alpha)$ by induction on $\mathcal{L}$:
(1) If α is a block, then $BKS(\alpha) = \{\alpha\}$.
(2) If α is $(\neg\beta)$, then $BKS(\alpha) = BKS(\beta)$.
(3) If α is $(\beta @ \gamma)$ for some @ in D, then

$$BKS(\alpha) = BKS(\beta) \cup BKS(\gamma).$$

1.2.10. Remark. $BKS(\alpha)$ is the collection of blocks that are used in α.

1.2.11. Lemma. Let S_1 and S_2 be two complete truth assignments. Let α be an expression of $\mathcal{L}$. Assume that $S_1(\mathtt{A}) = S_2(\mathtt{A})$ for every $\mathtt{A}$ in $BKS(\alpha)$.
Then $\bar{S}_1(\alpha) = \bar{S}_2(\alpha)$.

Proof: by induction on α.

(1) If α is a block, then $\bar{S}_1(\alpha) = S_1(\alpha) = S_2(\alpha) = \bar{S}_2(\alpha)$.
By unique readability, if α is not a block, then exactly one of the cases below must hold:

(2) If $\alpha = (\neg\beta)$ then by definition 1.2.9 $BKS(\beta) = BKS(\alpha)$,
so by induction, $\bar{S}_1(\beta) = \bar{S}_2(\beta)$
and thus by definition 1.2.2 $\bar{S}_1(\neg\beta) = \bar{S}_2(\neg\beta)$,
so $\bar{S}_1(\alpha) = \bar{S}_2(\alpha)$.

(3) If $\alpha = (\beta \wedge \gamma)$, then $BKS(\beta) \subseteq BKS(\alpha)$, so $\bar{S}_1(\mathtt{A}) = \bar{S}_2(\mathtt{A})$ for every $\mathtt{A}$ in $BKS(\beta)$,
and thus by induction, $\bar{S}_1(\beta) = \bar{S}_2(\beta)$.
Similarly, since $BKS(\gamma) \subseteq BKS(\alpha)$, we have $\bar{S}_1(\gamma) = \bar{S}_2(\gamma)$.
But from definition 1.2.2, this clearly implies $\bar{S}_1(\beta \wedge \gamma) = \bar{S}_2(\beta \wedge \gamma)$,
that is, $\bar{S}_1(\alpha) = \bar{S}_2(\alpha)$.

(4) If $\alpha = (\beta @ \gamma)$ for some other @, a similar argument works.

1.2.12. Corollary. Let α be in $\mathcal{L}$, and let S be a complete truth assignment. Then $\bar{S}(\alpha)$ depends only on the values of S on $BKS(\alpha)$.

1.2.13. Definition. Whenever S is a (not necessarily complete) truth assignment with dom(S) including $BKS(\alpha)$, for any complete truth assignments S_1, S_2 extending S, we have $\bar{S}_1(\alpha) = \bar{S}_2(\alpha)$. We define $\bar{S}(\alpha)$ to be this common value. Thus, $\bar{S}(\alpha)$ is well-defined whenever S is a (not necessarily complete) truth assignment that is defined on $BKS(\alpha)$.

We leave it to the reader to check that $\bar{S}(\alpha)$ satisfies all clauses of 1.2.2, whenever the dom(S) contains $BKS(\alpha)$.

1.2.14. Fact. α is a tautology iff for every truth assignment S on $BKS(\alpha)$, we have $\bar{S}(\alpha) = T$.

1.2.15. Definition. If S is a truth assignment defined on $BKS(\alpha)$, we say that S **satisfies** α if $\bar{S}(\alpha) = T$.
If A is a set of formulas, and S a truth assignment, then we say S satisfies A if S satisfies every $\alpha \in A$.

1.2.16. Definition. Let α be in $\mathcal{L}$. Then $|BKS(\alpha)|$ is the number of different blocks in $BKS(\alpha)$.

1.2.17. Corollary. Let α be in $\mathcal{L}$, and let $n = |BKS(\alpha)|$. Then there are 2^n truth assignments to check to determine whether α is a tautology.

Proof: By fact 1.2.14, we do not have to check all complete truth assignments to find out if α is a tautology, but only have to go through the truth assignments on $BKS(\alpha)$. We will show that there are 2^n such truth assignments, by induction on n.

(1) If $n = 1$, then there are two assignments, namely, if $BKS(\alpha) = \{A\}$, then $S_1(A) = T$ is one, and $S_2(A) = F$ is the second.

(2) If $n = k + 1$, let A be in $BKS(\alpha)$.
Then by the induction hypothesis, there are 2^k truth assignments over the remaining members of $BKS(\alpha) - \{A\}$.
For each such assignment, say S^*, we can create two assignments, say S_1^* and S_2^*, over *all* of $BKS(\alpha)$:
$S_1^*(B) = S^*(B)$ for all B in $BKS(\alpha) - \{A\}$, and $S_1^*(A) = T$;
$S_2^*(B) = S^*(B)$ for all B in $BKS(\alpha) - \{A\}$, and $S_2^*(A) = F$.
Therefore, the total number of truth assignments on $|BKS(\alpha)|$ is $2^k + 2^k = 2^{k+1} = 2^n$.

1.2.18. Definition. Let α be in $\mathcal{L}$. We define $SUB(\alpha)$ by induction:

(1) If α is a block, $SUB(\alpha) = \{\alpha\}$.
(2) If α is $(\neg\beta)$ then $SUB(\alpha) = \{\alpha\} \cup SUB(\beta)$.
(3) If α is $(\beta @ \gamma)$ for $@$ in $\{\wedge, \vee, \rightarrow, \leftrightarrow, |\}$, then $SUB(\alpha) = \{\alpha\} \cup SUB(\beta) \cup SUB(\gamma)$.

1.2.19. Remark. $SUB(\alpha)$ is the collection of expressions which are "subformulas" of α, where α is also considered a subformula of itself.

1.2.20. Example. Let α be the formula $((\neg A_1) \wedge (A_1 \vee A_2))$.
Then $SUB(\alpha) = \{\ A_1,\ (\neg A_1),\ A_2,\ (A_1 \vee A_2),\ ((\neg A_1) \wedge (A_1 \vee A_2))\ \}$.

1.2.21. Definition. If $f : A \rightarrow B$ is a function mapping A to B, and if $C \subseteq A$, then the **restriction** of f to C, denoted $f \upharpoonright C$, is the mapping from C to B defined for any c in C by:

$$(f \upharpoonright C)(c) = f(c).$$

1.2.22. Definition. Let α be in $\mathcal{L}$. The **truth table** of α is defined as

$$TABLE(\alpha) := \{f : \begin{array}{l} (1)\ f \text{ is a function with domain } SUB(\alpha), \\ \quad \text{range} \subseteq \{T, F\}, \text{ and} \\ (2)\ \text{for all } \beta \in SUB(\alpha),\ f(\beta) = \overline{f\lceil BKS(\beta)}(\beta) \end{array} \}$$

Condition (2) says that $f(\beta)$ is the truth value which is uniquely determined (using the rules in 1.2.2) from the restriction of f to the blocks of β.

We write a truth table by listing all functions in the table on separate lines. Each line first gives us a truth assignment on the blocks of the formula, and then describes how this truth assignment can be extended to all subformulas of the given formula.

1.2.23. Examples of Truth Tables.

(a) If α is a block, say A_1, then $SUB(\alpha) = \{A_1\}$, so $TABLE(\alpha)$ contains all functions from $\{A_1\}$ to $\{T, F\}$: (There are two such functions.)

Truth assignment	$\alpha = A_1$
S_1	T
S_2	F

(b) If α is $(A_1 \vee A_2)$, where both A_1 and A_2 are blocks, then $SUB(\alpha) = \{A_1, A_2, \alpha\}$, so its truth table contains all functions from $\{A_1, A_2, \alpha\}$ to $\{T, F\}$ that satisfy the coherence condition 1.2.22(2): (There are 4 such functions.)

Truth assignment	A_1	A_2	$\alpha = (A_1 \vee A_2)$
S_1	T	T	T
S_2	T	F	T
S_3	F	T	T
S_4	F	F	F

(c) If α is $((A_1 \wedge A_3) \vee A_{12})$
then its truth table is:

Truth assignment	A_1	A_3	A_{12}	$(A_1 \wedge A_3)$	α
S_1	T	T	T	T	T
S_2	T	T	F	T	T
S_3	T	F	T	F	T
S_4	T	F	F	F	F
S_5	F	T	T	F	T
S_6	F	T	F	F	F
S_7	F	F	T	F	T
S_8	F	F	F	F	F

From now on we will, for the sake of easier readability, omit parentheses whenever we trust the reader to deduce where they should be inserted.

Sometimes we also use other parentheses: $[\,]$, $\langle\,\rangle$, ..., where we officially should use $(\,)$.

We never omit a parenthesis immediately following a $\neg$ sign. Thus, $\neg A \vee B$ is to be read as $((\neg A) \vee B)$, whereas $\neg(A \vee B)$ is $(\neg(A \vee B))$.

1.2.24. Lemma. (1) Let $\varphi = \psi_1 \wedge \cdots \wedge \psi_n$, and let S be a truth assignment on $BKS(\varphi)$. Then $\bar{S}(\varphi) = T$ iff for ALL i in $\{1, \ldots, n\}$, we have $\bar{S}(\psi_i) = T$.

(2) Let $\varphi = \psi_1 \vee \cdots \vee \psi_n$, and let S be a truth assignment on $BKS(\varphi)$. Then $\bar{S}(\varphi) = T$ iff for SOME i in $\{1, \ldots, n\}$ we have $\bar{S}(\psi_i) = T$.

Proof: Exercise

1.2.25. Definition. A formula α is an n**-clause**, if α is of the form $\beta_1 \vee \ldots \vee \beta_n$, where each β_i is either A_i or $(\neg A_i)$.

We say that α is a clause, if α is an n-clause for some n.

1.2.26. Example. There are four 2-clauses:

$$(A_1 \vee A_2)$$
$$(\neg A_1 \vee A_2)$$
$$(A_1 \vee \neg A_2)$$
$$(\neg A_1 \vee \neg A_2).$$

1.2.27. Lemma. There are 2^n n-clauses.

1.2.28. Definition. A formula α is in **conjunctive normal form**, abbreviated **cnf** if (for some n, k)

$$\alpha = \gamma_1 \wedge \cdots \wedge \gamma_k$$

where the γ_i are different n-clauses. To emphasize that all clauses in α are n-clauses, we sometimes say that α is in n-cnf.

1.2.29. Examples. The following formulas are in cnf.

$\mathtt{A}_1$	1-cnf, consisting of one 1-clause
$\mathtt{A}_1 \wedge \neg\mathtt{A}_1$	1-cnf, using two 1-clauses
$(\mathtt{A}_1 \vee \mathtt{A}_2) \wedge (\neg\mathtt{A}_1 \vee \neg\mathtt{A}_2) \wedge (\mathtt{A}_1 \wedge \neg\mathtt{A}_2)$	2-cnf, using three 2-clauses
$(\mathtt{A}_1 \vee \neg\mathtt{A}_2 \vee \mathtt{A}_3) \wedge (\neg\mathtt{A}_1 \vee \neg\mathtt{A}_2 \vee \mathtt{A}_3)$	3-cnf, using two 3-clauses.

There is a natural correspondence between truth assignments on $\mathtt{A}_1, \ldots, \mathtt{A}_n$ and n-clauses:

1.2.30. Definition. Let S be a function from $\{\mathtt{A}_1, \ldots, \mathtt{A}_n\}$ to $\{T, F\}$, i.e., a truth assignment. We define the n-clause δ_S as

$$\delta_S := \beta_1 \vee \cdots \vee \beta_n$$

where for $i = 1, \ldots, n$

$$\beta_i := \begin{cases} \mathtt{A}_i & \text{if } S(\mathtt{A}_i) = F \\ \neg\mathtt{A}_i & \text{if } S(\mathtt{A}_i) = T \end{cases}$$

1.2.31. Example. Let S be the function that maps $\mathtt{A}_1$ to T and $\mathtt{A}_2$ to F. Then δ_S is the formula $(\neg\mathtt{A}_1) \vee \mathtt{A}_2$.

1.2.32. Lemma. Let S be a truth assignment on $\{\mathtt{A}_1, \ldots, \mathtt{A}_n\}$. Then
(1) $\bar{S}(\delta_S) = F$.
(2) For any n-clause $\alpha \neq \delta_S$, $\bar{S}(\alpha) = T$.

Proof: Let $\alpha = \beta_1 \vee \ldots \vee \beta_n$. If $\alpha = \delta_S$, then for all i we have $\bar{S}(\beta_i) = F$, so $\bar{S}(\alpha) = F$. But if $\alpha \neq \delta_S$, then there must be some i for which $\bar{S}(\beta_i) = T$, so $\bar{S}(\alpha) = T$.

1.2.33. Theorem. For each formula α that is not a tautology, there is a formula α' in cnf such that $\alpha \Leftrightarrow \alpha'$.

Proof: Let $BKS(\alpha) \subseteq \{\mathtt{A}_1, \ldots, \mathtt{A}_n\}$. We will find a formula α' in n-cnf.

Since α is not a tautology, there are truth assignments S on $\{\mathtt{A}_1, \ldots, \mathtt{A}_n\}$ with $\bar{S}(\alpha) = F$. Let $S_1, \ldots, S_k$ list all truth assignments on $\{\mathtt{A}_1, \ldots, \mathtt{A}_n\}$ with that property.
Let

$$\alpha' := \delta_{S_1} \wedge \cdots \wedge \delta_{S_k}$$

By construction, α' is in n-cnf. We have to show that $\alpha' \Leftrightarrow \alpha$.
So let S be any truth assignment on $\{\mathtt{A}_1, \ldots, \mathtt{A}_n\}$.
Case 1: $\bar{S}(\alpha) = F$. So $S = S_i$, for some i. Since $\bar{S}_i(\delta_{S_i}) = F$, also $\bar{S}_i(\alpha') = F$, so in this case $\bar{S}(\alpha) = \bar{S}(\alpha')$.
Case 2: $\bar{S}(\alpha) = T$. So for all $i \leq k$ we have $S \neq S_i$, so $\delta_S \neq \delta_{S_i}$, so by 1.2.32 we have $\bar{S}(\delta_{S_i}) = T$. Hence also $\bar{S}_i(\alpha') = T$, so also in this case we have $\bar{S}(\alpha) = \bar{S}(\alpha')$.

As an application of conjunctive normal forms, we give a proof of the "interpolation theorem (for sentential logic)."

1.2.34. Theorem. Assume that

$$\beta \Rightarrow \gamma,$$

and $BKS(\beta) \cap BKS(\gamma) = \{\mathtt{A}_1, \ldots, \mathtt{A}_m\}$, $(m > 0)$. Then there is a formula α with $BKS(\alpha) = \{\mathtt{A}_1, \ldots, \mathtt{A}_m\}$ such that

$$\beta \Rightarrow \alpha$$

$$\text{and} \qquad \alpha \Rightarrow \gamma$$

(The formula α is said to "interpolate" between β and γ.)

Proof: Assume that

$$BKS(\beta) = \{\mathtt{B}_1, \ldots, \mathtt{B}_k, \mathtt{A}_1, \ldots, \mathtt{A}_m\} \quad BKS(\gamma) = \{\mathtt{C}_1, \ldots, \mathtt{C}_l, \mathtt{A}_1, \ldots, \mathtt{A}_m\}$$

Our proof strategy is as follows: We will construct α in cnf. So α will be a conjunction of certain m-clauses. Which m-clauses should we take? Since we want "$\beta \Rightarrow \alpha$," and we have $\alpha \Rightarrow \varepsilon$, whenever ε is a clause appearing in α, we may only take clauses ε which satisfy "$\beta \Rightarrow \varepsilon$." To ensure that we also get $\alpha \Rightarrow \gamma$, we will take all such clauses ε. But first we have to consider the possibility that there are no such clauses:

Case 1: Assume there are no m-clauses ε (i.e., clauses built from $\mathtt{A}_1, \ldots, \mathtt{A}_m$) such that $\beta \Rightarrow \varepsilon$.
In this case let $\alpha := \neg(\mathtt{A}_1 \wedge \neg\mathtt{A}_1)$. So α is a tautology. In particular, we have $\beta \Rightarrow \alpha$. It remains to show $\alpha \Rightarrow \gamma$. We will show that also γ is a tautology. So let S_c be a truth assignment defined on $\{\mathtt{A}_1, \ldots, \mathtt{A}_m, \mathtt{C}_1, \ldots, \mathtt{C}_l\}$. We have to show that $\bar{S}_c(\gamma) = T$.
Let $S_a := S_c\lceil\{\mathtt{A}_1, \ldots, \mathtt{A}_m\}$. By our assumption, $\beta \not\Rightarrow \delta_{S_a}$.
So there is a truth assignment S_b on $BKS(\beta) \cup BKS(\delta_{S_a})$, i.e., on the set $\{\mathtt{B}_1, \ldots, \mathtt{B}_k, \mathtt{A}_1, \ldots, \mathtt{A}_m\}$, with $\bar{S}_b(\beta) = T$ and $\bar{S}_b(\delta_{S_a}) = F$. So by 1.2.32, $S_b\lceil\{\mathtt{A}_1, \ldots, \mathtt{A}_m\} = S_a = S_c\lceil\{\mathtt{A}_1, \ldots, \mathtt{A}_m\}$. So S_b and S_c agree on their common domain, so there is a (complete) truth assignment S extending both of them. We have $\bar{S}(\beta) = \bar{S}_b(\beta) = T$.
As $\beta \Rightarrow \gamma$, $\bar{S}(\gamma) = T$. So $\bar{S}_c(\gamma) = T$ which was what we wanted to show.
Case 2: Otherwise, let $\alpha_1, \ldots, \alpha_n$ be all m-clauses ε such that $\beta \Rightarrow \varepsilon$.
Let

$$\alpha := \alpha_1 \wedge \cdots \wedge \alpha_n.$$

Note that again $\beta \Rightarrow \alpha$. So we have to show $\alpha \Rightarrow \gamma$.
So let S_c be a truth assignment defined on $\{\mathtt{A}_1, \ldots, \mathtt{A}_m, \mathtt{C}_1, \ldots, \mathtt{C}_l\}$, with $\bar{S}_c(\alpha) = T$. We have to show that also $\bar{S}_c(\gamma) = T$.
Our strategy will be to extend this truth assignment to an assignment S such that $\bar{S}(\beta) = T$ and then use the fact that $\beta \Rightarrow \gamma$.
Let $S_a := S_c\lceil\{\mathtt{A}_1, \ldots, \mathtt{A}_m\}$. We have $\bar{S}_a(\alpha) = T$, so for all i, $\bar{S}_a(\alpha_i) = T$. By 1.2.32, this means that each α_i is different from δ_{S_a}. So by our choice of the α_i, we have $\beta \not\Rightarrow \delta_{S_a}$.
So there is a truth assignment S_b on $BKS(\beta) \cup BKS(\delta_{S_a})$, i.e., on the set $\{\mathtt{B}_1, \ldots, \mathtt{B}_l, \mathtt{A}_1, \ldots, \mathtt{A}_m\}$, with $\bar{S}_b(\beta) = T$ and $\bar{S}_b(\delta_{S_a}) = F$. So by 1.2.32, $S_b\lceil\{\mathtt{A}_1, \ldots, \mathtt{A}_m\} = S_a = S_c\lceil\{\mathtt{A}_1, \ldots, \mathtt{A}_m\}$. So S_b and S_c agree on their common domain, so there is a (complete) truth assignment S extending both of them. We have $\bar{S}(\beta) = \bar{S}_b(\beta) = T$.
As $\beta \Rightarrow \gamma$, $\bar{S}(\gamma) = T$. So $\bar{S}_c(\gamma) = T$ which was what we wanted to show.

1.2.35. Remark. Of course the interpolation theorem will still be true if $BKS(\beta) \cap BKS(\gamma)$ is not of the form $\{\mathtt{A}_1, \ldots, \mathtt{A}_n\}$. In general we have:
If $\beta \Rightarrow \gamma$, and $BKS(\beta) \cap BKS(\gamma)$ is not empty, then there is a formula α whose blocks are included in $BKS(\beta) \cap BKS(\gamma)$, such that $\beta \Rightarrow \alpha$ and $\alpha \Rightarrow \gamma$.
If $\beta \Rightarrow \gamma$, and $BKS(\beta) \cap BKS(\gamma) = \emptyset$, then at least one of $\neg\beta$ and γ is a tautology. (See exercise 10)

From our point of view, any two expressions α and β which are equivalent have the same logical meaning. With this in mind, it is natural to look for a small set of expressions such that for any α in $\mathcal{L}$, there is an expression β in the set such that $\alpha \Leftrightarrow \beta$. Some characterization of such sets should be very useful. In order to do this, we need the following:

1.2.36. Definition. A set $C \subseteq \mathcal{L}$ is **complete** if for every α in $\mathcal{L}$ there is a β in C such that $\alpha \Leftrightarrow \beta$.

1.2.37. Definition. Let $I \subseteq \{\neg, \wedge, \vee, \rightarrow, \leftrightarrow, |\}$. We define $\mathcal{L}_I$ to be the set of all expressions built up from the blocks by using only connectives in I. Clearly, $\mathcal{L}_I \subseteq \mathcal{L}$ and $\mathcal{L}_I$ is an inductive structure.
We say that I is complete iff $\mathcal{L}_I$ is complete.

1.2.38. Fact. Let @ be a connective and I a set of connectives, and assume that the language $\mathcal{L}_{I\cup\{@\}}$ is complete. Then also $\mathcal{L}_I$ will be complete, provided that:
Whenever $\alpha_1, \alpha_2 \in \mathcal{L}_I$, then there is $\beta \in \mathcal{L}_I$ such that $\alpha_1 @ \alpha_2 \Leftrightarrow \beta$.
Similarly, if the language $\mathcal{L}_{I\cup\{\neg\}}$ is complete, and
Whenever $\alpha \in \mathcal{L}_I$, then there is $\beta \in \mathcal{L}_I$ such that $\neg\alpha \Leftrightarrow \beta$,
then $\mathcal{L}_I$ will also be complete.

Proof: Using the inductive structure of $\mathcal{L}_{I\cup\{@\}}$ we can show that for every formula α in $\mathcal{L}_{I\cup\{@\}}$ there is a formula $\bar{\alpha} \in \mathcal{L}_I$ such that $\alpha \Leftrightarrow \bar{\alpha}$:

All blocks of $\mathcal{L}_{I\cup\{@\}}$ are also in $\mathcal{L}_I$.

If $\alpha = \beta\Delta\gamma$ for some $\Delta \in I$, then we already have $\bar{\beta}$ and $\bar{\gamma}$ in $\mathcal{L}_I$ which are equivalent to β and γ respectively. Hence $\bar{\beta}\Delta\bar{\gamma} \Leftrightarrow \alpha$.

Finally, if $\alpha = \beta @ \gamma$, then $\alpha \Leftrightarrow \bar{\beta} @ \bar{\gamma}$. By our assumption, $\bar{\beta} @ \bar{\gamma}$ is equivalent to an element of $\mathcal{L}_I$.

1.2.39. Theorem. The following are complete sets:

(1) $\mathcal{L}_{\{\neg,\wedge,\vee\}}$.
(2) $\mathcal{L}_{\{\neg,\wedge\}}$.
(3) $\mathcal{L}_{\{\neg,\vee\}}$.

Proof of (1): For any formula α, if α is not a tautology, then cnf of α is in $\mathcal{L}_{\{\neg,\wedge,\vee\}}$.

If α is a tautology, then the cnf of α is not defined, but it is easy to see that $\alpha \Leftrightarrow (\neg(A_1 \wedge (\neg A_1)))$.

Proof of (2): Since $\alpha_1 \vee \alpha_2 \Leftrightarrow \neg(\neg\alpha_1 \wedge \neg\alpha_2)$, we can apply fact 1.2.38.

Proof of (3): Similar.

1.2.40. Lemma.

(i) $\{\neg\}$ is not complete.

(ii) $\{\wedge, \vee, \rightarrow, \leftrightarrow\}$ is not complete.

Proof: (i) It is enough to show that no expression in $\mathcal{L}_{\{\neg\}}$ is a tautology. For this, we will show by induction that for every α in $\mathcal{L}_{\{\neg\}}$ there are assignments S_1 and S_2 such that $\bar{S_1}(\alpha) = T$ and $\bar{S_2}(\alpha) = F$.

(1) If α is a sentential symbol, then the result is obvious.

(2) If $\alpha = (\neg\beta)$, then by induction let S_1 and S_2 be such that $\bar{S_1}(\beta) = T$ and $\bar{S_2}(\beta) = F$. The $\bar{S_1}(\alpha) = \bar{S_1}(\neg\beta) = F$, and $\bar{S_2}(\alpha) = \bar{S_2}(\neg\beta) = T$,
so the proposition holds in this case.

(ii) Exercise.

1.2.41. Theorem. The language $\mathcal{L}_{|}$ is complete.

Proof: We already know that the language $\mathcal{L}_{\neg,\wedge}$ and hence also the language $\mathcal{L}_{|,\neg,\wedge}$ is complete. Since we have

$$\alpha_1 \wedge \alpha_2 \Leftrightarrow \neg(\alpha_1 | \alpha_2)$$

we can conclude from 1.2.38 than also $\mathcal{L}_{|,\neg}$ is complete. Since

$$\neg\alpha \Leftrightarrow (\alpha | \alpha)$$

we can again use 1.2.38 to get that $\mathcal{L}_{|}$ is complete.

Exercises

Truth Assignments

1. Find how many different assignments defined on the blocks $A_1, \ldots, A_n$ satisfy the following set of sentential formulas

$$\{\neg A_1 \vee A_2, \neg A_2 \vee A_3, \ldots, \neg A_i \vee A_{i+1}, \ldots, \neg A_{n-1} \vee A_n\}.$$

Tautologies

2. Prove 1.2.5.

3. Prove that de Morgan's laws are tautologies:

$$(\neg(\alpha \vee \beta)) \leftrightarrow ((\neg\alpha) \wedge (\neg\beta))$$
$$(\neg(\alpha \wedge \beta)) \leftrightarrow ((\neg\alpha) \vee (\neg\beta)).$$

4. Give the truth table for the following formulas. Which of them are tautologies?

(1) $((\mathtt{A} \leftrightarrow (\mathtt{B} \wedge \mathtt{C})) \vee ((\neg\mathtt{B}) \leftrightarrow (\neg\mathtt{A})))$.

(2) $(((\mathtt{A}_1 \leftrightarrow \mathtt{A}_2) \leftrightarrow \mathtt{A}_2) \leftrightarrow \mathtt{A}_2)$.

(3) $(((\neg\mathtt{A}_1) \leftrightarrow \mathtt{A}_2) \wedge (\neg(\mathtt{A}_1 \vee (\neg\mathtt{A}_2))))$.

(4) $((\mathtt{A}_1 \leftrightarrow ((\mathtt{A}_2 \leftrightarrow \mathtt{A}_3) \wedge (\mathtt{A}_3 \leftrightarrow \mathtt{A}_2))) \leftrightarrow ((\mathtt{A}_1 \wedge (\mathtt{A}_2 \wedge \mathtt{A}_3)) \vee ((\neg\mathtt{A}_1) \wedge ((\neg\mathtt{A}_2) \wedge (\neg\mathtt{A}_3)))))$.

5. Which of the following are true for all α, β in $\mathcal{L}$?

(1) If α and β are tautologies, then $\alpha \wedge \beta$ is a tautology.

(2) If α or β is a tautology, then $\alpha \vee \beta$ is a tautology.

(3) If $\alpha \wedge \beta$ is a tautology, then α is a tautology and β is a tautology.

(4) If $\alpha \vee \beta$ is a tautology, then α is a tautology or β is a tautology.

6. Show that for any formulas α, β, γ, the following formulas are tautologies:

(1) $((\alpha \vee \beta) \vee \gamma)) \leftrightarrow (\alpha \vee (\beta \vee \gamma))$.

(2) $((\alpha \wedge \beta) \wedge \gamma) \leftrightarrow (\alpha \wedge (\beta \wedge \gamma))$.

(3) $((\alpha \wedge \beta) \vee \gamma) \leftrightarrow ((\alpha \vee \gamma) \wedge (\beta \vee \gamma))$.

(4) $(\alpha \leftrightarrow (\neg(\neg\alpha)))$.

(5) $(\neg(\alpha \vee \beta)) \leftrightarrow (\neg\alpha \wedge \neg\beta)$.

7. Let α be a sentential formula such that the only logical connective that appears in α is: $\leftrightarrow$. Show that if each block that appears in α appears an even number of times, then α is a tautology. Is the converse of this claim also correct? Explain why!

(Hint: First show $(\alpha \leftrightarrow \beta) \leftrightarrow \gamma \Leftrightarrow \alpha \leftrightarrow (\beta \leftrightarrow \gamma)$. Then show by induction that: if $\mathtt{A}_{i_1}, \mathtt{A}_{i_2}, \ldots, \mathtt{A}_{i_k}$ are those variables which appear an odd number of times in α, then $\alpha \Leftrightarrow ((\cdots(\mathtt{A}_{i_1} \leftrightarrow \mathtt{A}_{i_2}) \leftrightarrow \cdots) \leftrightarrow \mathtt{A}_{i_k})$.)

The Relation $\Rightarrow$

8. Given a set A, we say that a relation $\mathcal{R}$ is:
(•) *reflexive on A* if $a\mathcal{R}a$ for all a in A.
(•) *symmetric* if $a\mathcal{R}b$ implies $b\mathcal{R}a$ for all a and b in A.
(•) *transitive* if $a\mathcal{R}b$ and $b\mathcal{R}c$ imply $a\mathcal{R}c$ for all a, b, and c in A. A relation $\mathcal{R}$ is an *equivalence* relation if it is reflexive, symmetric, and transitive. Show that the relation $\alpha \Leftrightarrow \beta$ (definition 1.2.6) is in fact an equivalence relation on $\mathcal{L}$.

9. Prove lemma 1.2.8: (1) If $\alpha \Rightarrow \beta$ and $\beta \Rightarrow \gamma$, then $\alpha \Rightarrow \gamma$.
(2) If α is a tautology, and $\alpha \Rightarrow \beta$, then also β is a tautology.

10. Show that if $\beta \Rightarrow \gamma$, and $BKS(\beta) \cap BKS(\gamma) = \emptyset$, then either γ or $(\neg\beta)$ (or both) must be a tautology.

11. Show that $\alpha \wedge \beta \Rightarrow \gamma$ iff $\alpha \Rightarrow (\beta \rightarrow \gamma)$.

12. (a) Show that $\alpha \wedge (\alpha \rightarrow \beta) \Leftrightarrow \alpha \wedge \beta$.
(b) Show that $\alpha \wedge \gamma \Rightarrow \beta$ iff $\gamma \Rightarrow (\alpha \rightarrow \beta)$.

13. Show that the following are equivalent:
(a) α is a tautology.
(b) for all β we have $\beta \Rightarrow \alpha$.
(c) for all β we have $\neg\alpha \Rightarrow \beta$.

14. Prove the following statements for all α, β, or disprove (i.e., find counterexamples).
(a) If $\Gamma \Rightarrow \alpha$ or $\Gamma \Rightarrow \beta$ then $\Gamma \Rightarrow (\alpha \vee \beta)$.
(b) If $\Gamma \Rightarrow \alpha$ and $\Gamma \Rightarrow \beta$ then $\Gamma \Rightarrow (\alpha \wedge \beta)$.
(c) If $\Gamma \Rightarrow (\alpha \vee \beta)$ then $\Gamma \Rightarrow \alpha$ or $\Gamma \Rightarrow \beta$.
(d) If $\Gamma \Rightarrow (\alpha \wedge \beta)$ then $\Gamma \Rightarrow \alpha$ and $\Gamma \Rightarrow \beta$.

Normal Forms

15. Prove lemma 1.2.24:
(1) Let $\varphi = \varphi_1 \wedge \cdots \wedge \varphi_n$, and let S be a truth assignment on $BKS(\varphi)$. Then $\bar{S}(\varphi) = T$ iff for ALL i in $\{1, \ldots, n\}$, we have $\bar{S}(\varphi_i) = T$.
(2) Let $\varphi = \varphi_1 \vee \cdots \vee \varphi_n$, and let S be a truth assignment on $BKS(\varphi)$. Then $\bar{S}(\varphi) = T$ iff for SOME i in $\{1, \ldots, n\}$ we have $\bar{S}(\varphi_i) = T$.

16. Prove that if α is composed of blocks and $\wedge$ only, then $\bar{S}(\alpha) = T$ iff for every block A_i in α, $\bar{S}(\mathsf{A}_i) = T$.

17. Formulate and prove a similar claim for the logical connective $\vee$.

18. Find an inductive structure with set of blocks $= \{A_1, \neg A_1\}$ and two unary operators that generates the set of all clauses.

19. Call a formula α a **dual** n**-clause**, if α is of the form $\beta_1 \wedge \ldots \wedge \beta_n$, where each β_i is either A_i or $(\neg A_i)$. A formula is in (n-)disjunctive normal form (dnf) iff it is of the form $\gamma_1 \vee \cdots \vee \gamma_k$, where each γ_j is a dual n-clause. Show that for each formula α: Either $\neg\alpha$ is a tautology, or there is a formula $\bar{\alpha}$ in disjunctive normal form with $\alpha \Leftrightarrow \bar{\alpha}$.

20. Find cnf and dnf (see previous exercise) of the following formulas:

$$A_1 \to A_2$$
$$A_1 \wedge A_2 \wedge A_3$$
$$A_1 \vee A_2 \vee A_3$$
$$A_1 \to (A_2 \to A_1).$$

21. Use fact 1.2.38 to show (directly, i.e., without using cnf or dnf) that the following sets are complete:

(1) $\{\neg, \vee, \to, \wedge, \leftrightarrow\}$.
(2) $\{\neg, \vee, \to, \wedge\}$.
(3) $\{\neg, \vee, \to\}$.
(4) $\{\neg, \to\}$.
(5) $\{\neg, \vee\}$.

Complete Languages

22. Finish the proof of theorem 1.2.39. That is, show that $\mathcal{L}_{\vee,\neg}$ and $\mathcal{L}_{\to,\neg}$ are complete languages.

23. Prove lemma 1.2.40(ii): $\{\wedge, \vee, \to, \leftrightarrow\}$ is not complete. (Hint: Consider the truth assignment S that assigns T to every block.)

24. Finish the proof of lemma 1.2.41.

Other Languages

25. Consider a modified language $\mathcal{L}'$ of sentential logic, where we have an additional connective ∇, with the interpretation $\alpha \nabla \beta =$ "neither α nor β."

(a) Give an inductive definition of $\mathcal{L}'$, and modify definition 1.2.2 appropriately.

(b) Define $\mathcal{L}_{\nabla}$ analogously to definition 1.2.37, and show that $\mathcal{L}_{\nabla}$ is complete.

26. Consider a modified language $\mathcal{L}''$ of sentential logic, where we have an additional connective $+$, with the interpretation $\alpha + \beta =$ "α or β, but not both." (This is sometimes called "exclusive or," as opposed to the "inclusive" $\vee$.) Also add two special blocks 0 and 1, and demand that for all truth assignments S we have $\bar{S}(0) = F$ and $\bar{S}(1) = T$.

(a) Give an inductive definition of $\mathcal{L}''$, and modify definition 1.2.2 appropriately.

(b) Show that for all α, β, γ in $\mathcal{L}''$:

(1) $\alpha \wedge \beta \Leftrightarrow \beta \wedge \alpha$.

(2) $\alpha + \beta \Leftrightarrow \beta + \alpha$.

(3) $\alpha \wedge (\beta \wedge \gamma) \Leftrightarrow (\alpha \wedge \beta) \wedge \gamma$.

(4) $\alpha + (\beta + \gamma) \Leftrightarrow (\alpha + \beta) + \gamma$.

(5) $\alpha \wedge (\beta + \gamma) \Leftrightarrow (\alpha \wedge \beta) + (\alpha \wedge \gamma)$.

(6) $\alpha + 0 \Leftrightarrow \alpha$.

(7) $\alpha \wedge 0 \Leftrightarrow 0$.

(8) $\alpha + 1 \Leftrightarrow \neg\alpha$.

(9) $\alpha \wedge 1 \Leftrightarrow \alpha$.

(c) Define $\mathcal{L}''_{1,+,\wedge}$ and show that it is a complete language.

1.3. First Order Logic

In this section we will develop a language which we can find implicit in all branches of mathematics. That means that we can use variations of this language to express facts about a group, or about sets. The difference will be the adequate parameters for, e.g., expressing the "multiplication" operation implicit in the concept of groups and the concept of membership implicit in set theory. In other words, we will provide an abstract study of a language that in one particular case is the language of groups, and in another case is the language of set theory, etc.

Mathematics talks about properties of objects in some underlying universe. A typical universe is, for instance, the set of natural numbers. The study of this universe is the branch of mathematics called number theory. Simple theorems in number theory are, for example

3 is less than 5.
7 is prime.

The first example states a binary relation between 3 and 5, namely the relation "is less than." The second example states a property about the number 7. This property also defines a relation, but this relation is unary.

Another example is the set of all points in the plane. In this universe we can consider the relations "P_1, P_2 and P_3 lie on one line," or "P_1, P_2 and P_3 form a triangle with a right angle at P_2." Another (4-place) relation is the following relation between four points: "P_1, P_2, P_3 and P_4 are the corners of a square."

This leads to the following

1.3.1. Definition. Let A be a set. Let n be a natural number. An **n-ary relation on A** is a set $R \subseteq A^n$, where $A^1 = A$, $A^2 = A \times A = \{(a,b) : a,b \in A\}$, $A^3 = (A \times A) \times A$, etc.

(Instead of "1-ary" we say "unary." "binary" is "2-ary," and "ternary" is "3-ary").

1.3.2. Example.

(a) $\{(n,m) : n \text{ is less than } m\}$ is a binary (=two-place) relation on $\mathbb{N}$.
(b) $\{(m,n,m+n) : m,n \in \mathbb{N}\}$ is a ternary (3-place) relation on $\mathbb{N}$.
(c) $\{n : n \text{ is prime}\}$ is a unary relation on $\mathbb{N}$.
(d) $\{n : n \text{ is a natural number}\}$ is a unary relation on $\mathbb{N}$ (a trivial one).
(e) $\{(p,q,r,s) : p,q,r,s \text{ are the corners of a square in the plane}\}$ is a 4-place relation on the sets of points in the plane.

Therefore, our language will have symbols for n-ary relations. We will also have to consider functions acting on the objects of the universe. When the universe is the set $\mathbb{N}$ of natural numbers, a typical example of a function is "+." This is a binary function like the multiplication operator in a group. This leads to the following

1.3.3. Definition. Let A be a set. Let n be a natural number. We say that F is an **n-ary function on A** if $F \subseteq A^{n+1}$ and satisfies

for every $a_1, \ldots, a_n$ in A, there is a unique b in A such that $(a_1, \ldots a_n, b) \in F$. In this case, we write $Fa_1 \ldots a_n = b$ or $F(a_1, \ldots, a_n) = b$.

For example, the set $\{(n, 2n) : n \in \mathbb{N}\} = \{(0,0), (1,2), (2,4), \dots\}$ is both a binary relation and a unary function.

In some universes there are distinguished objects. In number theory, for instance, "0" is a special object. For such objects we have special symbols in our language, called **constant symbols**, or sometimes **constants**.

Sometimes we want to establish a statement about several objects of the universe, but we don't want to talk about a *specific* object. For this purpose we use **variables**. A typical use of a variable is when we say

"Let n be a natural number. n can be even or odd ..."

Here, "n" is the variable. So we will include infinitely many variables in our language.

Finally, we want to make statements claiming that objects with a certain property exist in our universe, or that all objects of our universe have a certain property. A claim that a certain object exists is called an existential quantification, and a claim that all objects satisfy some property is called a universal quantification. An example using both types of quantification is

"**For every** natural number n, **there is** a natural number p such that p is a prime bigger than n."

Here, "for every" is a universal quantifier, and "there is" is an existential quantifier.

Now we are ready to define the symbols used in a first order language.

1.3.4. Definition. The following are the symbols (the "alphabet") used in a first order language:

(i) **Logical connectives:** $\neg$, $\wedge$, $\vee$, $\rightarrow$, $\leftrightarrow$ $|$.

(ii) **Variables:** $\mathtt{v}_1, \mathtt{v}_2, \mathtt{v}_3, \dots$. Usually, we will denote a variable with a letter from the end of the alphabet: $\mathtt{x}, \mathtt{y}, \mathtt{z}$. ($\mathtt{x}$ and $\mathtt{y}$ may denote the same variable $\mathtt{v}_n$).

(iii) **The equality symbol** $=$.

(iv) **Quantifiers:** $\forall$, $\exists$ ($\forall$ stands for "for every" and $\exists$ stands for "there is")

(v) **Constant symbols:** a set $\mathcal{C}$ of constants. Usually we will denote an element of $\mathcal{C}$ by $\mathtt{c}$.

(vi) **Function symbols:** a set $\mathcal{F}$ of function symbols. Usually, we will denote an element of $\mathcal{F}$ by $\mathtt{F}$. With each function symbol we associate a positive natural number, its "arity." For example, we say: "F is a 2-ary function symbol."

(vii) **Relation symbols:** a set $\mathcal{R}$ of relation symbols. Usually, we will denote an element of $\mathcal{R}$ by R. Again we associate an "arity" with each relation symbol.

1.3.5. Remark. Later we will consider a restricted language by only allowing the logical connectives $\neg$ and $\wedge$, and only the $\forall$ quantifier.

1.3.6. Remark. The symbols in 1.3.4(i), (ii), (iii) and (iv) are called **logical symbols.** They are part of the symbols of every first order language. 1.3.4(v), (vi), and (vii) are called **nonlogical symbols.** Which particular symbols we use depends on the specific branch of mathematics we are investigating.

1.3.7. Examples. We will give a few examples of first order languages. For each language we also include an example of a formula in that language, and we "translate" the formula into the usual mathematical language. A formal definition of what a formula "is" and what it "means" will be given later.

(i) The language of number theory:

Constant Symbols only "0"

Function Symbols S (successor function), "+" "$\cdot$", "$\uparrow$" (for exponentiation).

Relation Symbols "$<$"

We usually write $a < b$ instead of $<ab$, $a + b$ instead of $+ab$. Officially we use the symbol $\uparrow$ for exponentiation. Thus, "x to the yth power" would be written as $\uparrow(x, y)$. But in practice we almost always write x^y instead, conforming to common usage.

The formula $\exists x\, S(S(0)) \cdot x = y$ means: "y is an even number."

The formula $\exists y (S(S(0)))^y = x$ means: "x is a power of 2."

(ii) The poor man's language of number theory:

Constant Symbols no constant symbols.

Function Symbols "+", "$\cdot$"

Relation Symbols no relation symbols.

The formula $\exists x\, x + x = y$ means: "y is an even number."

(iii) The language of groups:

Constant Symbols "0"

Function Symbols one binary function symbol "+"

Relation Symbols no relation symbols.

The formula $(\forall \mathtt{x}\, (\mathtt{x} + \mathtt{x}) + \mathtt{x} = 0)$ means that every element of the group (i.e., the group this formula talks about) has order 3.

(Sometimes we write $\cdot$ for the binary operation and 1 instead of 0 for the constant symbol.)

(iv) The language of set theory:

Relation Symbols "$\in$"

Function Symbols no function symbols.

The formula $\exists \mathtt{y}\, \mathtt{y} \in \mathtt{x}$ says: "$\mathtt{x}$ is not the empty set (i.e., has some element)."

(v) The language of the real numbers:

Relation Symbols "$<$", "$\leq$"

Function Symbols "+", "−", "·", "/", "↑", "abs", "sin", "cos", "tan", "exp", ...

Constant Symbols "0", "1"

abs is the absolute value. We usually write $|\mathtt{x}|$ instead of abs($\mathtt{x}$). exp is the exponential function, $\exp(\mathtt{x}) = e^{\mathtt{x}}$.

The formula

$$\forall \mathtt{h} \Big[0 < \mathtt{h} \to \exists \mathtt{k} \big(0 < \mathtt{k} \wedge \forall \mathtt{z} : (|\mathtt{z}| < \mathtt{k} \to |\, ((\sin \mathtt{z})/\mathtt{z}) - 1| < \mathtt{h}) \big) \Big]$$

is usually written as

$$\lim_{x \to 0} \frac{\sin(x)}{x} = 1.$$

(vi) The (first order) language of calculus: The same as the language of the real numbers, but we need a few "generic" function symbols $\mathtt{f}, \mathtt{g}, \ldots$

For example, the formula

$$\forall \mathtt{a} \forall \mathtt{h} \Big[0 < \mathtt{h} \to \exists \mathtt{k} \big(0 < \mathtt{k} \wedge \forall \mathtt{x} (|\mathtt{x} - \mathtt{a}| < \mathtt{k} \to |\mathtt{f}(\mathtt{x}) - \mathtt{f}(\mathtt{a})| < \mathtt{h} \big) \Big]$$

says that $\mathtt{f}$ is continuous at every point.

(Remark: Usually calculus is developed using a more powerful language, but many concepts can be expressed in this first order language.)

As in the previous section, we will define the language in such a way that it will be an inductive structure. First we must define the blocks, which themselves will be given by an inductive process.

1.3.8. Definition (inductive). We define the concept of **term**. We say that τ is a **term** if one of the following conditions holds:

(i) τ is a constant

(ii) τ is a variable

(iii) there is an n-ary function symbol $\mathtt{F}$ and terms $\tau_1, \ldots, \tau_n$ such that $\tau = \mathtt{F}\tau_1 \ldots \tau_n$.

The set of terms is the inductive structure:

The blocks are the constant symbols and the variables.
The operators are the functions that map $\tau_1, \ldots, \tau_n$ to $\mathtt{F}\tau_1 \ldots \tau_n$. I.e., for each n-ary function symbol $\mathtt{F}$ we have an operator $K_{\mathtt{F}}$, defined by:
$K_{\mathtt{F}}(\tau_1, \ldots, \tau_n) = \mathtt{F}\tau_1 \cdots \tau_n$.

Constant symbols denote specific objects in the models we consider, and variables denote generic objects. Functions map objects to objects (if they are unary) or pairs of objects to objects (if they are binary), etc.

Thus, a term is the name of an object. For example, if $\mathbb{N}$ is the model of natural numbers, then the term $\mathtt{0}$ denotes the natural number 0. $\mathtt{S(0)}$ denotes the object obtained from the number 0 by applying the successor operation, i.e., the number 1.

We sometimes write a term $\mathtt{f}\tau_1 \cdots \tau_n$ as $\mathtt{f}(\tau_1, \ldots, \tau_n)$. E.g., if $\mathtt{f}$ is a binary function symbol, $\mathtt{x}$ and $\mathtt{y}$ are variables, we may write $\mathtt{f(x, f(y, y))}$ instead of $\mathtt{fxfyy}$. Similarly, we may write an atomic formula $\mathtt{R}\tau_1 \cdots \tau_n$ as $\mathtt{R}(\tau_1, \ldots, \tau_n)$.

In the language of number theory, $\mathtt{S(x)}$ is a term denoting the successor of the number denoted by $\mathtt{x}$. ($\mathtt{x}$ is a variable, so it does not correspond to any fixed object. But if we decide to assign the value 4 to $\mathtt{x}$, then $\mathtt{S(x)}$ must be assigned the value 5.)

1.3.9. Definition. We define the concept of **atomic formula**. We say that φ is an atomic formula if there is an n-ary relation symbol $\mathtt{R}$, and there are terms $\tau_1, \ldots, \tau_n$ such that φ is $\mathtt{R}\tau_1 \ldots \tau_n$, or if there are terms τ_1, τ_2 such that φ is $\tau_1 = \tau_2$.

Thus, atomic formulas correspond to the most basic statements one can make about a model. For example, if we work with the language of number theory, "$\mathtt{0} = \mathtt{0}$" is an atomic formula, $\mathtt{S(0)} < \mathtt{0}$ is an atomic formula, and $\mathtt{y} + \mathtt{0} = \mathtt{x}$ is an atomic formula.

We will construct general formulas from atomic formulas.

1.3.10. Definition (inductive). We define the concept of **formula**. We say that φ is a **formula** if one of the following conditions holds:

(i) φ is an atomic formula

(ii) there are formulas ψ and χ such that

φ is $(\neg\psi)$ or
φ is $(\psi \wedge \chi)$ or
φ is $(\psi \vee \chi)$ or
φ is $(\psi \rightarrow \chi)$ or
φ is $(\psi \leftrightarrow \chi)$ or
φ is $(\psi | \chi)$.

(iii) there is a formula ψ and a variable $\mathbf{x}$ such that φ is $(\forall \mathbf{x}\psi)$ or $(\exists \mathbf{x}\psi)$.

For example, again using the language of number theory, the following are formulas:

$$0 = 0$$
$$(0 = 0 \vee (y + 0) = x)$$
$$(\exists x \; y = (x + x)).$$

(We will omit parentheses if we trust the reader to insert them into the right places.)

The first order language $\mathcal{L}$ is the set of all formulas. Again, we need to start proving basic theorems about this language. The next goal is to show unique readability for $\mathcal{L}$. Since the proofs are very similar to the proofs following definition 1.1.25, we omit them.

1.3.11. Lemma. No term is an initial segment of a term.

Proof: Exercise

1.3.12. Lemma. For every term τ exactly one of the following holds:

(i) τ is a constant

(ii) τ is a variable

(iii) there are unique F and terms $\tau_1, \ldots, \tau_n$ such that $\tau = F\tau_1 \ldots \tau_n$.

1.3.13. Lemma. Let φ be an atomic formula. Then exactly one of the following holds:

(1) There exist unique $\mathtt{R}$ and $\tau_1, \ldots, \tau_n$ such that $\mathtt{R}$ is an n-ary relation symbol and $\tau_1, \ldots, \tau_n$ are terms and φ is $\mathtt{R}\tau_1 \ldots \tau_n$.

(2) There are unique terms τ_1, τ_2 such that φ is $\tau_1 = \tau_2$.

1.3.14. Theorem. (unique readability of $\mathcal{L}$) For every formula φ exactly one of the following holds:

(i) φ is atomic

(ii) there is a unique ψ such that φ is $(\neg\psi)$

(iii) there are unique ψ and χ and @ in $\{\wedge, \vee, \rightarrow, \leftrightarrow, |\}$ such that φ is $(\psi @ \chi)$

(iv) there is unique ψ such that φ is $\forall \mathtt{x}\psi$

(v) there is unique ψ such that φ is $\exists \mathtt{x}\psi$

Now that the definitions of "term" and "formula" have been given, the reader is advised to review again the examples in 1.3.7.

In order to give a semantics for a language $\mathcal{L}$, we have to fix a set called the universe. After this, we should interpret the constant symbols, the function symbols, and the relation symbols. The rest of the symbols will have a canonical interpretation.

1.3.15. Definition. Let $\mathcal{L}$ be a first order language. A **model** (or **structure**) M for $\mathcal{L}$ consists of the following:

(i) a set M, not empty, called the universe (sometimes written $|\mathcal{M}|$). (In general, if we use a calligraphic letter such as $\mathcal{A}$, $\mathcal{B}$, $\mathcal{M}$, $\mathcal{N}$, $\mathcal{X}$, etc., to denote a model, we use the corresponding italic letter A, B, M, N, X, etc. to denote the corresponding universe.)

(ii) For each relation symbol $\mathtt{R}$ in $\mathcal{R}$, a relation $\mathtt{R}^{\mathcal{M}}$ on M such that if $\mathtt{R}$ has arity k, then $\mathtt{R}^{\mathcal{M}} \subseteq M^k$. Thus, binary relation symbols are interpreted as binary relations, etc.

(iii) For each $\mathtt{F}$ in $\mathcal{F}$, a function $\mathtt{F}^{\mathcal{M}}$ with range in M such that if $\mathtt{F}$ has arity k, then $\mathtt{F}^{\mathcal{M}}$ is a function with domain M^k.

(iv) For each $\mathtt{c}$ in $\mathcal{C}$ an object $\mathtt{c}^{\mathcal{M}}$ in M.

For a specific language $\mathcal{L}$, with relation symbols $\mathtt{R}_1, \ldots$, function symbols $\mathtt{F}_1, \ldots$, and constant symbols $\mathtt{c}_1, \ldots$, we write, for a model $\mathcal{M}$,

$$\mathcal{M} = \langle M, \mathtt{R}_1^{\mathcal{M}}, \ldots, \mathtt{F}_1^{\mathcal{M}}, \ldots, \mathtt{c}_1^{\mathcal{M}}, \ldots \rangle.$$

For example, a group with underlying set X, group operation $*$, and neutral element ϕ will be written as $\langle X, *, \phi\rangle$.

1.3.16. Remark. The intention of definition 1.3.15 is the following:

(i) will give the universe.

(ii) will give, for each $\mathtt{R} \in \mathcal{R}$, an interpretation, namely, the relation $\mathtt{R}^{\mathcal{M}}$.

(iii) will give, for each $\mathtt{F} \in \mathcal{F}$, an interpretation, namely, the function $\mathtt{F}^{\mathcal{M}}$.

(iv) will give, for each $\mathtt{c} \in \mathcal{C}$, an interpretation, namely, the constant $\mathtt{c}^{\mathcal{M}}$.

Variables will be interpreted as generic objects of the universe. Logical connectives are interpreted as in section 2 and $\forall \mathtt{x}$ ... is interpreted as "every object in M ... " $\exists \mathtt{x}$ ... is interpreted as "there is an object in M ..."

1.3.17. Example.

(i) $\langle \mathbb{N}, +^{\mathbb{N}}, \cdot^{\mathbb{N}}, \rangle$ is a model for the poor man's language of number theory given in 1.3.7 where

$+^{\mathbb{N}} = \{\langle n, m, l\rangle : \langle n, m, l\rangle \text{ is in } \mathbb{N} \times \mathbb{N} \times \mathbb{N} \text{ and } n + m = l\}$

$\cdot^{\mathbb{N}} = \{\langle n, m, l\rangle : \langle n, m, l\rangle \text{ is in } \mathbb{N} \times \mathbb{N} \times \mathbb{N} \text{ and } n \cdot m = l\}$.

(ii) $\mathcal{Z} = \langle \mathcal{Z}, \cdot^{\mathcal{Z}}, 1^{\mathcal{Z}}\rangle$ is a model for the language of groups, where

$\cdot^{\mathcal{Z}} = \{\langle x, y, z\rangle \in \mathcal{Z} \times \mathcal{Z} \times \mathcal{Z} : x + y = z\}$

$1^{\mathcal{Z}}$ is the number 0.

Assume that $\mathcal{M}$ is a model for the language $\mathcal{L}$. Let φ be a formula in the language $\mathcal{L}$. We want to define the concept "φ is true in $\mathcal{M}$." For instance, let φ be the formula $\exists \mathtt{x}\mathtt{R}(\mathtt{x})$. If we want this formula to be "true" in $\mathcal{M}$, our intention is that, in fact, there is an element a in $\mathtt{R}^{\mathcal{M}}$.

We want to define truth by induction, so it seems that first we have to decide what the truth value of $\mathtt{R}(\mathtt{x})$ is. But $\mathtt{x}$ is only a variable and has no intrinsic connection to $\mathcal{M}$. To overcome this difficulty, we will generalize the definition of "terms" and "formulas" to "$\mathcal{M}$-terms" and "$\mathcal{M}$-formulas." The inductive definition is as follows:

1.3.18. Definition. Let $\mathcal{L}$ be a language, $\mathcal{M}$ a model for $\mathcal{L}$. We define the concept of **$\mathcal{M}$-term**. We say that τ is an **$\mathcal{M}$-term** if one of the following conditions hold:

(i) τ is a constant in the language $\mathcal{L}$.

(ii) τ is a variable.
(iii) there is an n-ary function symbol F in $\mathcal{L}$ and $\mathcal{M}$-terms $\tau_1, \ldots, \tau_n$ such that $\tau = F\tau_1 \ldots \tau_n$.
(iv) τ is an element of M.

Thus, $\mathcal{M}$-terms are defined like terms, but we also allow elements of the universe of $\mathcal{M}$ to be used as blocks.

To avoid ambiguities, always assume that no element in our language is also an element of any model that we consider.

$\mathcal{M}$-formulas are defined using the same definition as (plain) formulas, except that we also allow the expressions $\tau_1 = \tau_2$ (and $\mathtt{R}\tau_1 \cdots \tau_n$) as atomic $\mathcal{M}$-formulas whenever $\tau_1, \ldots, \tau_n$ are are $\mathcal{M}$-terms (and R is an n-ary relation symbol).

We say that an $\mathcal{M}$-term is **closed**, if it does not contain any variables.

To any closed $\mathcal{M}$-term τ we can associate a **value** $\tau^{\mathcal{M}}$, which will be an element of M. $\tau^{\mathcal{M}}$ is also called the interpretation of τ in $\mathcal{M}$.

1.3.19. Definition.

If τ is a constant symbol c, $\tau^{\mathcal{M}}$ is the element $\mathtt{c}^{\mathcal{M}}$.
If τ is an element of M, $\tau^{\mathcal{M}} := \tau$.
If τ is the term $\mathtt{F}(\tau_1, \ldots, \tau_n)$, then $\tau^{\mathcal{M}} := \mathtt{F}^{\mathcal{M}}(\tau_1^{\mathcal{M}}, \ldots, \tau_n^{\mathcal{M}})$, i.e., we compute $\tau^{\mathcal{M}}$ by first computing each $\tau_i^{\mathcal{M}}$, and then applying the function $\mathtt{F}^{\mathcal{M}}$.

1.3.20. Example. We will use the language of number theory (see 1.3.7) We let $\mathbb{N}$ be the model of natural numbers. The following are examples of $\mathbb{N}$-terms:

$$\mathtt{x}$$
$$\mathtt{S}(\mathtt{x} + \mathtt{0})$$
$$\mathtt{S}(3) + (2 \cdot 2).$$

The third term is a closed term. Its value (under interpretation in the natural numbers $\mathbb{N}$) is 8.
Note that there is a difference between the natural number 0 (which is an $\mathbb{N}$-term) and the constant symbol o (which is also an $\mathbb{N}$-term, and even a term). Both $\mathbb{N}$-terms have the same interpretation in the natural numbers: $\mathtt{0}^{\mathbb{N}} = 0 = 0^{\mathbb{N}}$.

Now we introduce the concept of **substitution**. If we have a term μ in which a variable x appears (for example, $\mathtt{S}(\mathtt{x})$), and we have another term

τ (for example, $0+0$), then we can construct the term "μ, with x replaced by τ" (in our example, $S(0+0)$).

If we interpret $S(x)$ as a general procedure for getting to a certain object $S(x)$ from a given object x in our model, then $S(0+0)$ is what we get if we start at $0+0$ instead of x.

1.3.21. Definition. Let $\mathcal{M}$ be a model.

(a) Let μ be an $\mathcal{M}$-term, x a variable, and τ an $\mathcal{M}$-term. We define $\mu(x/\tau)$ by induction on μ.

(i) If μ is a constant symbol c, or an element of $\mathcal{M}$, then $\mu(x/\tau)=\mu$. (We substitute only for variables.)

(ii) If μ is a variable y, then

$$\mu(x/\tau)=\begin{cases}\mu & \text{if } x\neq y\\ \tau & \text{if } x=y.\end{cases}$$

(iii) If $\mu = F\mu_1\ldots\mu_n$, then $\mu(x/\tau) = F\mu_1(x/\tau)\ldots\mu_n(x/\tau)$.

In other words, all occurrences of x are replaced by the term τ.

(b) Let φ be an $\mathcal{M}$-formula, x a variable, and τ an $\mathcal{M}$-term. We define $\varphi(x/\tau)$ by induction on φ.

(i) If $\varphi = R\mu_1\ldots\mu_n$, then $\varphi(x/\tau) = R\mu_1(x/\tau)\ldots\mu_n(x/\tau)$.

(ii) If $\varphi=(\neg\psi)$, then $\varphi(x/\tau)=(\neg\psi(x/\tau))$.

(iii) If $\varphi = (\psi\wedge\theta)$, then $\varphi(x/\tau) = (\psi(x/\tau)\wedge\theta(x/\tau))$. Similarly for the other connectives.

(iv) If $\varphi=\forall y\psi$ or $\exists y\psi$, where y is a variable different from x, then $\varphi(x/\tau)=\forall y\psi(x/\tau)$ (or $\exists y\psi(x/\tau)$). (We substitute only for free variables.)

(v) If $\varphi=\forall x\psi$ or $\exists x\psi$, then $\varphi(x/\tau)=\varphi$.

1.3.22. Example.

(1) If μ is the term $S(x_0)+(x_0\cdot x_1)$, and τ is the term x_0+x_2, then $\mu(x_0/\tau)$ is

$$S(x_0+x_2)+((x_0+x_2)\cdot x_1)$$

(2) If φ is the formula $\exists y(x+y = z)$, where x, y, z are distinct variables, and τ is the term $S(x)$, then $\varphi(x/\tau)$ is the formula

$$\exists y(S(x) + y = z)$$

and $\varphi(y/\tau)$ is the formula φ itself.

1.3.23. Lemma. If τ is an $\mathcal{M}$-term, then there are variables $y_1, \ldots, y_n$ and elements $m_1, \ldots, m_n$ of M, and there is term τ' such that $\tau = \tau'(y_1/m_1, \ldots, y_n/m_n)$.
A similar fact is true for $\mathcal{M}$-formulas.

Proof: By induction.

1.3.24. Definition. For every term $\tau \in \mathcal{L}$, we let $\mathrm{Var}(\tau)$ be the set of variables occurring in τ. That is, we define by induction
$\mathrm{Var}(c) = \emptyset$ for every constant c in $\mathcal{L}$
$\mathrm{Var}(x) = \{x\}$ for every variable x in $\mathcal{L}$
$\mathrm{Var}(F\tau_1 \ldots \tau_n) = \mathrm{Var}(\tau_1) \cup \cdots \cup \mathrm{Var}(\tau_n)$, whenever F is a function symbol in $\mathcal{L}$ and $\tau_1, \ldots, \tau_n$ are terms.

1.3.25. Definition. For any formula φ we define the sets $\mathrm{Var}(\varphi)$ (the set of variables occurring in φ), $\mathrm{Bound}(\varphi)$ (the set of bound variables occurring in φ) and $\mathrm{Free}(\varphi)$ (the set of free variables occurring in φ).

(i) If φ is atomic, say φ is $\tau_1 = \tau_2$ or $R\tau_1 \ldots \tau_n$, then $\mathrm{Bound}(\varphi) = \emptyset$ and $\mathrm{Var}(\varphi) = \mathrm{Free}(\varphi) = \mathrm{Var}(\tau_1) \cup \mathrm{Var}(\tau_2)$ or $\mathrm{Var}(\tau_1) \cup \cdots \cup \mathrm{Var}(\tau_n)$, respectively.

(ii) If φ is $(\neg\psi)$, then $\mathrm{Var}(\varphi) = \mathrm{Var}(\psi)$, $\mathrm{Bound}(\varphi) = \mathrm{Bound}(\psi)$, and $\mathrm{Free}(\varphi) = \mathrm{Free}(\psi)$.

(iii) If φ is $(\varphi_1 @ \varphi_2)$ (with $@ \in \{\wedge, \vee, \rightarrow, \leftrightarrow, |\}$), then $\mathrm{Var}(\varphi) = \mathrm{Var}(\varphi_1) \cup \mathrm{Var}(\varphi_2)$, $\mathrm{Bound}(\varphi) = \mathrm{Bound}(\varphi_1) \cup \mathrm{Bound}(\varphi_2)$, and $\mathrm{Free}(\varphi) = \mathrm{Free}(\varphi_1) \cup \mathrm{Free}(\varphi_2)$.

(iv) If φ is $\forall x\psi$ or $\exists x\psi$, then $\mathrm{Var}(\varphi) = \mathrm{Var}(\psi) \cup \{x\}$, $\mathrm{Bound}(\varphi) = \mathrm{Bound}(\psi) \cup \{x\}$, and $\mathrm{Free}(\varphi) = \mathrm{Free}(\psi) - \{x\}$. (Note that x thus changes from a free variable to a bound variable. We say that $\forall$ "binds" x.)

1.3.26. Example. If φ is the formula

$$((\forall x \exists y R(x, y, z)) \wedge \exists y R(x, y, y)$$

(where x, y, z are distinct variables) then $\mathrm{Var}(\varphi) = \{x, y, z\}$, $\mathrm{Free}(\varphi) = \{x, z\}$, and $\mathrm{Bound}(\varphi) = \{x, y\}$. Note that x is both a bound and a free variable in φ. More specifically we can label each occurrence of each of the variables as follows:

bound bound free free bound

$$((\forall x\, \exists y\, R(x, y, z)) \wedge \exists y\, R(x, y, y))$$

bound bound bound bound

1.3.27. Definition. We say that a formula (or an $\mathcal{M}$-formula) is a **closed formula** or a **sentence** if it contains no free variables.

Why do we only replace the free variables? Consider the formula $\exists y\, x + y = z$. (with x, y, z distinct variables). The formula φ can be interpreted to mean

"there is a number y such that $x + y = z$." (*)

This is a property that the numbers x and z may have or not have, but our choice of the particular variable y is irrelevant. For example, let v be another variable, then the formula φ_1: $\exists v\, x + v = z$ says

"there is a number v such that $x + v = z$." (**)

which is really the same thing as (*). In fact, both (*) and (**) can be expressed as

"there is a number which we can add to x to obtain z"

So the formula φ does not "talk" about the variable y at all, so when we perform a substitution we pretend that the variable y does not occur in the formula at all.

Since $\mu(x/\tau)$ is again a term (call it μ_1), we can apply another substitution to this term, for example we can consider $\mu_1(y/\sigma)$, where y is some variable (not necessarily distinct from x) and σ is a term. Instead of $\mu_1(y/\sigma)$ we can also write

$$\big(\mu(x/\tau)\big)(y/\sigma) \quad \text{or} \quad \mu(x/\tau)(y/\sigma).$$

Sometimes we want to replace several variables simultaneously:

1.3.28. Definition.

(a) Let μ be a term, $x_1, \ldots, x_k$ different variables, and $\tau_1, \ldots, \tau_k$ terms. We define $\mu(x_1/\tau_1, \ldots, x_k/\tau_k)$ by induction on μ.

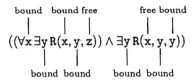

(i) If μ is a constant symbol $\mathbf{c}$, then $\mu(\mathrm{x}_1/\tau_1, \ldots, \mathrm{x}_k/\tau_k) = \mu$.

(ii) If μ is a variable y, then

$$\mu(\mathrm{x}_1/\tau_1, \ldots, \mathrm{x}_k/\tau_k) = \begin{cases} \tau_i & \text{if } \mathrm{y} = \mathrm{x}_i, \text{ some } i \in \{1, \ldots, k\} \\ \mu & \text{otherwise.} \end{cases}$$

(iii) If $\mu = \mathbf{F}\mu_1 \ldots \mu_n$, then $\mu(\mathrm{x}_1/\tau_1, \ldots, \mathrm{x}_k/\tau_k) = \mathbf{F}\mu_1(\mathrm{x}_1/\tau_1, \ldots, \mathrm{x}_k/\tau_k) \ldots \mu_n(\mathrm{x}_1/\tau_1, \ldots, \mathrm{x}_k/\tau_k)$.

In other words, all occurrences of x_1 are replaced by the term τ_1, etc.

(b) Let φ be a formula, $\mathrm{x}_1, \ldots, \mathrm{x}_k$ different variables, and $\tau_1, \ldots, \tau_k$ terms. We define $\varphi(\mathrm{x}_1/\tau_1, \ldots, \mathrm{x}_k/\tau_k)$ by induction on φ.

(i) If $\varphi = \mathbf{R}\mu_1 \ldots \mu_n$, then $\varphi(\mathrm{x}_1/\tau_1, \ldots, \mathrm{x}_k/\tau_k) = \mathbf{R}\mu_1(\mathrm{x}_1/\tau_1, \ldots, \mathrm{x}_k/\tau_k) \ldots \mu_n(\mathrm{x}_1/\tau_1, \ldots, \mathrm{x}_k/\tau_k)$.

(ii) If $\varphi = (\neg\psi)$, then $\varphi(\mathrm{x}_1/\tau_1, \ldots, \mathrm{x}_k/\tau_k) = (\neg\psi(\mathrm{x}_1/\tau_1, \ldots, \mathrm{x}_k/\tau_k))$.

(iii) If $\varphi = (\psi \wedge \theta)$, then

$$\varphi(\mathrm{x}_1/\tau_1, \ldots, \mathrm{x}_k/\tau_k) = (\psi(\mathrm{x}_1/\tau_1, \ldots, \mathrm{x}_k/\tau_k) \wedge \theta(\mathrm{x}_1/\tau_1, \ldots, \mathrm{x}_k/\tau_k))$$

Similarly for the other connectives.

(iv) If $\varphi = \forall \mathrm{y}\psi$, and y is not among $\mathrm{x}_1, \ldots, \mathrm{x}_k$, then $\varphi(\mathrm{x}_1/\tau_1, \ldots, \mathrm{x}_k/\tau_k) = \forall \mathrm{y}\psi(\mathrm{x}_1/\tau_1, \ldots, \mathrm{x}_k/\tau_k)$.

(v) If $\varphi = \forall \mathrm{x}_i\psi$ or $\exists \mathrm{x}_i\psi$ for some $i \in \{1, \ldots, k\}$, then $\varphi(\mathrm{x}_1/\tau_1, \ldots, \mathrm{x}_k/\tau_k)$ is

$$\forall \mathrm{x}_i(\psi(\mathrm{x}_1/\tau_1, \ldots, \mathrm{x}_{i-1}/\tau_{i-1}, \mathrm{x}_{i+1}/\tau_{i+1}, \ldots, \mathrm{x}_k/\tau_k)$$

(so we substitute for all the x_j except for x_i).

In other words, all **free** occurrences of x_1 are replaced by the term τ_1, etc.

1.3.29. Example. Let μ be the term $(\mathrm{x} + \mathrm{y})$ (where x and y are distinct variables). Let τ_1 be the term $(\mathrm{x} \cdot \mathrm{y})$, and let τ_2 be the term $\mathbf{z}$ (a variable different from x and y). Then

$$\begin{aligned}
\mu(\mathrm{x}/\tau_1, \mathrm{y}/\tau_2) &= ((\mathrm{x} \cdot \mathrm{y}) + \mathrm{z}) \\
\mu(\mathrm{y}/\tau_2, \mathrm{x}/\tau_1) &= ((\mathrm{x} \cdot \mathrm{y}) + \mathrm{z}) \\
\mu(\mathrm{x}/\tau_1) &= ((\mathrm{x} \cdot \mathrm{y}) + \mathrm{y}) \\
\mu(\mathrm{x}/\tau_1)(\mathrm{y}/\tau_2) &= ((\mathrm{x} \cdot \mathrm{y}) + \mathrm{y})(\mathrm{y}/\tau_2) = ((\mathrm{x} \cdot \mathrm{z}) + \mathrm{z}) \\
\mu(\mathrm{y}/\tau_2) &= (\mathrm{x} + \mathrm{z}) \\
\mu(\mathrm{y}/\tau_2)(\mathrm{x}/\tau_1) &= (\mathrm{x} + \mathrm{z})(\mathrm{x}/\tau_1) = ((\mathrm{x} \cdot \mathrm{y}) + \mathrm{z}).
\end{aligned}$$

1.3.30. Example. Let φ be the formula $(\exists \mathtt{u}(\mathtt{x}+\mathtt{u})=\mathtt{y})$ (where x, y, u and z are distinct variables). Then

$$\varphi(\mathtt{x}/\mathtt{x}\cdot\mathtt{u},\mathtt{y}/\mathtt{z}) = (\exists \mathtt{u}((\mathtt{x}\cdot\mathtt{u})+\mathtt{u})=\mathtt{z})$$
$$\varphi(\mathtt{x}/\mathtt{x}\cdot\mathtt{y},\mathtt{y}/\mathtt{z}) = (\exists \mathtt{u}((\mathtt{x}\cdot\mathtt{y})+\mathtt{u})=\mathtt{z})$$

As we shall see later, the first substitution is not "allowed."

1.3.31. Remark. If μ is a term in which the variable x does not appear, then $\mu(\mathtt{x}/\tau)=\mu$.
Similarly, if the variable x does not appear **freely** in the formula φ (see 1.3.25), then $\varphi(\mathtt{x}/\tau)=\varphi$.

Proof: By induction on μ (or by induction on φ).

1.3.32. Definition. A formula φ is called a prime formula iff φ is atomic or φ is of the form $\exists \mathtt{x}\psi$ or $\forall \mathtt{x}\psi$.

1.3.33. Fact. The set of formulas in the first order language $\mathcal{L}$ is an inductive structure with the prime formulas as blocks, and the functions $F_\neg$, $F_\wedge$, $F_\vee$, $F_\leftrightarrow$, $F_\rightarrow$, $F_|$ as operators. (See 1.1.12.)

1.3.34. Definition. A map s from the prime formulas into the set $\{T,F\}$ will be called a **sentential** truth assignment. If s is a sentential truth assignment, then s can be extended to a function $\bar{s}$ defined on all formulas as follows:

If φ is a prime formula, then $\bar{s}(\varphi)=s(\varphi)$.
If φ is the formula $\psi_1 @ \psi_2$ with @ in $\wedge, \vee, \rightarrow, \leftrightarrow, |$, then $\bar{s}(\varphi) = G_@(\bar{s}(\psi_1), \bar{s}(\psi_2))$, where $G_@$ is the function described in 1.2.3.
If φ is the formula $(\neg\psi)$, then $\bar{s}(\varphi)=G_\neg(\bar{s}(\psi))$.

1.3.35. Definition. A formula φ is called a **tautology**, if for all sentential truth assignments s we have $\bar{s}(\varphi)=T$.

(It is clear that when we talk about $\bar{s}$, we need the values of s only on the prime formulas occurring in φ.)

See exercise 8 for a reformulation of this definition.

1.3.36. Example.

(1) $\forall x\, P(x) \wedge \exists y\, Q(x) \rightarrow \forall x P(x)$ is a tautology.

(2) $\forall x\, P(x) \rightarrow \forall y\, P(y)$ is not a tautology.

Proof of (1): Let s be a sentential truth assignment, and let φ be the formula $\forall x\, P(x) \wedge \exists y\, Q(x) \rightarrow \forall x P(x)$. $s(\forall x P(x))$ can be equal to T or F. If it is equal to T, then also $\bar{s}(\varphi) = T$, because φ is of the form $\psi \rightarrow \forall x P(x))$. If it is equal to F, then also

$$\bar{s}(\forall x P(x) \wedge (\exists y Q(x))) \ = F$$

so again $\bar{s}(\varphi) = T$.

Proof of (2): There is a sentential truth assignment s that assigns T to $\forall x\, P(x)$ and F to $\forall y\, P(y)$.

1.3.37. Definition (The satisfaction relation). If φ is a **closed $\mathcal{M}$-formula** we define

$$\mathcal{M} \models \varphi \qquad \text{(read: "}\varphi\text{ is valid in }\mathcal{M}\text{" or "}\mathcal{M}\text{ satisfies }\varphi\text{")}$$

as follows:

If φ is the $\mathcal{M}$-formula $\tau_1 = \tau_2$ (where τ_1 and τ_2 are closed $\mathcal{M}$-terms), then we let $\mathcal{M} \models \varphi$ be true iff $\tau_1^{\mathcal{M}}$ and $\tau_2^{\mathcal{M}}$ are the same element of M.

If φ is $R(\tau_1, \ldots, \tau_n)$, where R is a relation symbol, and τ_1, $\ldots$, τ_n are closed $\mathcal{M}$-terms, then we let $\mathcal{M} \models \varphi$ be true iff the n-tuple $(\tau_1^{\mathcal{M}}, \ldots, \tau_n^{\mathcal{M}})$ is in $R^{\mathcal{M}}$.

If $\varphi = (\psi_1 \wedge \psi_2)$, then we we let $\mathcal{M} \models \varphi$ be true iff both $\mathcal{M} \models \psi_1$ and $\mathcal{M} \models \psi_2$ are true.

If $\varphi = (\psi_1 \vee \psi_2)$, then we we let $\mathcal{M} \models \varphi$ be true iff either $\mathcal{M} \models \psi_1$ or $\mathcal{M} \models \psi_2$ (or both) are true.

If $\varphi = (\psi_1 | \psi_2)$, then we we let $\mathcal{M} \models \varphi$ be true iff not both $\mathcal{M} \models \psi_1$ and $\mathcal{M} \models \psi_2$ are true (i.e., if at most one of these statements is true).

If $\varphi = (\psi_1 \rightarrow \psi_2)$, then we we let $\mathcal{M} \models \varphi$ be true iff either $\mathcal{M} \models \psi_1$ is false or $\mathcal{M} \models \psi_2$ is true (or both).

If $\varphi = (\psi_1 \leftrightarrow \psi_2)$, then we we let $\mathcal{M} \models \varphi$ be true iff either both or neither of the statements $\mathcal{M} \models \psi_1$ and $\mathcal{M} \models \psi_2$ are true.

If $\varphi = (\neg\psi)$, we let $\mathcal{M} \models \varphi$ be true iff $\mathcal{M} \models \psi$ is false.

If $\varphi = (\forall x \psi)$, then we let $\mathcal{M} \models \varphi$ be true, iff for all elements m of M, $\mathcal{M} \models \psi(x/m)$ is true. (This definition is legal, because the only free variable of ψ (if any) is $\mathtt{x}$, so $\psi(\mathtt{x}/m)$ is again a closed $\mathcal{M}$-formula.)

If $\varphi = \exists \mathtt{x}\psi$, then we let $\mathcal{M} \models \varphi$ be true, iff for some element m of M, $\mathcal{M} \models \psi(\mathtt{x}/m)$ is true.

1.3.38. Definition. If the free variables of φ are $\mathtt{y}_1, \ldots, \mathtt{y}_n$, then we call the formula $\forall \mathtt{y}_1 \cdots \forall \mathtt{y}_n \varphi$ the **universal closure** of φ.

[It will turn out that it does not matter in which order we write the quantifiers. To be specific, we fix some order among all variables of our language, and we define the universal closure of φ to be the formula produced by writing the quantifiers in the corresponding order.

Thus, the universal closure of $\mathtt{R}(\mathtt{x}, \mathtt{y})$ will be $\forall \mathtt{x} \forall \mathtt{y}\, \mathtt{R}(\mathtt{x}, \mathtt{y})$ or $\forall \mathtt{y} \forall \mathtt{x}\, \mathtt{R}(\mathtt{x}, \mathtt{y})$, according to whether the variable $\mathtt{x}$ is before or after the variable $\mathtt{y}$ in our fixed order. (Of course, if $\mathtt{x}$ and $\mathtt{y}$ are the same variable, then the universal closure of $\mathtt{R}(\mathtt{x}, \mathtt{y})$ is just $\forall \mathtt{x} \mathtt{R}(\mathtt{x}, \mathtt{y})$.)]

1.3.39. Definition and Remarks.

(1) If φ is not closed, we let ψ be the universal closure of φ, and we write $\mathcal{M} \models \varphi$ for $\mathcal{M} \models \psi$.

(2) In other words, assuming that the free variables of φ are the (distinct) variables $\mathtt{x}_1, \ldots, \mathtt{x}_n$, $\mathcal{M} \models \varphi$ means that for all n-tuples $(m_1, \ldots, m_n)$ from M we have $\mathcal{M} \models \varphi(\mathtt{x}_1/m_1, \ldots, \mathtt{x}_n/m_n)$.

(3) If the free variables of φ are **among** $\mathtt{x}_1, \ldots, \mathtt{x}_n$, then we also have $\mathcal{M} \models \varphi$ iff for all n-tuples $(m_1, \ldots, m_n)$ from M we have $\mathcal{M} \models \varphi(\mathtt{x}_1/m_1, \ldots, \mathtt{x}_n/m_n)$.

(4) If Γ is a set of ($\mathcal{M}$-)formulas, we write $\mathcal{M} \models \Gamma$ iff for all $\varphi \in \Gamma$, $\mathcal{M} \models \varphi$. We read this as "$\mathcal{M}$ models Γ," or "$\mathcal{M}$ thinks that Γ is true," or "$\mathcal{M}$ is a model of Γ."

(5) We write $\mathcal{M} \not\models \varphi$ if it is not the case that $\mathcal{M} \models \varphi$. (If φ is closed, then this is equivalent to $\mathcal{M} \models \neg\varphi$.) We let $\mathcal{M} \not\models \Gamma$ mean that $\mathcal{M} \models \Gamma$ is not true, i.e., there is $\psi \in \Gamma$ such that $\mathcal{M} \not\models \psi$.

(6) If τ and τ' are closed $\mathcal{M}$-terms with the same interpretation, i.e., satisfying $\tau^{\mathcal{M}} = \tau'^{\mathcal{M}}$, then for any formula φ we have $\mathcal{M} \models \varphi(\mathtt{x}/\tau)$ iff $\mathcal{M} \models \varphi(\mathtt{x}/\tau')$.

Proof:

(2) A formal proof would use induction on n. The fact is clear for $n = 0$ and $n = 1$, and for $n = k + 1 > 1$ we have to use the fact that the $\mathcal{M}$-formula

$\varphi(\mathrm{x}_1/m_1,\ldots,\mathrm{x}_k/m_k,\mathrm{x}_{k+1}/m_{k+1})$ is the same as

$$\varphi(\mathrm{x}_1/m_1,\ldots,\mathrm{x}_k/m_k)(\mathrm{x}_{k+1}/m_{k+1}).$$

(3) If the free variables of φ are, for example, x_1, ..., x_k, and the variables x_{k+1}, ..., x_n are not free in φ, then it is easy to see that the formula $\varphi(\mathrm{x}_1/m_1,\ldots,\mathrm{x}_k/m_k,\mathrm{x}_{k+1}/m_{k+1},\ldots,\mathrm{x}_n/m_n))$ is the same as the formula $\varphi(\mathrm{x}_1/m_1,\ldots,\mathrm{x}_k/m_k)$.
(6) By induction on φ.

1.3.40. Definition. If Γ is a set of **closed** formulas, and φ a (not necessarily closed) formula, then we write

$$\Gamma \models \varphi$$

iff

For all models $\mathcal{M}$: If $\mathcal{M} \models \Gamma$, then $\mathcal{M} \models \varphi$

or in other words:

There is no model $\mathcal{M}$ that satisfies both $\mathcal{M} \models \Gamma$ and $\mathcal{M} \not\models \varphi$

If also φ is closed, then this is equivalent to: "There is no model of $\Gamma \cup \{\neg\varphi\}$."

1.3.41. Definition. $\models \varphi$ means $\emptyset \models \varphi$, i.e., $\mathcal{M} \models \varphi$ for all models $\mathcal{M}$ (of our language).
We say that φ is **valid** iff $\models \varphi$.

Note that we do not define $\Gamma \models \varphi$ if there are free variables in some formula of Γ.

1.3.42. Definition. If φ, ψ are formulas, we write $\psi \models \varphi$ iff $\models \psi \to \varphi$. (Remember that if the free variables of φ and ψ are x_1, ..., x_n, then this means $\models \forall\mathrm{x}_1 \cdots \forall\mathrm{x}_n\,(\psi \to \varphi)$. We write $\psi \mathrel{=\!\!|\!\!\models} \varphi$ (and we say that φ is **elementarily equivalent** to ψ) iff $\psi \models \varphi$ and $\varphi \models \psi$.

1.3.43. Remark. If τ is a closed $\mathcal{M}$-term, $\tau^{\mathcal{M}} = m$, then $\mathcal{M} \models \varphi(\mathrm{x}/\tau)$ iff $\mathcal{M} \models \varphi(\mathrm{x}/m)$. We leave the proof as an exercise.

1.3.44. Lemma.

(1) For any $\mathcal{M}$-formulas φ and ψ:

If $\mathcal{M} \models \varphi \to \psi$ and $\mathcal{M} \models \varphi$, then $\mathcal{M} \models \psi$

(2) If $\Gamma \cup \{\varphi\}$ is a set of closed formulas, and ψ is any formula, then

$$\Gamma \cup \{\varphi\} \models \psi \quad \text{iff} \quad \Gamma \models \varphi \to \psi$$

Proof: (1) Let $x_1, \ldots, x_n$ (all distinct) be the free variables of $\varphi \to \psi$, let $a_1, \ldots, a_n$ be elements of $\mathcal{M}$, and write φ' for $\varphi(x_1/a_1, \ldots, x_n/a_n)$, similarly for ψ. So the free variables of ψ are among $x_1, \ldots, x_n$.

Since $\mathcal{M} \models \varphi' \to \psi'$, by 1.3.37 we have either $\mathcal{M} \not\models \varphi'$ or $\mathcal{M} \models \psi'$ (or both). Since $\mathcal{M} \models \varphi$, we must also have $\mathcal{M} \models \varphi'$ by 1.3.39. So the first alternative is impossible, and therefore the second possibility must hold: $\mathcal{M} \models \psi'$.
(2) Exercise.

1.3.45. Lemma.
(i) If $\varphi_1 \models \psi_1$ and $\varphi_2 \models \psi_2$, then

$$\begin{aligned} \varphi_1 \wedge \varphi_2 &\models \psi_1 \wedge \psi_2 \\ \varphi_1 \vee \varphi_2 &\models \psi_1 \vee \psi_2 \\ \neg\psi_1 &\models \neg\varphi_1 \\ \forall x\, \varphi_1 &\models \forall x\, \psi_1 \\ \exists x\, \varphi_1 &\models \exists x\, \psi_1 \end{aligned}$$

(ii) If $\varphi_1 \mathrel{=\!\mid\models} \psi_1$ and $\varphi_2 \mathrel{=\!\mid\models} \psi_2$, then

$$\begin{aligned} \varphi_1 \wedge \varphi_2 &\mathrel{=\!\mid\models} \psi_1 \wedge \psi_2 \\ \varphi_1 \vee \varphi_2 &\mathrel{=\!\mid\models} \psi_1 \vee \psi_2 \\ \neg\varphi_1 &\mathrel{=\!\mid\models} \neg\psi_1 \\ \forall x\, \varphi_1 &\mathrel{=\!\mid\models} \forall x\, \psi_1 \\ \exists x\, \varphi_1 &\mathrel{=\!\mid\models} \exists x\, \psi_1 \\ \varphi_1 \to \varphi_2 &\mathrel{=\!\mid\models} \psi_1 \to \psi_2 \\ \varphi_1 \leftrightarrow \varphi_2 &\mathrel{=\!\mid\models} \psi_1 \leftrightarrow \psi_2 \\ \varphi_1 | \varphi_2 &\mathrel{=\!\mid\models} \psi_1 | \psi_2 \end{aligned}$$

The proof is left as an exercise.

We will prove several theorems by induction on formulas. To show that

all formulas φ have the property P

we have to show that all atomic formulas have the property P, and that whenever φ and ψ have the property P, $\neg\varphi$, $\varphi \wedge \psi$, $\varphi \vee \psi$, ..., $\varphi|\psi$, $\exists x\varphi$, and $\forall x\varphi$ have the property P also.

In order to make these proofs shorter, we will restrict ourselves to a sublanguage that uses only $\neg$ and $\wedge$ as connectives and only one quantifier. But before we can do that, we have to show that for every formula there is an equivalent formula (in the sense of $\dashv\models$) in this sublanguage.

1.3.46. Definition. $\mathcal{L}_{\wedge,\neg,\forall}$ is the sublanguage of the language of first order logic that does not use the connectives $\vee$, $\rightarrow$, $\leftrightarrow$, $|$, and does not use the quantifier $\exists$. (I.e., we use definitions 1.3.4, 1.3.8, ..., but allow only $\wedge$ and $\neg$ in 1.3.4(i), and only $\forall$ in 1.3.4(iv).)

1.3.47. Definition. For each formula φ in $\mathcal{L}$ we define a formula $\tilde{\varphi}$, the "translation" of φ, as follows:

If φ is atomic, then $\tilde{\varphi} = \varphi$.
If $\varphi = (\neg\psi)$, then $\tilde{\varphi} = (\neg\tilde{\psi})$.
If $\varphi = (\psi_1 \wedge \psi_2)$, then $\tilde{\varphi} = (\tilde{\psi}_1 \wedge \tilde{\psi}_2)$.
If $\varphi = (\psi_1 \vee \psi_2)$, then $\tilde{\varphi} = (\neg((\neg\tilde{\psi}_1) \wedge (\neg\tilde{\psi}_2)))$.
If $\varphi = (\psi_1 \rightarrow \psi_2)$, then $\tilde{\varphi} = (\neg(\tilde{\psi}_1 \wedge (\neg\tilde{\psi}_2)))$.
If $\varphi = (\psi_1 \leftrightarrow \psi_2)$, we let

$$\tilde{\varphi} = \left[\neg((\neg\tilde{\psi}_1) \wedge \tilde{\psi}_2)\right] \wedge \left[\tilde{\psi}_1 \wedge (\neg\tilde{\psi}_2)\right].$$

If $\varphi = (\psi_1|\psi_2)$, then $\tilde{\varphi} = \neg(\tilde{\psi}_1 \wedge \tilde{\psi}_2)$.
If $\varphi = \forall x\psi$, then $\tilde{\varphi} = \forall x\tilde{\psi}$.
If $\varphi = \exists x\psi$, then $\tilde{\varphi} = \neg(\forall x(\neg\tilde{\psi}))$.

1.3.48. Lemma.

(1) For every formula φ in $\mathcal{L}$ the formula $\tilde{\varphi}$ is in $\mathcal{L}_{\neg,\wedge,\forall}$.
(2) For every formula φ in $\mathcal{L}$ we have $\tilde{\varphi} \dashv\models \varphi$.

Proof: Exercise.

1.3.49. Notation. From now on we will exclusively work in the language $\mathcal{L}_{\neg,\wedge,\forall}$. Whenever we write a formula φ in $\mathcal{L}$ that seems to be outside the language $\mathcal{L}_{\neg,\wedge,\forall}$, we are using φ as an ABBREVIATION for $\tilde{\varphi}$.

For example, $(\mathrm{x} = \mathrm{y} \rightarrow \mathrm{y} = \mathrm{x})$ is an abbreviation for

$$(\neg(\mathrm{x} = \mathrm{y} \wedge (\neg \mathrm{y} = \mathrm{x})))$$

and $\exists \mathrm{x}\ \mathrm{x} = \mathrm{y}$ abbreviates

$$(\neg(\forall \mathrm{x}(\neg\, \mathrm{x} = \mathrm{y}))).$$

Now we will give a list of valid formulas. These formulas will be the axioms used in the next section.

1.3.50. Examples of Valid Formulas.

(i) Tautologies.

(ii) Let φ and ψ be formulas. Then $\forall \mathrm{x}(\varphi \rightarrow \psi) \rightarrow (\forall \mathrm{x}\varphi \rightarrow \forall \mathrm{x}\psi)$ is a valid formula.

(iii) Let φ be a formula. Assume that x is not in Free(φ). Then $\varphi \rightarrow \forall \mathrm{x}\varphi$ is a valid formula.

(iv) The following are valid formulas:

(a) $\mathrm{x} = \mathrm{x}$.

(b) $\mathrm{x} = \mathrm{y} \rightarrow \mathrm{y} = \mathrm{x}$.

(c) $\mathrm{x} = \mathrm{y} \wedge \mathrm{y} = \mathrm{z} \rightarrow \mathrm{x} = \mathrm{z}$.

(d) $\mathrm{x}_1 = \mathrm{y}_1 \wedge \cdots \wedge \mathrm{x}_k = \mathrm{y}_k \rightarrow (\mathrm{R}\mathrm{x}_1 \ldots \mathrm{x}_k \leftrightarrow \mathrm{R}\mathrm{y}_1 \ldots \mathrm{y}_k)$, where R is a k-ary relation symbol.

(e) $\mathrm{x}_1 = \mathrm{y}_1 \wedge \cdots \wedge \mathrm{x}_k = \mathrm{y}_k \rightarrow \mathrm{F}\mathrm{x}_1 \ldots \mathrm{x}_k = \mathrm{F}\mathrm{y}_1 \ldots \mathrm{y}_k$, where F is a k-ary function symbol.

Proof: (i) Let φ be a tautology. Let $\mathcal{M}$ be a model. Assume that the free variables of φ are $\{\mathrm{x}_1, \ldots, \mathrm{x}_n\}$, and let $m_1, \ldots, m_n$ be elements of M. Let $\mathcal{P}$ be the set of prime formulas ψ with free variables among $\mathrm{x}_1, \ldots, \mathrm{x}_n$, and let $\mathcal{L}_0$ be the inductive structure obtained from the blocks in $\mathcal{P}$ using the operators $F_\wedge$ and $F_\neg$ only. Note that φ is in $\mathcal{L}_0$.

Now define a sentential truth assignment s on $\mathcal{L}_0$ by

$$\forall \psi \in \mathcal{P}\ s(\psi) = \begin{cases} T & \text{if } \mathcal{M} \models \psi(\mathrm{x}_1/m_1, \ldots, \mathrm{x}_k/m_k) \\ F & \text{if } \mathcal{M} \not\models \psi(\mathrm{x}_1/m_1, \ldots, \mathrm{x}_k/m_k) \end{cases}$$

and extend it to a function $\bar{s}$ as in 1.3.34. Since φ is a tautology, we have $\bar{s}(\varphi) = T$.

Now it is easy to show by induction that for all $\psi \in \mathcal{L}_0$ we have

$$\bar{s}(\psi) = T \quad \leftrightarrow \quad \mathcal{M} \models \psi(\mathrm{x}_1/m_1, \ldots, \mathrm{x}_k/m_k)$$

Hence, $\mathcal{M} \models \varphi(\mathrm{x}_1/m_1, \ldots, \mathrm{x}_k/m_k)$.

Proof: (ii) Assume there is a model $\mathcal{M}$ such that $\mathcal{M} \not\models \forall\mathrm{x}(\varphi \to \psi) \to ((\forall\mathrm{x}\varphi) \to (\forall\mathrm{x}\psi))$. So, assuming that the free variables of this formula are $\mathrm{x}_1, \ldots, \mathrm{x}_k$, there are elements $m_1, \ldots, m_k$ of M such that (writing φ' for $\varphi(\mathrm{x}_1/m_1, \ldots, \mathrm{x}_k/m_k)$, etc.), we have

$$\mathcal{M} \models \neg \left(\forall\mathrm{x}(\varphi' \to \psi') \to ((\forall\mathrm{x}\varphi') \to (\forall\mathrm{x}\psi'))\right)$$

(a) $\mathcal{M} \models \forall\mathrm{x}(\varphi' \to \psi')$.
(b) $\mathcal{M} \models \forall\mathrm{x}\varphi'$.
(c) $\mathcal{M} \models \neg\forall\mathrm{x}\psi'$.

By (c), there is an element $m \in M$, such that $\mathcal{M} \models \neg\psi'(\mathrm{x}/m)$. But by (a) and (b) we have that $\mathcal{M} \models \varphi'(\mathrm{x}/m) \to \psi'(\mathrm{x}/m)$ and $\mathcal{M} \models \varphi'(\mathrm{x}/m)$. This contradicts the definition 1.3.37 of $\models$.

The other formulas are also left as an exercise to for the reader.

Let φ_1 be $\neg\forall\mathrm{x}(\mathrm{x} = \mathrm{y})$, and let φ_2 be $\neg\forall\mathrm{z}(\mathrm{z} = \mathrm{y})$, where x, y, z are distinct variables. Both formulas say that y is not the unique element of the universe. $\varphi_2(\mathrm{y}/\mathrm{x})$ says that x is not the unique element of the universe. But $\varphi_1(\mathrm{y}/\mathrm{x})$ does not say anything about x, because x is not free any more. This example leads to the following definition.

1.3.51. Definition. Let φ be an $\mathcal{M}$-formula, τ an $\mathcal{M}$-term, and x a variable. We define *allow*$(\varphi, \tau, \mathrm{x})$ ("τ may be substituted for x in φ") by induction on φ.

(i) If φ is atomic, then *allow*$(\varphi, \tau, \mathrm{x})$.
(ii) If $\varphi = (\neg\psi)$, then *allow*$(\varphi, \tau, \mathrm{x})$ iff *allow*(ψ, τ, x).
(iii) if $\varphi = (\psi \wedge \theta)$, then *allow*$(\varphi, \tau, \mathrm{x})$ iff *allow*(ψ, τ, x) and *allow*$(\theta, \tau, \mathrm{x})$.
(iv) if $\varphi = \forall\mathrm{y}\psi$, then *allow*$(\varphi, \tau, \mathrm{x})$ iff: either $\mathrm{x} \notin \mathrm{Fr}(\varphi)$, or *allow*$(\psi, \tau, \mathrm{x})$ and y is not in $\mathrm{Var}(\tau)$. (In particular, this will be true if y is the variable x, since then x is not free in φ.)

In other words, *allow*$(\varphi, \tau, \mathrm{x})$ if after substituting τ for all free occurrences x in φ no free variable of τ becomes bound.

1.3.52. Example. Let φ be the formula $(\forall x\exists y R(x,y,z)) \to \exists y R(x,y,z)$, where x, y, z are distinct variables. Then we have *allow*$(\varphi, g(x,z), x)$, but not *allow*$(\varphi, f(y), x)$.

1.3.53. Fact. Each of the following conditions implies *allow*(φ, τ, x):

(1) The sets $\mathrm{Var}(\tau)$ and $\mathrm{Bound}(\varphi)$ are disjoint. (In particular,this will be true if τ is a closed term.)

(2) x is not in $\mathrm{Fr}(\varphi)$.

Proof: Exercise.

We have seen in 1.3.29 that in general for an $\mathcal{M}$-term μ we do not have $\mu(x/\tau_1)(y/\tau_2) = \mu(y/\tau_2)(x/\tau_1)$. However, if τ_2 is a closed term, then it is easy to see that $\mu(x/\tau_1)(y/\tau_2) = \mu(y/\tau_2)(x/\tau_1')$, where $\tau_1' = \tau_1(y/\tau_2)$. This is a special case of the following lemma:

1.3.54. Lemma. Assume that $\mathcal{M}$ is a model, $m_1, \ldots, m_k, m$ are elements of M (so in particular, they are closed $\mathcal{M}$-terms), x, y_1, ..., y_k are distinct variables, τ and μ are $\mathcal{M}$-terms. Then

$$\mu(x/\tau)(y_1/m_1, \ldots, y_k/m_k, x/m) = \mu'(x/\tau') \qquad (*)$$

where $\mu' = \mu(y_1/m_1, \ldots, y_k/m_k)$ and $\tau' = \tau(y_1/m_1, \ldots, y_k/m_k, x/m)$.

Proof: To make the proof more legible, we will write $\overline{y}/\overline{m}$ as an abbreviation for $y_1/m_1, \ldots, y_k/m_k$.

We prove this by induction on the term μ. The induction step is trivial, so we only have to consider the case where μ is a variable or a constant or an element of M. If μ is a constant symbol or an element of M, or a variable not among y_1, ..., y_k, x, then both sides in $(*)$ are equal to μ.

If μ is the variable x, then $\mu' = x$, so the right hand side in $(*)$ is equal to τ'. The left hand side is equal to

$$\mu(x/\tau)(\overline{y}/\overline{m}, x/m) = \tau(\overline{y}/\overline{m}, x/m) = \tau'.$$

If μ is one of the variables y_i, then μ' as well as $\mu'(x/\tau')$ are equal to m_i, and we also have

$$\mu(x/\tau)(\overline{y}/\overline{m}, x/m) = y_i(\overline{y}/\overline{m}, x/m) = m_i.$$

Now we can prove a similar lemma for formulas:

1.3.55. Lemma. Assume that $\mathcal{M}$ is a model, $m_1, \ldots, m_k$ are elements of M, x, y_1, ..., y_k are distinct variables, τ is an $\mathcal{M}$-term and φ is an $\mathcal{M}$-formula, and *allow*(φ, τ, x). Then

$$\varphi(\mathrm{x}/\tau)(\mathrm{y}_1/m_1,\ldots,\mathrm{y}_k/m_k,\mathrm{x}/m) = \varphi'(\mathrm{x}/\tau')$$

where $\varphi' = \varphi(\mathrm{y}_1/m_1,\ldots,\mathrm{y}_k/m_k)$ and $\tau' = \tau(\mathrm{y}_1/m_1,\ldots,\mathrm{y}_k/m_k,\mathrm{x}/m)$.

Proof: By induction. (Again we will write $\overline{\mathrm{y}}/\overline{m}$ instead of $\mathrm{y}_1/m_1,\ldots,\mathrm{y}_k/m_k$.)

For atomic formulas, we just use the previous lemma.
The induction step where $\varphi = \psi_1 \wedge \psi_2$ or $\varphi = \neg\psi$ is easy.
So we only have to deal with the case where φ is obtained from a formula ψ using universal quantification. We have to consider several cases:
Case 1: $\varphi = [\forall \mathrm{z}\psi]$, where z is a variable different from x and also different from $\mathrm{y}_1, \ldots, \mathrm{y}_k$. Then

$$\begin{aligned}
\varphi(\mathrm{x}/\tau)(\overline{\mathrm{y}}/\overline{m},\mathrm{x}/m) &= [\forall \mathrm{z}\,\psi](\mathrm{x}/\tau)(\overline{\mathrm{y}}/\overline{m},\mathrm{x}/m)\\
&= [\forall \mathrm{z}\,(\psi(\mathrm{x}/\tau)(\overline{\mathrm{y}}/\overline{m},\mathrm{x}/m))]\\
\text{(by induction hypothesis)} \quad &= [\forall \mathrm{z}\,(\psi(\overline{\mathrm{y}}/\overline{m})(\mathrm{x}/\tau'))]\\
&= [\forall \mathrm{z}\,\psi](\overline{\mathrm{y}}/\overline{m})(\mathrm{x}/\tau')\\
&= \varphi(\overline{\mathrm{y}}/\overline{m})(\mathrm{x}/\tau').
\end{aligned}$$

Case 2: $\varphi = [\forall \mathrm{x}\,\psi]$. Since x is not free in $[\forall \mathrm{x}\,\psi]$, we have

$$\begin{aligned}
\varphi(\mathrm{x}/\tau)(\overline{\mathrm{y}}/\overline{m},\mathrm{x}/m) &= [\forall \mathrm{x}\,\psi](\mathrm{x}/\tau)(\overline{\mathrm{y}}/\overline{m},\mathrm{x}/m)\\
&= [\forall \mathrm{x}\,\psi](\overline{\mathrm{y}}/\overline{m})\\
&= [\forall \mathrm{x}\,\psi](\overline{\mathrm{y}}/\overline{m})(\mathrm{x}/\tau')\\
&= \varphi(\overline{\mathrm{y}}/\overline{m})(\mathrm{x}/\tau')
\end{aligned}$$

Case 3: $\varphi = [\forall \mathrm{y}_i\,\psi]$, for some i. To simplify the notation, we assume that $\varphi = [\forall \mathrm{y}_1\,\psi]$. Since we have $\mathit{allow}(\varphi,\tau,\mathrm{x})$, y_1 is not free in τ. So $\tau' = \tau(\mathrm{y}_2/m_2,\ldots,\mathrm{y}_k/m_k,\mathrm{x}/m)$. So we get

$$\begin{aligned}
\varphi(\mathrm{x}/\tau)(\overline{\mathrm{y}}/\overline{m},\mathrm{x}/m) &= [\forall \mathrm{y}_1\,\psi](\mathrm{x}/\tau)(\mathrm{y}_1/m_1,\mathrm{y}_2/m_2,\ldots,\mathrm{y}_k/m_k,\mathrm{x}/m)\\
&= [\forall \mathrm{y}_1\,\psi(\mathrm{x}/\tau)(\mathrm{y}_2/m_2,\ldots,\mathrm{y}_k/m_k,\mathrm{x}/m)]\\
\text{(by induction hypothesis)} \quad &= [\forall \mathrm{y}_1\,\psi(\mathrm{y}_2/m_2,\ldots,\mathrm{y}_k/m_k)(\mathrm{x}/\tau')]\\
&= [\forall \mathrm{y}_1\,\psi](\mathrm{y}_2/m_2,\ldots,\mathrm{y}_k/m_k)(\mathrm{x}/\tau')\\
&= [\forall \mathrm{y}_1\,\psi](\mathrm{y}_1/m_1,\mathrm{y}_2/m_2,\ldots,\mathrm{y}_k/m_k)(\mathrm{x}/\tau')\\
&= \varphi(\overline{\mathrm{y}}/\overline{m})(\mathrm{x}/\tau').
\end{aligned}$$

1.3.56. Lemma. Let φ be a formula, τ a term, and x a variable. Assume $\mathit{allow}(\varphi,\tau,\mathrm{x})$. Then the formula

$$(\forall \mathrm{x}\, \varphi) \to \varphi(\mathrm{x}/\tau)$$

is valid.

Proof: First we will prove this lemma under the additional assumption that the formula $(\forall \mathrm{x}\, \varphi) \to \varphi(\mathrm{x}/\tau)$ is closed, and then we will show how to reduce the general case to this special case.

So assume that the formula $(\forall \mathrm{x}\, \varphi) \to \varphi(\mathrm{x}/\tau)$ has no free variables. So φ has no free variables other than x. Let $b := \tau^{\mathcal{M}}$ be the interpretation of τ in $\mathcal{M}$. (If x is free in φ, τ has to be a closed term, since $\varphi(\mathrm{x}/\tau)$ is closed. If x is not free in φ, then $\tau^{\mathcal{M}}$ may not be well-defined. In that case we let b be an arbitrary element of M.)
If $\mathcal{M} \not\models \forall \mathrm{x}\, \varphi$, then by the definition of $\models$ we get that $\mathcal{M} \models (\forall \mathrm{x}\, \varphi) \to \varphi(\mathrm{x}/\tau)$.
So assume $\mathcal{M} \models \forall \mathrm{x}\, \varphi$. Again by the definition of $\models$ we get that $\mathcal{M} \models \varphi(\mathrm{x}/b)$. But as we have remarked above (1.3.39(6)), this means the same as $\mathcal{M} \models \varphi(\mathrm{x}/\tau)$, so again we get $\mathcal{M} \models (\forall \mathrm{x} \varphi) \to \varphi(\mathrm{x}/\tau)$.

Before we prove the theorem in its full generality, we remark that the proof so far really shows

$$\mathcal{M} \models (\forall \mathrm{x} \varphi) \to \varphi(\mathrm{x}/\tau) \qquad (*)$$

for all closed $\mathcal{M}$-formulas rather than only for all closed formulas.

Now let $(\forall \mathrm{x} \varphi) \to \varphi(\mathrm{x}/\tau)$ be any formula (or indeed any $\mathcal{M}$-formula), not necessarily closed. Assume that x, $\mathrm{y}_1, \ldots, \mathrm{y}_n$ are distinct variables, and that the free variables of $\forall \mathrm{x} \varphi \to \varphi(\mathrm{x}/\tau)$ are among x, $\mathrm{y}_1, \ldots, \mathrm{y}_n$. (Note that x may or may not be a free variable of $\varphi(\mathrm{x}/\tau)$.)

Let $\mathcal{M}$ be a model, and let $a, m_1, \ldots, m_n$ be elements of M. We have to check that

$$\mathcal{M} \models (\forall \mathrm{x} \varphi \to \varphi(\mathrm{x}/\tau))(\mathrm{x}/a, \mathrm{y}_1/m_1, \ldots, \mathrm{y}_n/m_n) \qquad (**)$$

holds. Again abbreviate $\mathrm{y}_1/m_1, \ldots, \mathrm{y}_n/m_n$ by $\overline{\mathrm{y}}/\overline{m}$. Write ψ for the $\mathcal{M}$-formula $\varphi(\overline{\mathrm{y}}/\overline{m})$.
First note that the formula $(\forall \mathrm{x}\, \varphi)(\mathrm{x}/a, \overline{\mathrm{y}}/\overline{m})$ is really the formula $\forall \mathrm{x}\, \psi$.
Secondly, we note that by 1.3.55,

$$\varphi(\mathrm{x}/\tau)(\mathrm{x}/a, \overline{\mathrm{y}}/\overline{m}) = \varphi(\overline{\mathrm{y}}/\overline{m})(\mathrm{x}/\tau')$$

for some closed $\mathcal{M}$-term τ'.
Finally, since we have already established $(*)$ whenever the formula in question was a **closed** $\mathcal{M}$-formula, we can conclude that

$$\mathcal{M} \models \forall \mathrm{x}\, \psi \to \psi(\mathrm{x}/\tau')$$

which is the same as $(**)$ above.

We call a language "purely relational" if there are no constant symbols and no function symbols in the language. For example, we can construct a purely relational language for groups by not using a binary function for the group operation, but instead the ternary relation $\{(x, y, z) : x.y = z\}$. It might also be convenient to have a unary relation (i.e. a subset of the group) whose only element is the neutral element of the group.

In general, we use the following definition:

1.3.57. Definition. Let $\mathcal{L}$ be a language of first order logic. We define the associated purely relational language $\mathcal{L}^{\text{rel}}$ as follows:

$\mathcal{L}^{\text{rel}}$ has no function symbols and no constant symbols.
$\mathcal{L}^{\text{rel}}$ has all relation symbols from $\mathcal{L}$.
For every constant symbol $\mathtt{c}$ in $\mathcal{L}$ there is a unary relation symbol $\mathtt{R}_{\mathtt{c}}$ in $\mathcal{L}^{\text{rel}}$.
For every n-ary function symbol $\mathtt{f}$ in $\mathcal{L}$ there is an $n+1$-ary relation symbol $\mathtt{R}_{\mathtt{f}}$ in $\mathcal{L}^{\text{rel}}$.
$\mathcal{L}^{\text{rel}}$ has all the variables of the language $\mathcal{L}$, and it additionally has (distinct) variables $\mathtt{t}_\tau$ for every term τ in $\mathcal{L}$.

1.3.58. Definition. If $\mathcal{M}$ is a model for $\mathcal{L}$, we define the **corresponding relational structure** $\mathcal{M}^{\text{rel}}$ for $\mathcal{L}^{\text{rel}}$ as follows:

M^{rel}, the universe of $\mathcal{M}^{\text{rel}}$, is identical to M, the universe of $\mathcal{M}$.
For any relation symbol $\mathtt{R}$ in $\mathcal{L}$,

$$\mathtt{R}^{\mathcal{M}^{\text{rel}}} = \mathtt{R}^{\mathcal{M}}.$$

For any (n-ary) function symbol $\mathtt{f}$ in $\mathcal{L}$,

$$\mathtt{R}_{\mathtt{f}}^{\mathcal{M}^{\text{rel}}} = \{(m_1, \ldots, m_n, \mathtt{f}^{\mathcal{M}}(m_1, \ldots, m_n) : m_1, \ldots, m_n \in M^{\text{rel}}\}.$$

For any constant symbol $\mathtt{c}$ in $\mathcal{L}$,

$$\mathtt{R}_{\mathtt{c}}^{\mathcal{M}^{\text{rel}}} = \{\mathtt{c}^{\mathcal{M}}\}.$$

1.3.59. Lemma. Assume $\mathcal{N}$ is a model for $\mathcal{L}^{\text{rel}}$ such that the following are true:

(1) For every n-ary function symbol $\mathtt{f}$ in $\mathcal{L}$,

$$\mathcal{N} \models \forall \mathtt{x}_1 \cdots \forall \mathtt{x}_n \, \exists \mathtt{y} \, \mathtt{R}_{\mathtt{f}}^{\mathcal{N}}(\mathtt{x}_1, \ldots, \mathtt{x}_n, \mathtt{y})$$
$$\mathcal{N} \models \forall \mathtt{x}_1 \cdots \forall \mathtt{x}_n \forall \mathtt{y} \forall \mathtt{y}' \, \left[\mathtt{R}_{\mathtt{f}}^{\mathcal{N}}(\mathtt{x}_1, \ldots, \mathtt{x}_n, \mathtt{y}) \wedge \mathtt{R}_{\mathtt{f}}^{\mathcal{N}}(\mathtt{x}_1, \ldots, \mathtt{x}_n, \mathtt{y}') \rightarrow \mathtt{y} = \mathtt{y}'\right].$$

(2) For every constant symbol $\mathtt{c}$ in $\mathcal{L}$

$$\mathcal{N} \models \exists \mathtt{y}\, \mathtt{R}_{\mathtt{c}}^{\mathcal{N}}(\mathtt{y})$$
$$\mathcal{N} \models \forall \mathtt{y} \forall \mathtt{y}' \left[\mathtt{R}_{\mathtt{c}}^{\mathcal{N}}(\mathtt{y}) \wedge \mathtt{R}_{\mathtt{c}}^{\mathcal{N}}(\mathtt{y}') \to \mathtt{y} = \mathtt{y}'\right].$$

Then there is a unique structure $\mathcal{M}$ for the language $\mathcal{L}$ such that $\mathcal{M}^{\mathrm{rel}} = \mathcal{N}$.

Proof: Exercise.

Now that we have a purely relational structure $\mathcal{M}^{\mathrm{rel}}$ for each model $\mathcal{M}$, we can translate formulas about $\mathcal{M}$ into formulas about $\mathcal{M}^{\mathrm{rel}}$. For example, if $(G, \cdot, 1)$ is a group, and the corresponding relational structure is $(G, R_{\cdot}, R_1)$, a possible translation of $\mathtt{x} \cdot \mathtt{x} = 1$ would be

$$\exists \mathtt{y}(\mathtt{R}_{\cdot}(\mathtt{x}, \mathtt{x}, \mathtt{y}) \wedge \mathtt{R}_1(\mathtt{y})).$$

In general, we first have to define a formula φ_τ for each term τ of $\mathcal{L}$ which insures that the variable $\mathbf{t}_\tau$ is an interpretation of τ.

1.3.60. Definition.

(a) For any term τ in $\mathcal{L}$ we define a formula φ_τ in $\mathcal{L}^{\mathrm{rel}}$ as follows (by induction on τ).

(1) If τ is a variable $\mathtt{x}$, the φ_τ is the formula $\mathbf{t}_\tau = \mathtt{x}$.

(2) If τ is a constant $\mathtt{c}$, then φ_τ is the formula $\mathtt{R}_{\mathtt{c}}(\mathbf{t}_\tau)$.

(3) If τ is a term $\mathtt{F}\tau_1 \cdots \tau_n$ then φ_τ is the formula

$$\exists \mathbf{t}_{\tau_1} \cdots \exists \mathbf{t}_{\tau_n} (\varphi_{\tau_1} \wedge \cdots \varphi_{\tau_n} \wedge \mathtt{R}_{\mathtt{F}} \mathbf{t}_{\tau_1} \cdots \mathbf{t}_{\tau_n} \mathbf{t}_\tau)$$

(b) If τ is an $\mathcal{M}$-term, then τ is of the form $\tau'(\mathtt{x}_1/m_1, \ldots, \mathtt{x}_k/m_k)$ for some term τ'. In this case we let φ_τ be the $\mathcal{M}^{\mathrm{rel}}$-formula

$$\varphi_{\tau'}(\mathtt{x}_1/m_1, \ldots, \mathtt{x}_k/m_k).$$

1.3.61. Lemma. If $\mathcal{M}$ is a model for the language $\mathcal{L}$, and $\mathcal{M}^{\mathrm{rel}}$ is the corresponding purely relational model, then

Whenever τ is a closed $\mathcal{M}$-term in $\mathcal{L}$ and

$$\mathcal{M} \models \tau = m$$

then

$$\mathcal{M}^{\mathrm{rel}} \models \varphi_\tau \leftrightarrow \mathbf{t}_\tau = m.$$

Proof: This is clear if τ is a constant symbol or an element of M. If τ is a variable $\mathtt{x}$, then $\mathcal{M} \models \mathtt{x} = m$ implies

$$\mathcal{M} \models \mathtt{t_x} = \mathtt{x} \leftrightarrow \mathtt{t_x} = m.$$

Now assume τ is $\mathtt{f}\tau_1 \cdots \tau_n$. Let $\mathcal{M} \models \tau = m$, $\mathcal{M} \models \tau_1 = m_1, \ldots,$ $\mathcal{M} \models \tau_n = m_n$. So we get $\mathtt{f}^{\mathcal{M}}(m_1, \ldots, m_n) = m$.

By induction hypothesis we have $\mathcal{M} \models \varphi_{\tau_i}$, so

$$\begin{aligned}\mathcal{M}^{\mathrm{rel}} \models \varphi_\tau &\leftrightarrow \exists \mathtt{t}_{\tau_1} \cdots \exists \mathtt{t}_{\tau_n} (\mathtt{t}_{\tau_1} = m_1 \wedge \cdots \mathtt{t}_{\tau_n} = m_n \wedge \mathtt{R_f} \mathtt{t}_{\tau_1} \cdots \mathtt{t}_{\tau_n}) \\ &\leftrightarrow (\mathtt{R_f} m_1 \cdots m_n \mathtt{t}_\tau) \leftrightarrow \mathtt{t}_\tau = m.\end{aligned}$$

Now we can translate formulas:

1.3.62. Definition. For each formula φ in $\mathcal{L}$, we define a corresponding formula φ^{rel} in $\mathcal{L}^{\mathrm{rel}}$ by induction on φ:

(1) If φ is $\tau_1 = \tau_2$, then φ^{rel} is the formula

$$\exists \mathtt{t}_{\tau_1} \exists \mathtt{t}_{\tau_2} \varphi_{\tau_1} \wedge \varphi_{\tau_2} \wedge \mathtt{t}_{\tau_1} = \mathtt{t}_{\tau_2}.$$

(2) If φ is $\mathtt{R}\tau_1 \ldots \tau_n$, then φ^{rel} is the formula

$$\exists \mathtt{t}_{\tau_1} \cdots \exists \mathtt{t}_{\tau_n} \varphi_{\tau_1} \wedge \cdots \wedge \varphi_{\tau_n} \wedge \mathtt{R} \mathtt{t}_{\tau_1} \cdots \mathtt{t}_{\tau_n}.$$

(3) If φ is $\psi_1 \wedge \psi_2$, then φ^{rel} is $\psi_1^{\mathrm{rel}} \wedge \psi_2^{\mathrm{rel}}$, similarly for $\neg$.

(4) If φ is $\forall \mathtt{x} \psi$, then φ^{rel} is $\forall \mathtt{x} \psi^{\mathrm{rel}}$.

If φ is an $\mathcal{M}$-formula, say $\varphi = \varphi'(\mathtt{x}_1/m_1, \ldots)$, then we let $\varphi^{\mathrm{rel}} = \varphi'^{\mathrm{rel}}(\mathtt{x}_1/m_1, \ldots)$.

1.3.63. Lemma. For any closed $\mathcal{M}$-formula φ:

$$\mathcal{M} \models \varphi \quad \text{iff} \quad \mathcal{M}^{\mathrm{rel}} \models \varphi^{\mathrm{rel}}$$

Proof: Exercise (by induction on φ).

In mathematics it is very common to ask when two distinct structures are essentially the same, or "isomorphic." Now we will study isomorphisms between two models for a fixed language.

1.3.64. Definition. Let $\mathcal{M}$ and $\mathcal{N}$ be two models for the language $\mathcal{L}$. We say that a function $\Phi : M \to N$ is an **isomorphism** if Φ is a 1-1 function from M onto N, and

(i) for every $\mathtt{R} \in \mathcal{R}$ we have

$$\mathcal{M} \models \mathtt{R}(a_1 \dots a_n) \quad \text{iff} \quad \mathcal{N} \models \mathtt{R}(\Phi(a_1) \dots \Phi(a_n)),$$

(ii) for every $\mathtt{F} \in \mathcal{F}$ we have

$$\mathcal{M} \models \mathtt{F}(a_1 \dots a_n) = b \quad \text{iff} \quad \mathcal{N} \models F(\Phi(a_1) \dots \Phi(a_n)) = \Phi(b),$$

(ii) for all c, constant symbols in the language, $\Phi(\mathtt{c}^{\mathcal{M}}) = \mathtt{c}^{\mathcal{N}}$.

1.3.65. Theorem. Let $\Phi : M \to N$ be an isomorphism from M onto N. Let $\psi(\mathtt{x}_1, \dots, \mathtt{x}_n)$ be a formula. Then

$$\mathcal{M} \models \psi(a_1, \dots, a_n) \quad \text{iff} \quad \mathcal{N} \models \psi(\Phi(a_1) \dots \Phi(a_n)).$$

Proof: By induction.

1.3.66. Definition. We say that $\Phi : M \to M$ is an **automorphism** if Φ is an isomorphism from M onto M.

1.3.67. Definition. We say that a set $Q \subseteq M^n$ is **definable in M** if there is a formula $\varphi(\mathtt{x}_1, \dots, \mathtt{x}_n)$ such that

$$(a_1, \dots, a_n) \in Q \quad \text{iff} \quad \mathcal{M} \models \varphi(a_1, \dots, a_n).$$

1.3.68. Fact. If Q_1 and Q_2 (subsets of M^n) are both definable, then so are

$$Q_1 \cup Q_2 \quad Q_1 \cap Q_2 \quad M^n - Q_1 \quad Q_1 - Q_2.$$

Proof: Assume that Q_1 and Q_2 are defined by φ_1 and φ_2. Then
$Q_1 \cup Q_2$ is defined by the formula $\varphi_1 \vee \varphi_2$,
$Q_1 \cap Q_2$ is defined by the formula $\varphi_1 \wedge \varphi_2$,

$M^n - Q_1$ is defined by the formula $(\neg\varphi_1)$, and
$Q_1 - Q_2$ is defined by the formula $\varphi_1 \wedge (\neg\varphi_2)$

1.3.69. Examples.

(a) Let $\langle \mathbb{N}, +, \cdot \rangle$ be the model of natural numbers. Then the following sets are definable:

(i) $\{1\} = \{\mathtt{x} : (\forall \mathtt{y})(\mathtt{y} \cdot \mathtt{x} = \mathtt{y})\}$.

(ii) $\{5\} = \{\mathtt{x} : (\exists \mathtt{y})(\forall \mathtt{z}\, \mathtt{y} \cdot \mathtt{z} = \mathtt{z} \wedge \mathtt{y} + \mathtt{y} + \mathtt{y} + \mathtt{y} + \mathtt{y} = \mathtt{x})\}$.

(iii) $\{p : p \text{ is prime}\} \cup \{1\} = \{\mathtt{x} : (\forall \mathtt{y})(\forall \mathtt{z})(\mathtt{y} \cdot \mathtt{z} = \mathtt{x} \rightarrow \mathtt{y} = \mathtt{x} \vee \mathtt{z} = \mathtt{x})\}$.

(iv) $\{p : p \text{ is prime}\}$, using (i) and (iii).

(b) Let $\mathcal{M} = \langle \mathbf{R}, \{+^{\mathbf{R}}, \cdot^{\mathbf{R}}\}, 0 \rangle$. (where $\mathbf{R}$ is the set of real numbers, and $+^{\mathbf{R}}$, $\cdot^{\mathbf{R}}$ are the usual operations). Then the following sets are definable.

(i) $\mathbf{R}^+ = \{\mathtt{x} : (\exists \mathtt{y})(\mathtt{y} \cdot \mathtt{y} = \mathtt{x})\}$.

(ii) $\{\mathtt{x}\}$, for any algebraic real number $\mathtt{x}$.

1.3.70. Lemma. Let $\Phi : M \rightarrow M$ be an automorphism. Let $Q \subseteq M^n$ be definable. Then Q is **invariant under** Φ, that is,

$$\{(\Phi(a_1), \ldots, \Phi(a_n)) : (a_1, \ldots, a_n) \in Q\} = Q.$$

Proof: Assume $(a_1, \ldots, a_n) \in M^n$, and let Q be defined by $\psi(\mathtt{x}_1, \ldots, \mathtt{x}_n)$. Then

$$(a_1, \ldots, a_n) \in Q \text{ iff } \mathcal{M} \models \psi(a_1, \ldots, a_n)$$
$$\mathcal{M} \models \psi(\Phi(a_1), \ldots, \Phi(a_n)) \text{ iff } (\Phi(a_1), \ldots, \Phi(a_n)) \in Q$$

using 1.3.65.

1.3.71. Example. We will show that the relation $Q = \{\langle m_1, m_2, m_3 \rangle : m_1 + m_2 = m_3\}$ is not definable in $\langle \mathbb{N}, \{\cdot^{\mathbb{N}}\} \rangle$.

Proof: The fundamental theorem of arithmetic says that each natural number can be written in a unique way as the product of powers of prime numbers. So, if n is a natural number, let

$$n = p_1^{k_1} \cdots p_l^{k_l}.$$

Now we define the automorphism $\Phi : \mathbb{N} \rightarrow \mathbb{N}$ that switches 3 and 5 in the prime representation of n. For example,

$$\begin{aligned}\Phi(3) &= 5\\ \Phi(3\cdot 5) &= 5\cdot 3 = 15\\ \Phi(5^5) &= 3^5\\ \Phi(2^2) &= 2^2\,.\end{aligned}$$

Then by lemma 1.3.70, we have that if Q is definable, then

$$Q = \{(\Phi(a_1), \ldots, \Phi(a_n)) : (a_1, \ldots, a_n) \in Q\}.$$

Let us consider $(3, 2, 5) \in Q$. Then $(\Phi(3), \Phi(2), \Phi(5)) = (5, 2, 3) \in Q$, a contradiction.

1.3.72. Notation. In the rest of the book will sometimes employ a less exact but more convenient notation. If φ is a formula with only free variable $\mathbf{x}$, then we may write $\varphi(\mathbf{x})$ instead of φ to emphasize that $\mathbf{x}$ is the free variable of φ.
Instead of $\varphi(\mathbf{x}/\tau)$ we can then write $\varphi(\tau)$.
Similarly, if φ has several free variables, we may write this as $\varphi(\mathbf{x}, \mathbf{y}, \ldots)$, thus establishing an order on the free variables of φ. This makes it possible to simplify the formula $\varphi(\mathbf{x}/\sigma, \mathbf{y}/\tau, \ldots)$ to $\varphi(\sigma, \tau, \ldots)$.

Exercises

Examples

1. Find a formula φ (with free variable $\mathbf{x}$) in the "poor man's language of number theory" (1.3.7) such that $\mathbb{N} \models \varphi(\mathbf{x}/n)$ iff n is a power of 2. (Hint: What property do all divisors of n have?)

2. Give an example of a first order language capable of describing a model which is

(a) a group
(b) a ring
(c) a field
(d) the set of natural numbers
(e) a graph
(f) a tree

3. For each one of the following sets of formulas give an example of a model that satisfies this set of formulas. Try to describe all finite models satisfying this formula.

(a) (1) $R(x_1, x_2) \wedge R(x_2, x_3) \rightarrow R(x_1, x_3)$.
(2) $R(x_1, x_2) \wedge R(x_1, x_3) \wedge R(x_2, x_4) \wedge R(x_3, x_4) \rightarrow R(x_2, x_3) \vee R(x_3, x_2) \vee (x_2 = x_3)$.

(b) (1) $R(x_1, x_2) \wedge R(x_2, x_3) \rightarrow R(x_1, x_3)$.
(2) $R(x_1, x_3) \wedge R(x_2, x_3) \rightarrow R(x_1, x_2) \vee R(x_2, x_1) \vee (x_1 = x_2)$.
(3) $R(c_1, x_1)$.

4. In each one of the following cases find an appropriate first order language and formula such that there are models that satisfy the formula, and every model that satisfies the formula has the property that:

(a) the model is a finite set with exactly n elements (for a given n).
(b) the model is a dense linear ordering (*e.g.*, the rational numbers).
(c) the model is a field.
(d) the model is a field with characteristic 3.

Terms, Formulas, and Substitution

5. Give an explicit definition by induction of $\mathcal{M}$-formulas.

6. Identify all the free variables in the following formulas and match the appropriate quantifier with each variable that is not free.

(a) $\forall x(\exists y P(x, y, z, w) \wedge \forall z \exists x R(y, x, z, w))$
(b) $\exists x \forall y(\forall z Q(x, z, w, s) \wedge \forall x R(x, z, w, t))$

7. If x and y are distinct variables, σ and τ **closed** $\mathcal{M}$-terms, and μ is any $\mathcal{M}$-term, show that

$$\mu(x/\sigma)(y/\tau) = \mu(x/\sigma, y/\tau) = \mu(y/\tau)(x/\sigma).$$

8. Let $A_1, \ldots, A_n$ be sentential symbols, and let σ be a map from $\{A_1, \ldots, A_n\}$ into the set of all formulas of first order logic. By induction, we can extend σ to a map $\bar{\sigma}$ defined on all sentential formulas built only from $\{A_1, \ldots, A_n\}$ as follows:

(1) $\bar{\sigma}(A_i) = \sigma(A_i)$.

(2) $\bar{\sigma}(\alpha @ \beta) = \sigma(\alpha) @ \sigma(\beta)$, for $@ \in \{\wedge, \rightarrow, \leftrightarrow, \vee\}$ and $\bar{\sigma}(\neg\alpha) = \neg\bar{\sigma}(\alpha)$. Show that a first order formula φ is a tautology if and only if there is a sentential tautology α and a map σ from $BKS(\alpha)$ into the set of first order formulas such that $\bar{\sigma}(\alpha) = \varphi$.

9. If $\mathcal{M} \not\models \varphi$, does it follow that $\mathcal{M} \models \neg\varphi$?

10. Find formulas φ and ψ with free variable $\mathbf{x}$ such that $\forall \mathbf{x}\, \varphi \models \forall \mathbf{x}\, \psi$, but NOT $\varphi \models \psi$.

11. Let $\mathbf{x}_1, \ldots, \mathbf{x}_n$ be distinct variables and let φ, ψ be formulas with free variables among $\mathbf{x}_1, \ldots, \mathbf{x}_n$. Show that $\varphi \mathrel{=\!\mid\!\models} \psi$ iff

for all models $\mathcal{M}$, for all (not necessarily distinct) elements $a_1, \ldots, a_n$ in M,

$$M \models \varphi(\mathbf{x}_1/a_1, \ldots \mathbf{x}_n/a_n) \text{ iff } \mathcal{M} \models \psi(\mathbf{x}_1/a_1, \ldots \mathbf{x}_n/a_n)$$

i.e., the "meaning" of φ and ψ is the same in all models.

12. Prove 1.3.44(2): If $\Gamma \cup \{\varphi\}$ is a set of closed formulas, then

$$\Gamma \cup \{\varphi\} \models \psi \quad \text{iff} \quad \Gamma \models \varphi \rightarrow \psi$$

<u>The Restricted Language</u>

13. Prove 1.3.45

14. Prove lemma 1.3.48.

15. Complete the proof of 1.3.50.

16. Prove fact 1.3.53, and find an example where *allow*$(\varphi, \tau, \mathbf{x})$ holds, but neither of the conditions in 1.3.53 is satisfied.

17. Suppose that the substitutions $\varphi(\mathbf{x}/\mathbf{y})$ and $\varphi(\mathbf{x}/\mathbf{y})(\mathbf{y}/\mathbf{x})$ are permitted.

(1) Is it true that $\varphi = \varphi(\mathbf{x}/\mathbf{y})(\mathbf{y}/\mathbf{x})$? Explain why!

(2) If your answer is no, provide a necessary and sufficient condition for equality.

18. Let $\mathbf{x}$, $\mathbf{y}$ be distinct variables, and let τ be the term $\mathbf{y}$, and $\mathbf{c}$ a constant symbol. Let $\tau' = \tau(\mathbf{x}/\mathbf{y}, \mathbf{y}/\mathbf{c})$, i.e., $\tau' = \mathbf{c}$. Find a formula φ such that

$$\varphi(\mathbf{x}/\tau)(\mathbf{y}/\mathbf{c}) \neq \varphi(\mathbf{y}/\mathbf{c})(\mathbf{x}/\tau').$$

19. Prove lemma 1.3.23: If τ is an $\mathcal{M}$-term, then there are variables $\mathtt{y}_1$, …, $\mathtt{y}_n$ and elements m_1, …, m_n of M, and there is term τ' such that $\tau = \tau'(\mathtt{y}_1/m_1, \ldots, \mathtt{y}_n/m_n)$.

20. Prove 1.3.43: If $\tau^{\mathcal{M}} = a$, then $\mathcal{M} \models \varphi(x/\tau)$ iff $\mathcal{M} \models \varphi(x/a)$.

21. Prove lemma 1.3.59.

22. Prove lemma 1.3.63: For any closed $\mathcal{M}$-formula φ:

$$\mathcal{M} \models \varphi \quad \text{iff} \quad \mathcal{M}^{\text{rel}} \models \varphi^{\text{rel}}.$$

23. Find a purely relational language for groups, and translate the group axioms into this language.

Orders

24. $\mathtt{R}(\mathtt{u},\mathtt{v})$ is an **order relation** in a model $\mathcal{M}$ if the following formulas are valid in the model:

(i) $\neg\mathtt{R}(\mathtt{u},\mathtt{u})$.
(ii) $\mathtt{R}(\mathtt{u},\mathtt{v}) \rightarrow \neg\mathtt{R}(\mathtt{v},\mathtt{u})$.
(iii) $\mathtt{R}(\mathtt{u}_1,\mathtt{u}_2) \wedge \mathtt{R}(\mathtt{u}_2,\mathtt{u}_3) \rightarrow \mathtt{R}(\mathtt{u}_1,\mathtt{u}_3)$.

Given a formula $\varphi(\mathtt{x}_1)$, we define a subset $A_\varphi \subseteq M$ by

$$A_\varphi := \{a \in M : \mathcal{M} \models \varphi(\mathtt{x}/a)\}.$$

Similarly, if we have a formula with two free variables $\varphi(\mathtt{x}_1, \mathtt{x}_2)$, we define $A_\varphi \subseteq M \times M$, a subset of the set of pairs from M, by

$$A_\varphi := \{\langle a, b\rangle : a \in M, b \in M, \mathcal{M} \models \varphi(\mathtt{x}_1/a, \mathtt{x}_2/b)\}.$$

We call A the set characterized by φ. What are the sets characterized by the following formulas?

(a) $\mathtt{R}$ is an order relation in the model $\mathcal{M}$ and $\varphi(\mathtt{u},\mathtt{v}) = \neg\mathtt{R}(\mathtt{u},\mathtt{v})$.
(b) $\mathtt{R}$ is an order relation in the model $\mathcal{M}$ and $\varphi(\mathtt{u},\mathtt{v}) = \neg\mathtt{R}(\mathtt{u},\mathtt{v}) \wedge \neg\mathtt{R}(\mathtt{v},\mathtt{u})$.
(c) $\mathtt{R}$ is an order relation in the model $\mathcal{M}$ which is a tree, (*i.e.*, the following formula is valid in the model: $\mathtt{R}(\mathtt{u}_1,\mathtt{u}_2) \wedge \mathtt{R}(\mathtt{u}_1,\mathtt{u}_3) \rightarrow \mathtt{R}(\mathtt{u}_2,\mathtt{u}_3) \vee \mathtt{R}(\mathtt{u}_3,\mathtt{u}_2) \vee (\mathtt{u}_2 = \mathtt{u}_3)$)

and
$\varphi(u, v) = \neg R(v, u)$.

25. Describe all the models that satisfy the statement: $\forall x \forall y \exists z (R(x, y) \rightarrow R(x, z) \wedge R(z, y))$, where R is an order relation.

26. Let $\mathcal{L}$ be a first order language **without the equality symbol** and with exactly n unary relation symbols. For every statement φ in $\mathcal{L}$, let K_φ be the minimal number of elements in any model satisfying φ. Find a statement φ for which K_φ is maximal (*i.e.*, for every ψ in $\mathcal{L}$, $K_\varphi \geq K_\psi$). What is K_φ?

Validity

27. (a) Show $\models \exists u(\varphi \vee \psi) \leftrightarrow \exists u\varphi \vee \exists u\psi$.

(b) Assume that u is not free in φ. Show $\models \exists u(\varphi \wedge \psi) \leftrightarrow \varphi \wedge \exists u\psi$ and $\models \forall u(\varphi \vee \psi) \leftrightarrow \varphi \vee \forall u\psi$.

28. In a language that has one unary relation symbol find:

(a) formulas φ, ψ for which
$\forall u(\varphi \vee \psi)$ is not equivalent (i.e., in the sense of $\dashv\models$) to
$\forall u\varphi \vee \forall u\psi$.

(b) formulas φ, ψ for which
$\exists u(\varphi \wedge \psi)$ is not equivalent to $\exists u\varphi \wedge \exists u\psi$.

29. For each of the following formulas prove that the formula is valid or give a model to show that it is not valid.

(a) $\forall u(P(u) \rightarrow R(u)) \rightarrow (\forall uP(u) \rightarrow \forall uR(u))$.
(b) $(\forall uP(u) \rightarrow \forall uR(u)) \rightarrow \forall u(P(u) \rightarrow R(u))$.

30. Which of the following formulas are always valid? Prove your answer.

(a) $\forall u(\varphi \rightarrow \psi) \rightarrow (\forall u\varphi \rightarrow \forall u\psi)$.
(b) $(\forall u\varphi \rightarrow \forall u\psi) \rightarrow \forall u(\varphi \rightarrow \psi)$,
where u is not free in ψ.
(c) $(\forall u\varphi \rightarrow \forall u\psi) \rightarrow \forall u(\varphi \rightarrow \psi)$,
where u is not free in φ.

31. Let $\varphi(u, v_1, \ldots, v_n)$ be a formula in the language $\mathcal{L}$ with free variables $u, v_1, \ldots, v_n$ only, and let x, y be two variables that do not appear in φ. Prove or disprove the validity of each one of the following formulas:

(a) $\forall u\varphi(u, v_1, \ldots, v_n) \rightarrow \exists u\varphi(u, v_1, \ldots, v_n)$.
(b) $\forall u\varphi(u, v_1, \ldots, v_n) \rightarrow \forall x\forall y\varphi(F(x, y), v_1, \ldots, v_n)$, where F is a function symbol in $\mathcal{L}$.
(c) $\exists u\varphi(u, v_1, \ldots, v_n) \rightarrow \exists x\exists y\varphi(F(x, y), v_1, \ldots, v_n)$, where F is a function symbol in $\mathcal{L}$.

32. Let $\mathcal{L}_1 = \{P_1, F_1, c_1\}$.

(1) Find two different interpretations for $\mathcal{L}_1$ in a model whose universe is the set of natural numbers, $P_1(x, y)$ has the interpretation $x \leq y$, and such that the following is valid in both interpretations:
$\forall x(P_1(F_1(x, c_1), x) \wedge P_1(x, F(x, c_1)))$.
(2) Does the following sentence hold in both interpretations?
$\forall x\forall y(F_1(x, y) = F_1(y, x))$.
(3) If your answer to (2) was yes, find a third interpretation for which the sentence does not hold. If your answer was no, find a third interpretation for which it does.

33. Let φ be the following formula

$$[\forall xP(x, x) \wedge \forall x\forall y\forall z(P(x, y) \wedge P(y, z) \rightarrow P(x, z)) \wedge \forall x\forall y(P(x, y) \vee P(y, x))] \rightarrow [\exists x\forall yP(x, y)].$$

Prove that any finite model (a model with a finite universe) is a model of φ but there exist infinite models that do not satisfy φ.

1.4. Proof Systems

In this section we will study the concept of proof. The mathematics of this century have been strongly influenced by Euclid's axiomatic method. Typical examples are geometry, number theory, and set theory. An axiomatic system provides a rigorous way to establish the basic principles from which mathematicians develop a subject.

In mathematical logic we also use an axiomatic system. The only difference is that we also formalize the concept of proof, defining a **proof system**, also called a **deductive system**. Schematically, we will have a *set of axioms* and a *set of rules*. Given this we will obtain the theorems by applying the rules of our proof system to the axioms and to the previously proved theorems.

What shall we choose as the set of axioms for our logic? This is not an easy question. Usually it takes a long time to find an adequate set of axioms which satisfies the taste of mathematicians. Our only criterion is that the axioms be simple. People also usually want them to be "economical" in the sense that the chosen set of axioms should be a minimal set.

We are interested in finding a set of axioms able to capture "logical truth." This means that if Γ is the collection of all formulas $\varphi \in \mathcal{L}$ that we consider "logically true" then φ should be a consequence of our axioms, and conversely, every consequence of the axioms should be a member of Γ. Of course, every axiom must be a member of $\mathcal{L}$.

"Logical truth" has a simple definition in view of our discussion in section 1.3: We say that a formula φ is logically true iff φ is valid, i.e., φ is valid in every model for the language $\mathcal{L}$. We write this as $\models \varphi$.

Thus, the simplest set of logical axioms would be

$$\Gamma := \{\varphi \in \mathcal{L} : \varphi \text{ is valid}\}$$

and we could say that φ is a theorem iff φ is an axiom. That is, we would have an inductive system where Γ is the set of blocks, and there are no operators.

This is a very trivial deduction system, and it clearly satisfies our requirement, but the only method to find out whether a formula is an axiom is to check the validity of this formula in every model.

The main objection to this axiomatic system is that it does not give a **syntactic** method that we can use to find out if a formula is true. A "syntactic method" only considers the formulas themselves, as finite strings of symbols, and does not have to check through an infinity of possible models. In this section we will develop a syntactic logical deduction system. Every formula that can be derived in this system will be true (i.e., valid in all models). The main theorem of the next chapter will show that our system is the right one, i.e., that every true formula can be derived in our deductive system.

In the previous sections we showed that we can restrict our language to only the connectives $\{\neg, \wedge\}$ and the quantifier $\forall$ and still be able to express every other formula of the language of first order logic in an elementarily equivalent way. Therefore, in order to make induction on formulas easier, we will work with this restricted language. (When we use other logical connectives or the quantifier $\exists$, it should be understood that they are used as abbreviations only, e.g. $\exists x\varphi$ is an abbreviation for $\neg\forall x\neg\varphi$, $\varphi \rightarrow \psi$ abbreviates $\neg(\varphi \wedge \neg\psi)$, etc.)

For this section, we fix an arbitrary first order language $\mathcal{L}$.

The axioms will be classified into several groups:

First we will give a subset of the axioms, the "pure" axioms. Then we will define a logical axiom to be any universal quantification of a pure axiom.

There will be seven groups of "pure" axioms, defined as follows:

1.4.1. Axiom Group I — Tautology axioms. Let φ be a formula. If φ is a tautology (see 1.3.35) then φ is a pure axiom.

1.4.2. Axiom Group II — Distributivity axioms. If φ and ψ are formulas, and $\mathbf{x}$ a variable, then

$$\forall \mathbf{x}(\varphi \to \psi) \to (\forall \mathbf{x}\, \varphi \to \forall \mathbf{x}\, \psi)$$

is a pure axiom.

1.4.3. Axiom Group III — Substitution axioms. If φ is a formula, $\mathbf{x}$ a variable and τ a term, and <u>*allow*$(\varphi, \tau, \mathbf{x})$ holds</u>, then

$$\forall \mathbf{x}\, \varphi \to \varphi(\mathbf{x}/\tau)$$

is a pure axiom.

1.4.4. Axiom Group IV — Generalization axioms. If φ is a formula, $\mathbf{x}$ is a variable, and <u>$\mathbf{x}$ is not in Free(φ)</u>, then

$$\varphi \to \forall \mathbf{x}\, \varphi$$

is a pure axiom.

1.4.5. Axiom Group V — Equality axioms. If $\mathbf{x}$, $\mathbf{y}$ and $\mathbf{z}$ are variables (not necessarily distinct), then

$$\mathbf{x} = \mathbf{x}$$

$$\mathbf{x} = \mathbf{y} \to \mathbf{y} = \mathbf{x}$$

$$\mathbf{x} = \mathbf{y} \wedge \mathbf{y} = \mathbf{z} \to \mathbf{x} = \mathbf{z}$$

are pure axioms.

1.4.6. Axiom Group VI — Equivalence axioms for relations. Assume that R is a k-ary relation symbol and $x_1, \dots, x_k$ and $y_1, \dots, y_k$ are variables. Then

$$x_1 = y_1 \wedge \cdots \wedge x_k = y_k \rightarrow (R(x_1, \dots, x_k) \leftrightarrow R(y_1, \dots, y_k))$$

is a pure axiom.

1.4.7. Axiom Group VII — Equivalence axioms for functions . Assume that F is a k-ary function symbol and $x_1, \dots, x_k$ and $y_1, \dots, y_k$ are variables. Then

$$x_1 = y_1 \wedge \cdots \wedge x_k = y_k \rightarrow F(x_1, \dots, x_k) = F(y_1, \dots, y_k)$$

is a pure axiom.

1.4.8. Definition. We say that a formula φ is a **logical axiom**, if there is pure axiom ψ and variables $y_1, \dots, y_n$ such that

$$\varphi = \forall y_1 \cdots \forall y_n \, \psi.$$

Here we also allow $n = 0$, that is, the pure axioms are special cases of logical axioms.

We list examples of pure axioms in the table on the next page.

Our next task is to define the deductive system which allows us to prove logical theorems from axioms. There will be only one rule. This rule is called "modus ponens." An example of everyday use of modus ponens is the following:

(1) We know that lightning is always followed by thunder.
(2) We see lightning.
(3) Therefore, we know there will be thunder.

Formally, we can give the following definition.

1.4.9. Definition. Assume φ_1 and φ_2 are formulas, and ψ is a formula. Then

ψ is derived from φ_1 and φ_2 by **modus ponens (MP)**

just means

φ_1 is of the form $\varphi_2 \rightarrow \psi$

Pure Axioms

In the following table, we use as an example a language $\mathcal{L}$ that has a one place relation symbol $\mathtt{R}$, a two place relation symbol $\mathtt{Q}$, a one place function symbol $\mathtt{f}$, and a constant symbol $\mathtt{c}$. $\mathtt{v}, \mathtt{x}, \mathtt{y}, \mathtt{z}$ are distinct variables.

Group	Axiom Name	Example
I	Tautology	$[\mathtt{R}(\mathtt{x}) \wedge \forall \mathtt{y}\mathtt{Q}(\mathtt{x},\mathtt{y})] \rightarrow \mathtt{R}(\mathtt{x})$
II	Distributivity	$\forall \mathtt{x}(\mathtt{R}(\mathtt{x}) \rightarrow \mathtt{x} = \mathtt{c}) \rightarrow [(\forall \mathtt{x}\, \mathtt{R}(\mathtt{x})) \rightarrow (\forall \mathtt{x}\, \mathtt{x} = \mathtt{c})]$
III	Substitution	$\forall \mathtt{x}(\mathtt{R}(\mathtt{x}) \rightarrow \exists \mathtt{y}\mathtt{Q}(\mathtt{x},\mathtt{y})) \rightarrow$ $\rightarrow (\mathtt{R}(\mathtt{f}(\mathtt{c})) \rightarrow \exists \mathtt{y}\mathtt{Q}(\mathtt{f}(\mathtt{c}),\mathtt{y}))$
IV	Generalization	$\exists \mathtt{y}\mathtt{Q}(\mathtt{x},\mathtt{y}) \rightarrow \forall \mathtt{z}\, \exists \mathtt{y}\mathtt{Q}(\mathtt{x},\mathtt{y})$
V	Equality 1	$\mathtt{x} = \mathtt{x}$
	Equality 2	$\mathtt{x} = \mathtt{y} \rightarrow \mathtt{y} = \mathtt{x}$
	Equality 3	$\mathtt{x} = \mathtt{y} \wedge \mathtt{y} = \mathtt{z} \rightarrow \mathtt{x} = \mathtt{z}$
VI	Equiv. (relations)	$\mathtt{x} = \mathtt{y} \wedge \mathtt{z} = \mathtt{v} \rightarrow (\mathtt{Q}(\mathtt{x},\mathtt{z}) \leftrightarrow \mathtt{Q}(\mathtt{y},\mathtt{v}))$
VII	Equiv. (functions)	$\mathtt{x} = \mathtt{y} \rightarrow \mathtt{f}(\mathtt{x}) = \mathtt{f}(\mathtt{y})$

In other words, whenever we have the formulas φ_2 and $\varphi_2 \to \psi$, we can say that ψ can be derived from these two formulas by MP.

In the previous example, if we write φ_2 for "there is lightning," and ψ for "there will be thunder," we have

(1) If φ_2, then ψ.
(2) φ_2.
(3) Therefore, ψ.

Now we are ready to introduce the fundamental concept of this section:

1.4.10. Definition. Let $\langle \varphi_1, \ldots, \varphi_n \rangle = \bar{\varphi}$ be a sequence of formulas. We say that $\bar{\varphi}$ is a **derivation** if for each j, $1 \leq j \leq n$, (at least) one of the following conditions holds

(1) φ_j is a logical axiom.
(2) there exists $i, k < j$ such that φ_j is derived from φ_i and φ_k by MP. (I.e., φ_i is the formula $\varphi_k \to \varphi_j$.)

1.4.11. Definition. Let φ be a formula. We say that φ is a **logical theorem** if there is a derivation $\langle \varphi_1, \ldots, \varphi_n \rangle$ with $\varphi_n = \varphi$. We write this as $\vdash \varphi$.

In other words, a formula is a theorem if it can be obtained from the axioms using the rule of modus ponens. We can also express this fact in the following way:

1.4.12. Remark. The set of theorems is an inductive structure: The blocks are given by the axioms, and the only operator is the function MP, defined by

$$MP(\varphi, \psi) = \begin{cases} \chi, & \text{if } \psi = \varphi \to \chi \text{ (so } MP(\varphi, \varphi \to \chi) = \chi) \\ \psi & \text{otherwise.} \end{cases}$$

When we prove something "by induction on the derivation of φ," we mean that we use induction on this structure.

In mathematics we usually work in some specific branch, like number theory. Each specific branch of mathematics has its own set of axioms, call it Γ. We can say that a theorem is obtained from Γ by applying some rules of inference which are accepted by the mathematical community. In mathematical logic we establish explicitly how to prove theorems from Γ. The following is our definition of the formal concept of proof. This will lead us to the definition of theorem.

1.4.13. Definition. Let $\langle \varphi_1, \ldots, \varphi_n \rangle = \bar{\varphi}$ be a sequence of formulas. We say that $\bar{\varphi}$ is a *derivation from* Γ if for each j, $1 \leq j \leq n$, (at least) one of the following conditions holds

(1) φ_j is a logical axiom.
(2) there exists $i, k < j$ such that $\varphi_i = \varphi_k \to \varphi_j$.
(3) φ_j is in Γ. (In this case, we call φ_j a nonlogical axiom.)

1.4.14. Definition. Let Γ be a set of formulas and let φ be a formula. We say that φ is a theorem of Γ, in symbols

$$\Gamma \vdash \varphi$$

if there is a derivation $\langle \varphi_1, \ldots, \varphi_n \rangle$ from Γ such that $\varphi = \varphi_n$.

1.4.15. Corollary.

(1) Every formula of Γ is a theorem of Γ.
(2) Every logical axiom is a theorem of Γ.
(3) If $\Gamma \subseteq \Gamma'$, then every theorem of Γ is also a theorem of Γ'.

1.4.16. Corollary. Let Γ be a set of formulas. Then $\Gamma \vdash \varphi$ iff there is a finite $\Gamma_0 \subseteq \Gamma$ such that $\Gamma_0 \vdash \varphi$.

Proof: Each derivation step uses at most one nonlogical axiom from Γ, and there are only finitely many steps.

Formally, let $\langle \varphi_1, \ldots, \varphi_n \rangle$ be a derivation of φ from Γ. Let ψ_0 be an element of Γ. (If Γ is empty, or even finite, then there is nothing to prove.) To each $i \leq n$ we assign a formula $\psi_i \in \Gamma$ as follows:

(1) If φ_i is in Γ, then $\psi_i := \varphi_i$.
(2) If φ_i is not in Γ, then $\psi_i := \psi_0$.

Note that in case 2, φ is not a nonlogical axiom, so either it must be a logical axiom, or derived by modus ponens from two previous formulas. So $\langle \varphi_1, \ldots, \varphi_n \rangle$ is really a derivation from $\{\psi_1, \ldots, \psi_n\}$.

1.4.17. Remark. Assume that $\Gamma \vdash \varphi$ and that $\varphi \Rightarrow \psi$. (I.e., $\varphi \to \psi$ is a tautology, see 1.3.35.) Then $\Gamma \vdash \psi$. In this case we say that we obtained $\Gamma \vdash \psi$ from $\Gamma \vdash \varphi$ by "tautological reasoning."

Proof: Since $\varphi \to \psi$ is a tautology, it is an axiom. So $\Gamma \vdash \varphi \to \psi$. By modus ponens we can conclude $\Gamma \vdash \psi$.

1.4.18. Remark. Instead of saying "$\langle\varphi_1, \ldots, \varphi_n\rangle$ is a derivation from Γ," we write

$$\begin{array}{ll} 1 & \Gamma \vdash \varphi_1 \\ & \ldots \\ n & \Gamma \vdash \varphi_n. \end{array}$$

1.4.19. Example. We will show how to derive $\{\forall x(\varphi \to \psi)\} \vdash \exists x\varphi \to \exists x\psi$.

(Remember that $\exists x$ is really an abbreviation for $\neg\forall x\neg$.)

1	$\forall x(\varphi \to \psi) \vdash \forall x((\varphi \to \psi) \to (\neg\psi \to \neg\varphi))$	(Group I)
2	$\forall x(\varphi \to \psi) \vdash \forall x((\varphi \to \psi) \to (\neg\psi \to \neg\varphi)) \to$ $\to (\forall x(\varphi \to \psi) \to \forall x(\neg\psi \to \neg\varphi))$	(Group II)
3	$\forall x(\varphi \to \psi) \vdash (\forall x(\varphi \to \psi) \to \forall x(\neg\psi \to \neg\varphi))$	(modus ponens)
4	$\forall x(\varphi \to \psi) \vdash \forall x(\varphi \to \psi)$	(nonlogical axiom)
5	$\forall x(\varphi \to \psi) \vdash \forall x(\neg\psi \to \neg\varphi)$	(modus ponens)
6	$\forall x(\varphi \to \psi) \vdash \forall x(\neg\psi \to \neg\varphi) \to (\forall x\neg\psi \to \forall x\neg\varphi)$	(Group II)
7	$\forall x(\varphi \to \psi) \vdash \forall x\neg\psi \to \forall x\neg\varphi$	(modus ponens)
8	$\forall x(\varphi \to \psi) \vdash (\forall x\neg\psi \to \forall x\neg\varphi) \to (\neg\forall x\neg\varphi \to \neg\forall x\neg\psi)$	(Group I)
9	$\forall x(\varphi \to \psi) \vdash \neg\forall x\neg\varphi \to \neg\forall x\neg\psi$	(modus ponens)

(Note: By 1.4.23 below, this implies $\emptyset \vdash \forall x[\varphi \to \psi] \to [\exists x\varphi \to \exists x\psi]$.)

1.4.20. Proposition ("Soundness"). Let Γ be a set of closed formulas. Assume that $\mathcal{M}$ is a model such that $\mathcal{M} \models \Gamma$. Let φ be a formula such that $\Gamma \vdash \varphi$. Then $\mathcal{M} \models \varphi$.
(In other words: If $\Gamma \vdash \varphi$, then $\Gamma \models \varphi$.)

Proof: By induction on the derivation, i.e., by induction on the inductive structure described in 1.4.12:

(i) If φ is in Γ, then by assumption $\mathcal{M} \models \varphi$.
(ii) If φ is a logical axiom, then $\mathcal{M} \models \varphi$, because φ is valid.
(iii) If φ is derived from $\psi \to \varphi$ and ψ, then by induction hypothesis $\mathcal{M} \models \psi \to \varphi$ and $\mathcal{M} \models \psi$, so by 1.3.44 $\mathcal{M} \models \varphi$.

1.4.21. Corollary. If φ is a logical theorem (1.4.11) and $\mathcal{M}$ is a model, then $\mathcal{M} \models \varphi$.

1.4.22. Corollary.

(1) If $\Gamma \vdash \varphi \to \psi$ and $\Gamma \vdash \varphi$, then $\Gamma \vdash \psi$.
(2) If $\Gamma \vdash \varphi \to \psi$, then $\Gamma \cup \{\varphi\} \vdash \psi$.

The following theorem shows that this last implication can be reversed:

1.4.23. Theorem (Deduction). If $\Gamma \cup \{\varphi\} \vdash \psi$, then $\Gamma \vdash \varphi \to \psi$.

Proof: Let

$$\Gamma \cup \{\varphi\} \vdash \psi_1$$
$$\cdots$$
$$\Gamma \cup \{\varphi\} \vdash \psi_n$$

be a derivation of $\Gamma \cup \{\varphi\} \vdash \psi$ $(= \psi_n)$. We will produce a derivation of $\Gamma \vdash \varphi \to \psi_n$ by replacing each line $\Gamma \cup \{\varphi\} \vdash \psi_k$ by one or three lines of the form $\Gamma \vdash \ldots$, ending in $\Gamma \vdash \varphi \to \psi_k$.

If ψ_k was a logical axiom, or a nonlogical axiom in Γ, replace $\Gamma \cup \{\varphi\} \vdash \psi_k$ by

$$\begin{array}{ll} \Gamma \vdash \psi_k & \text{(logical or nonlogical axiom)} \\ \Gamma \vdash \psi_k \to (\varphi \to \psi_k) & \text{(group I (tautology))} \\ \Gamma \vdash \varphi \to \psi_k & \text{(modus ponens)} \end{array}$$

If ψ_k is the nonlogical axiom φ, then $\varphi \to \psi_k$ is really the tautology $\varphi \to \varphi$, so we can replace $\Gamma \cup \{\varphi\} \vdash \psi_k$ by

$$\Gamma \vdash \varphi \to \psi_k \quad \text{(Group I)}$$

If ψ_k was obtained by modus ponens, say from $\Gamma \cup \{\varphi\} \vdash \psi_i \to \psi_k$ and $\Gamma \cup \{\varphi\} \vdash \psi_i$, then these two lines have already been replaced by lines that contain the lines

$$\Gamma \vdash \varphi \to (\psi_i \to \psi_k)$$

and

$$\Gamma \vdash \varphi \to \psi_i.$$

Replace $\Gamma \cup \{\varphi\} \vdash \psi_k$ by

$$\begin{array}{ll} \Gamma \vdash (\varphi \to (\psi_i \to \psi_k)) \to ((\varphi \to \psi_i) \to (\varphi \to \psi_k)) & \text{(group I)} \\ \Gamma \vdash (\varphi \to \psi_i) \to (\varphi \to \psi_k) & \text{(modus ponens)} \\ \Gamma \vdash \varphi \to \psi_k & \text{(modus ponens)} \end{array}$$

After having replaced all lines $\Gamma \cup \{\varphi\} \vdash \psi_k$ of the original derivation we get a derivation of $\Gamma \vdash \varphi \to \psi$.

See exercise 6 for a reformulation of this proof.

1.4.24. Definition. We say that Γ is **inconsistent** if there is φ such that $\Gamma \vdash \varphi$ and $\Gamma \vdash \neg\varphi$. Γ is **consistent** if it is not inconsistent.

1.4.25. Corollary. Γ is inconsistent iff $\Gamma \vdash \psi$ for *every* ψ.

Proof: Note that $\neg\varphi \to (\varphi \to \psi)$ is a tautology. Hence, if $\Gamma \vdash \varphi$ and $\Gamma \vdash \neg\varphi$, then by applying modus ponens twice we get $\Gamma \vdash \psi$.

1.4.26. Theorem. $\Gamma \vdash \varphi$ iff $\Gamma \cup \{\neg\varphi\}$ is inconsistent.

Proof: Assume that $\Gamma \vdash \varphi$. Then also

$$\Gamma \cup \{\neg\varphi\} \vdash \varphi \quad \text{(since already } \Gamma \vdash \varphi\text{)}$$

and

$$\Gamma \cup \{\neg\varphi\} \vdash \neg\varphi. \quad \text{(nonlogical axiom)}$$

Therefore $\Gamma \cup \{\neg\varphi\}$ is inconsistent.

Conversely, assume that $\Gamma \cup \{\neg\varphi\}$ is inconsistent. Then by 1.4.25, $\Gamma \cup \{\neg\varphi\} \vdash \varphi$. By the deduction theorem,

$$\Gamma \vdash \neg\varphi \to \varphi$$

Using the tautology $(\neg\varphi \to \varphi) \to \varphi$, we obtain $\Gamma \vdash \varphi$.

1.4.27. Theorem (Generalization). Let Γ be a set of formulas and let x be a variable such that for every ψ in Γ the variable x is not in Free(ψ). Assume that $\Gamma \vdash \varphi$. Then $\Gamma \vdash \forall x\varphi$.

Proof: By induction on the length of the proof.
Case (1): If φ is a logical axiom then $\forall x\varphi$ is a logical axiom.
Case (2): If φ is in Γ, then x does not belong to Free(φ). Therefore, $\varphi \to \forall x\varphi$ is a logical axiom (Group IV). Therefore $\Gamma \vdash \varphi \to \forall x\varphi$ and $\Gamma \vdash \varphi$, so $\Gamma \vdash \forall x\varphi$.
Case (3): There is a formula ψ such that

$$\Gamma \vdash \psi \text{ and } \Gamma \vdash \psi \to \varphi.$$

Therefore $\Gamma \vdash \forall x\psi$ and $\Gamma \vdash \forall x(\psi \to \varphi)$.

But $\forall x(\psi \to \varphi) \to (\forall x\psi \to \forall x\varphi)$ is a logical axiom (Group II). Hence by applying MP,

$$\Gamma \vdash \forall x\psi \to \forall x\varphi$$

and by applying MP again

$$\Gamma \vdash \forall x\varphi.$$

1.4.28. Example. It is impossible to avoid the assumption that x is not free in the formulas of Γ. The following exemplifies this.

We know that

$$x = y \vdash x = y.$$

Applying the WRONG principle we get

$$x = y \vdash \forall y(x = y).$$

By the deduction theorem, this would imply

$$\vdash x = y \rightarrow \forall y(x = y)$$

and by 1.4.27 this would imply

$\vdash \forall y(x = y \rightarrow \forall y(x = y))$	(Generalization)
$\vdash \forall y(x = y \rightarrow \forall y(x = y)) \rightarrow (x = x \rightarrow \forall y(x = y))$	(Group III)
$\vdash x = x \rightarrow \forall y(x = y))$	(modus ponens)
$\vdash x = x$	(Group V)
$\vdash \forall y(x = y)$	(modus ponens)

hence (again by 1.4.27), $\vdash \forall x \forall y(x = y)$. But this clearly contradicts 1.4.20.

1.4.29. Remark. 1.4.27 corresponds to the familiar line of reasoning: "From the hypotheses Γ, we have concluded that x has the property φ. But as x was arbitrary, *every* x has the property φ."

1.4.30. Lemma (Introduction of $\forall$). Assume that φ and ψ are formulas, x is a variable, and τ is a term. Then

(1) IF *allow*(ψ, τ, x), and if

$$\Gamma \vdash \psi(x/\tau) \rightarrow \varphi$$

THEN

$$\Gamma \vdash \forall x \psi \rightarrow \varphi.$$

(2) IF x is not free in ψ and x is not free in any formula of Γ, and if

$$\Gamma \vdash \psi \rightarrow \varphi,$$

THEN

$$\Gamma \vdash \psi \rightarrow \forall x \varphi$$

Proof of (1):

1 $\Gamma \vdash \psi(x/\tau) \to \varphi$	(By assumption)
2 $\Gamma \vdash \forall x\, \psi \to \psi(x/\tau)$	(Substitution axiom)
3 $\Gamma \vdash (\forall x\, \psi \to \psi(x/\tau)) \to$ $\to [(\psi(x/\tau) \to \varphi) \to (\forall x\, \psi \to \varphi)]$	(tautology)
4 $\Gamma \vdash (\psi(x/\tau) \to \varphi) \to (\forall x\, \psi \to \varphi)$	(modus ponens from 2,3)
5 $\Gamma \vdash \forall x\, \psi \to \varphi$	(modus ponens from 1,4)

Proof of (2):

1 $\Gamma \cup \{\psi\} \vdash \psi \to \varphi$	(by assumption)
2 $\Gamma \cup \{\psi\} \vdash \psi$	(nonlogical axiom)
3 $\Gamma \cup \{\psi\} \vdash \varphi$	(MP)
4 $\Gamma \cup \{\psi\} \vdash \forall x\, \psi$	(Generalization theorem)
5 $\Gamma \vdash \psi \to \forall x\, \psi$	(Deduction theorem)

1.4.31. Lemma (Introduction of $\exists$). Assume that φ and ψ are formulas, x is a variable, and τ is a term. Then

(1) IF τ may be substituted for x in φ (i.e., *allow*(φ, τ, x)) and if

$$\Gamma \vdash \varphi \to \psi(x/\tau)$$

THEN

$$\Gamma \vdash \varphi \to \exists x \psi.$$

(2) IF x is not free in ψ and x is not free in any formula of Γ, and if

$$\Gamma \vdash \varphi \to \psi$$

THEN

$$\Gamma \vdash \exists x \varphi \to \psi.$$

Proof: (1) follows (using tautological reasoning) from

$$\vdash \forall x \neg \psi \to (\neg \psi)(x/\tau).$$

For the proof of (2), note $\Gamma \vdash \varphi \to \psi$ is tautologically equivalent to $\Gamma \vdash \neg\psi \to \neg\varphi$. So by the previous lemma we get

$$\Gamma \vdash \neg\psi \to \forall x \neg\varphi$$

and hence again by tautological reasoning $\Gamma \vdash \exists x \varphi \to \psi$.

1.4.32. Example. We will prove that

$$\vdash \exists x \forall y \varphi \to \forall y \exists x \varphi$$

(if x and y are distinct variables).

Proof:

$\vdash \varphi \to \varphi$	(Tautology)
$\vdash \varphi \to \exists x \varphi$	(by 1.4.31(1), since x may be substituted for x)
$\vdash \forall y \varphi \to \exists x \varphi$	(by 1.4.30(1), since y may be substituted for y)
$\vdash \forall y \varphi \to \forall y \exists x \varphi$	(by 1.4.30(2), since y is not free in $\forall y \varphi$)
$\vdash \exists x \forall y \varphi \to \forall y \exists x \varphi$	(by 1.4.31(2), since x is not free in $\forall y \exists x \varphi$.)

1.4.33. Definition. We write $\varphi \dashv\vdash \psi$ ("φ is equivalent to ψ") for $\vdash \varphi \leftrightarrow \psi$. (Later we will see that $\varphi \dashv\vdash \psi$ iff φ and ψ are elementarily equivalent.)

1.4.34. Remark. $\varphi \dashv\vdash \psi$ iff $\varphi \vdash \psi$ and $\psi \vdash \varphi$.

Proof: Exercise.

1.4.35. Fact.

(1) $\dashv\vdash$ is an equivalence relation.

(2) If $\varphi \dashv\vdash \psi$, then $\neg\varphi \dashv\vdash \neg\psi$, $\forall x \varphi \dashv\vdash \forall x \psi$ and $\exists x \varphi \dashv\vdash \exists x \psi$. Also, if $\varphi_1 \dashv\vdash \psi_1$ and $\varphi_2 \dashv\vdash \psi_2$, then $\varphi_1 \wedge \varphi_2 \dashv\vdash \psi_1 \wedge \psi_2$, $\varphi_1 \vee \varphi_2 \dashv\vdash \psi_1 \vee \psi_2$, and so on for the other connectives.

(3) If x and y are variables, and y is not free in φ, and *allow*(φ, y, x), then

$$\vdash \forall x \varphi \to \forall y \varphi(x/y).$$

Proof: Assume $\varphi \dashv\vdash \psi$. Then $\vdash \varphi \to \psi$. Then by 1.4.27, also $\vdash \forall x(\varphi \to \psi)$, so using an axiom from group II and MP we get $\vdash \forall x \varphi \to \forall x \psi$. Similarly we get $\vdash \forall x \psi \to \forall x \varphi$, and using MP and tautology axioms, we get $\forall x \varphi \dashv\vdash \forall x \psi$. The remainder of (1) and (2) are left to the reader.

Proof of (3):

$\forall x \varphi \vdash \varphi(x/y)$	(Group III+MP)
$\forall x \varphi \vdash \forall y \varphi(x/y)$	(By theorem 1.4.27)
$\vdash \forall x \varphi \to \forall y \varphi(x/y)$	(By theorem 1.4.23)

(Remark: The above is a mathematical proof, not a derivation as defined in 1.4.13. However, using theorems 1.4.27 and 1.4.23, the above three lines can easily be converted to a formal derivation.)

When we investigate logical formulas, it is sometimes convenient to replace a formula φ by an equivalent formula φ^* that has a special structure.

1.4.36. Definition. A formula φ is in **prenex normal form**, if it is of the form

$$Q_1 y_1 Q_2 y_2 \cdots Q_n y_n \psi,$$

where $n \geq 0$, $Q_1, \ldots, Q_n$ are quantifiers ($\exists$ or $\forall$), $y_1, \ldots, y_n$ are variables, and ψ is quantifier-free. (For $n = 0$ this just means that $\varphi = \psi$ is quantifier-free.)

Later in this section we will show that every formula is equivalent to a formula in prenex normal form.

Recall the definition of free and bound variables (1.3.25).

1.4.37. Definition. We will define by induction when a formula φ is "simple":

(i) Every atomic formula is "simple."

(ii) $\varphi = (\neg\psi)$ is "simple" iff ψ is.

(iii) $\varphi = (\varphi_1 \wedge \varphi_2)$ is "simple," iff $\mathrm{Bound}(\varphi_1) \cap \mathrm{Var}(\varphi_2) = \mathrm{Var}(\varphi_1) \cap \mathrm{Bound}(\varphi_2) = \emptyset$, and both φ_1 and φ_2 are "simple."

(iv) $\varphi = \forall y \psi$ is "simple" iff ψ is "simple" and $y \notin \mathrm{Bound}(\psi)$.

Clearly, a formula φ is "simple" if all its bound variables are distinct from each other and distinct from any free variable, i.e., whenever $\forall x$ or $\exists x$ occurs in φ, neither $\forall x$ nor $\exists x$ occurs anywhere else in φ, nor is x free in φ.

1.4.38. Lemma. Let φ be a formula. For every finite set A of variables, there is a "simple" formula $\varphi^A \dashv\vdash \varphi$ such that no bound variable in φ^A is in A, and φ and φ^A have the same free variables.

Proof: We will prove this by induction on φ (for all sets A simultaneously).

If φ is atomic, then we can let $\varphi^A = \varphi$.

If $\varphi = \neg\psi$, we can let $\varphi^A = \neg(\psi^A)$.

If $\varphi = \forall x \psi$, let y be a variable neither in $\mathrm{Var}(\psi^A)$ nor in A. As $\psi \dashv\vdash \psi^A$, $\forall x \psi \dashv\vdash \forall x \psi^A$. By 1.4.35, $\forall x \psi^A \dashv\vdash \forall y \psi^A(x/y)$. Now it is easy to see that $\varphi^A = \forall y \psi^A(x/y)$ satisfies all requirements.

If $\varphi = \psi_1 \wedge \psi_2$, let

$$B = A \cup \mathrm{Var}(\psi_2)$$
$$C = A \cup \mathrm{Var}(\psi_1^B)$$
$$\varphi^A = \psi_1^B \wedge \psi_2^C$$

It is clear that $\psi_1^B \wedge \psi_2^C \dashv\vdash \varphi$. We have to check that this formula is "simple": $\mathrm{Var}(\psi_1^B) \cap \mathrm{Bound}(\psi_2^C) \subseteq C \cap \mathrm{Bound}(\psi_2^C) = \emptyset$. What about $\mathrm{Bound}(\psi_1^B) \cap \mathrm{Var}(\psi_2^C)$? We can write $\mathrm{Var}(\psi_2^C)$ as the union of $\mathrm{Bound}(\psi_2^C)$ and a $\mathrm{Free}(\psi_2^C)$, so we have to check that $\mathrm{Bound}(\psi_1^B) \cap \mathrm{Free}(\psi_2^C) = \emptyset$ and $\mathrm{Bound}(\psi_1^B) \cap \mathrm{Bound}(\psi_2^C) = \emptyset$:

$$\mathrm{Bound}(\psi_1^B) \cap \mathrm{Free}(\psi_2^C) = \mathrm{Bound}(\psi_1^B) \cap \mathrm{Free}(\psi_2) \subseteq \mathrm{Bound}(\psi_1^B) \cap B = \emptyset$$

$$\mathrm{Bound}(\psi_1^B) \cap \mathrm{Bound}(\psi_2^C) \subseteq C \cap \mathrm{Bound}(\psi_2^C) = \emptyset$$

1.4.39. Corollary. Every formula is equivalent (i.e., $\dashv\vdash$) to a "simple" formula.

1.4.40. Lemma. If $\mathbf{y}$ does not occur in ψ_2, then

$$\begin{array}{lcl} (\forall \mathbf{y}\psi_1) \wedge \psi_2 & \dashv\vdash & \forall \mathbf{y}(\psi_1 \wedge \psi_2) \\ (\forall \mathbf{y}\psi_1) \vee \psi_2 & \dashv\vdash & \forall \mathbf{y}(\psi_1 \vee \psi_2) \\ (\exists \mathbf{y}\psi_1) \wedge \psi_2 & \dashv\vdash & \exists \mathbf{y}(\psi_1 \wedge \psi_2) \\ (\exists \mathbf{y}\psi_1) \vee \psi_2 & \dashv\vdash & \exists \mathbf{y}(\psi_1 \vee \psi_2). \end{array}$$

Similarly, if $\mathbf{z}$ does not occur in ψ_1, then
$(\forall \mathbf{z}\psi_1) \wedge \psi_2 \dashv\vdash \forall \mathbf{z}(\psi_1 \wedge \psi_2)$, etc.
(Note that in particular, if $(\forall \mathbf{y}\, \psi_1) \wedge \psi_2$ is "simple," then $\mathbf{y}$ does not occur in ψ_2.)

Proof: (Sketch, for the first equivalence only) $\varphi \wedge \psi \to \varphi$ and $\varphi \wedge \psi \to \psi$ are tautology axioms. From this it is easy to get

$$(\forall \mathbf{y}\psi_1) \wedge \psi_2 \vdash \forall \mathbf{y}\psi_1$$

and

$$(\forall \mathbf{y}\psi_1) \wedge \psi_2 \vdash \psi_2.$$

Using the axiom $\vdash (\forall \mathbf{y}\psi_1 \to \psi_1)$ and MP, we get

$$(\forall \mathbf{y}\psi_1) \wedge \psi_2 \vdash \psi_1.$$

Using the tautology $\varphi \to (\psi \to \varphi \wedge \psi)$ and MP twice, we get

$$(\forall \mathbf{y}\psi_1) \wedge \psi_2 \vdash (\psi_1 \wedge \psi_2).$$

By the generalization theorem 1.4.27, this implies

$$(\forall \mathbf{y}\psi_1) \wedge \psi_2 \vdash \forall \mathbf{y}(\psi_1 \wedge \psi_2).$$

1.4.41. Lemma. Assume that $\varphi = (Q_1\mathbf{y}_1 \dots Q_n\mathbf{y}_n\psi_1) \wedge (R_1\mathbf{z}_1 \dots R_m\mathbf{z}_m\psi_2)$ is a "simple" formula (where the Q_i's and R_i's are quantifiers). Then $(Q_1\mathbf{y}_1 \dots Q_n\mathbf{y}_n\psi_1) \wedge (R_1\mathbf{z}_1 \dots R_m\mathbf{z}_m\psi_2)$ is equivalent (in the sense of $\dashv\vdash$) to $Q_1\mathbf{y}_1 \dots Q_n\mathbf{y}_n R_1\mathbf{z}_1 \dots R_m\mathbf{z}_m(\psi_1 \wedge \psi_2)$, and similarly if we replace $\wedge$ by $\vee$.

Proof: Use induction, and 1.4.40 to get

$$\psi_1 \wedge (R_1\mathbf{z}_1 \dots R_m\mathbf{z}_m\psi_2) \dashv\vdash R_1\mathbf{z}_1 \dots R_m\mathbf{z}_m(\psi_1 \wedge \psi_2),$$

then induction and 1.4.35 to get $Q_1\mathbf{y}_1 \dots Q_n\mathbf{y}_n(\psi_1 \wedge (R_1\mathbf{z}_1 \dots R_m\mathbf{z}_m\psi_2)) \dashv\vdash Q_1\mathbf{y}_1 \dots Q_n\mathbf{y}_n R_1\mathbf{z}_1 \dots R_m\mathbf{z}_m(\psi_1 \wedge \psi_2)$. Then we use again 1.4.40 and induction to get $Q_1\mathbf{y}_1 \dots Q_n\mathbf{y}_n(\psi_1 \wedge (R_1\mathbf{z}_1 \dots R_m\mathbf{z}_m\psi_2)) \dashv\vdash (Q_1\mathbf{y}_1 \dots Q_n\mathbf{y}_n\psi_1) \wedge (R_1\mathbf{z}_1 \dots R_m\mathbf{z}_m\psi_2)$, and the theorem follows.

1.4.42. Theorem. Every formula is equivalent (i.e., $\dashv\vdash$) to a formula in prenex normal form.

Proof: By 1.4.39, it is enough to prove it for "simple" formulas. We will prove by induction the (apparently) stronger statement:

For all "simple" formulas φ:

There are "simple" formulas φ^* and $\bar{\varphi}$ in prenex normal form with the same bound and free variables as φ, such that

$$\varphi \dashv\vdash \varphi^* \text{ and } \neg\varphi \dashv\vdash \bar{\varphi}.$$

For atomic φ we can take $\varphi^* = \varphi$ and $\bar{\varphi} = \neg\varphi$.

If $\varphi = \neg\psi$, let $\varphi^* = \bar{\psi} \dashv\vdash \neg\psi = \varphi$ and $\bar{\varphi} = \psi^* \dashv\vdash \psi \dashv\vdash \neg\varphi$.

If $\varphi = \forall\mathbf{x}\psi$, then by induction hypothesis we already have formulas $\psi^* \dashv\vdash \psi$ and $\bar{\psi} \dashv\vdash \neg\psi$ in prenex normal form. Clearly $\forall\mathbf{x}\psi^*$ and $\exists\mathbf{x}\bar{\psi}$ are in prenex normal form, and

$$\forall\mathbf{x}\psi^* \dashv\vdash \forall\mathbf{x}\psi = \varphi$$

and

$$\exists\mathbf{x}\bar{\psi} \dashv\vdash \exists\mathbf{x}\neg\psi \dashv\vdash \neg\forall\mathbf{x}\psi = \neg\varphi.$$

If $\varphi = \psi_1 \wedge \psi_2 \dashv\vdash \psi_1^* \wedge \psi_2^*$, then applying 1.4.41 to $\psi_1^* \wedge \psi_2^*$ will yield φ^*, and applying 1.4.41 to $\bar{\psi}_1 \vee \bar{\psi}_2 \dashv\vdash \neg\psi_1 \vee \neg\psi_2 \dashv\vdash \neg(\psi_1 \wedge \psi_2)$ will yield $\bar{\varphi}$.

1.4.43. Definition. A Σ_0-formula ($= \Pi_0$-formula) is a formula with no quantifiers.

For $n \geq 0$, a Σ_{n+1}-formula is a formula φ of the form $\exists y_1 \cdots \exists y_k \psi$, where $k \geq 0$ and ψ is a Π_n-formula. (In particular, taking $k = 0$, every Π_n-formula is a Σ_{n+1}-formula.)
For $n \geq 0$, a Π_{n+1}-formula is a formula φ of the form $\forall y_1 \cdots \forall y_k \psi$, where $k \geq 0$ and ψ is a Σ_n-formula. (In particular, taking $k = 0$, every Σ_n-formula is a Π_{n+1}-formula.)

1.4.44. Corollary. For every formula φ there is some n such that φ is equivalent ($\dashv\vdash$) to some Σ_n-formula.

1.4.45. Example. $\forall x(\exists z P(x,z) \to \exists y Q(x,y))$ is equivalent to the Π_2-formula

$$\forall x \forall z \exists y (P(x,z) \to Q(x,y)),$$

and it is also equivalent to the Π_3-formula

$$\forall x \exists y \forall z (P(x,z) \to Q(x,y)).$$

We are usually interested in the smallest n such that φ is a Σ_n (or Π_n)-formula. This measures, in some sense, the "complexity" of the formula φ.

1.4.46. Fact. If ψ_1 and ψ_2 are equivalent to Σ_n-formulas, then so are $\psi_1 \wedge \psi_2$, $\psi_1 \vee \psi_2$ and $\exists x \psi_1$.
A similar fact is true for Π_n-formulas (with $\exists$ replaced by $\forall$.)
φ is equivalent to a Σ_n-formula iff $\neg\varphi$ is equivalent to a Π_n-formula.

Exercises

Axioms

1. For each of the following formulas say whether it is an instance of an axiom. Explain why!
   (1) $((\forall x P(x) \to \forall y P(y)) \to P(z)) \to (\forall x P(x) \to (\forall y P(y) \to P(z)))$.
   (2) $\forall y(\forall x(P(x) \to P(x)) \to (P(z) \to P(z)))$.
   (3) $\forall x \exists y P(x,y) \to \exists y P(y,y)$.
   (4) $\forall x(\forall y P(x,y) \to \forall y P(x,y))$.
   (5) $\forall x(\forall y P(x,y) \to \forall z P(x,z))$.

Derivations

2. Using the axioms, give a complete derivation for each of the following formulas.

(1) $\forall x(\varphi \to \psi) \to (\varphi \to \forall x\psi)$
where x is not free in φ.

(2) $\varphi(t) \to \exists x\varphi(x)$
where t is free for x in φ.

(3) $\forall x\varphi \to \exists x\varphi$.

(4) $\forall x(\varphi \wedge \psi) \leftrightarrow (\forall x\varphi \wedge \forall x\psi)$.

3. Which of the following are true for all φ, ψ?

(1) If $\vdash \varphi \wedge \psi$, then $\vdash \varphi$ and $\vdash \psi$.

(2) If $\vdash \varphi \vee \psi$, then $\vdash \varphi$ or $\vdash \psi$.

(3) If $\vdash \varphi$ and $\vdash \psi$, then $\vdash \varphi \wedge \psi$.

(4) If $\vdash \varphi$ or $\vdash \psi$, then $\vdash \varphi \vee \psi$.

4. Prove or disprove: (here you may use the generalization theorem, deduction theorem, etc.)

(1) $\vdash \exists x(\varphi \to \forall x\varphi)$.

(2) $\vdash (\forall x\varphi \vee \forall x\psi) \to \forall x(\varphi \vee \psi)$.

(3) $\vdash (\exists x\varphi \wedge \exists x\psi) \to \exists x(\varphi \wedge \psi)$.

(4) $\{\varphi(x), \forall y(\varphi(y) \to \forall z\psi(z))\} \vdash \forall x\psi(x)$.

(5) $\vdash \exists x\forall y\varphi \to \forall y\exists x\varphi$.

5. Suppose that $\Gamma \vdash \varphi$ and let P be a relation symbol that does not appear in Γ or in φ. Is there a proof of φ from Γ in which P does not appear? Is there a proof of φ from Γ that includes only relation symbols that appear both in φ and Γ? Prove your answers!

6. Formulate 1.4.23 as a proof by induction on the inductive structure defined in 1.4.12.

7. Prove 1.4.34: $\varphi \dashv\vdash \psi$ iff $\varphi \vdash \psi$ and $\psi \vdash \varphi$.

Prenex Normal Form

8. Suppose that x is not free in ψ. Prove that:

(1) $\forall x(\varphi(x) \to \psi) \dashv\vdash \exists x\varphi(x) \to \psi$
(2) $\exists x(\varphi(x) \to \psi) \dashv\vdash \forall x\varphi(x) \to \psi$
(3) $\forall x(\psi \to \varphi(x)) \dashv\vdash \psi \to \forall x\varphi(x)$
(4) $\exists x(\psi \to \varphi(x)) \dashv\vdash \psi \to \exists x\varphi(x)$

9. For each of the following formulas, find an equivalent formula in *prenex normal form.*

(1) $(\forall x\, R(x)) \wedge (\forall x\, Q(x))$
(2) $(\exists x\, R(x)) \vee (\exists x\, Q(x))$
(3) $(\forall x\, R(x)) \vee Q(y)$ where x and y are different variables.
(4) $(\exists x\, R(x)) \wedge Q(y)$ where x and y are different variables.

10. For each of the following formulas, find an equivalent formula in *prenex normal form.* Use the previous exercise.

(1) $\forall xP(x) \to \forall xR(x)$
(2) $\exists xP(x) \to \exists xR(x)$
(3) $\forall xP(x) \vee \forall xR(x)$
(4) $\exists xP(x) \wedge \exists xR(x)$
(5) $\exists x_1Q(x_1, x) \to \forall y_1(\exists zR(x, y, y_1, z) \leftrightarrow \exists zR(x_1, y, y_1, z))$
(6) $\forall x(P(x) \to Q(x, y)) \to (\exists yP(y) \to \exists zQ(y, z))$
(7) $\exists xQ(x, y) \to (P(x) \to \neg\exists uQ(x, u))$

11. Give a procedure that yields for every given formula φ an equivalent formula ψ in *prenex normal form.*

Chapter 2

Completeness

2.1. Enumerability

2.1.1. Definition. A set X is **enumerable** if it is finite or if there is an onto function $f : \mathbb{N} \to X$. We then call $\{f(0), f(1), \ldots\}$ an **enumeration** of X.

2.1.2. Remark. The empty set is finite, so by definition, it is also enumerable. We also have:

(1) If X is enumerable, and $f : X \to Y$ is onto, then Y is enumerable.

(2) If X and Y are each enumerable and infinite, then there is a one-to-one onto function $f : X \to Y$.

2.1.3. Examples.

(1) Clearly, $\mathbb{N}$ is enumerable.

(2) The set of even numbers,

$$E = \{m \in \mathbb{N} : \exists n \in \mathbb{N}(m = 2n)\}$$

is enumerable. Define $f(n) = 2n$.

(3) The set $\mathbb{Z}$ of integers is enumerable:

$$\mathbb{Z} = \{\ldots, -2, -1, 0, 1, 2, \ldots\} = \{0, 1, -1, 2, -2, \ldots\}.$$

2.1.4. Example.
$\mathbb{N}$ can be written as the union of infinitely many disjoint infinite sets, each of which is enumerable.

Let

$$\begin{aligned}
T_0 &= \{0, 1, 3, 5, 7, 9, \ldots\} = \{0\} \cup \{2k+1 : k \in \mathbb{N}\} \\
T_1 &= \{2, 6, 10, 14, 18, \ldots\} \\
T_2 &= \{4, 12, 20, 28, 36, \ldots\} \\
&\vdots \\
T_n &= \{2^n \cdot m : m \in T_0 \text{ and } m \neq 0\} \\
&\vdots
\end{aligned}$$

To see that the T_n are disjoint, suppose $x \in T_i \cap T_j$. Without loss of generality, $i < j$, and clearly $i > 0$. Then $x = 2^i \cdot m = 2^j \cdot n$, for some $m, n \in T_0 \setminus \{0\}$. But then, $m = 2^{j-i} \cdot n$, which implies that m is an even number. But $m \in T_0 \setminus \{0\}$.

The proof that every natural number is in some T_n is by induction and is left as an exercise.

2.1.5. Lemma.
(i) Every subset of an enumerable subset is enumerable.
(ii) The union of enumerably many enumerable sets is enumerable.
(iii) The cross product of finitely many enumerable sets is enumerable.

Proof:
(i) Exercise.

(ii) Let $\{A_i : i \in \mathbb{N}\}$ be an enumerable set of enumerable sets. We assume each A_i is infinite, and leave the finite case as an exercise. To show that $A = \bigcup\{A_i : i \in \mathbb{N}\}$ is enumerable, we will associate each A_i with the T_i of the previous example. Given $i \in \mathbb{N}$, both T_i and A_i are enumerable and infinite, so let $f_i : T_i \to A_i$ be a one-to-one onto function. Now we have

$$A_0 = \{f_0(0), f_0(1), f_0(3), \ldots\}$$

$$A_1 = \{f_1(2), f_1(6), \ldots\}$$
$$\vdots$$

Now it is clear how to define an onto function $f : \mathbb{N} \to A$. Given $n \in \mathbb{N}$, there is a unique i such that $n \in T_i$. Hence n is in the domain of f_i, so define

$$f(n) = f_i(n).$$

Then $A = \{f(0), f(1), f(2), \ldots\}$.

(iii) Let X and Y be enumerable infinite sets. To show that $X \times Y = \{(x, y) : x \in X \text{ and } y \in Y\}$ is enumerable, we first write $X \times Y$ as

$$\{(x_0, y) : y \in Y\} \cup \{(x_1, y) : y \in Y\} \cup \cdots ,$$

where $\{x_0, x_1, \ldots\}$ is an enumeration of X. Clearly each $\{(x_i, y) : y \in Y\}$ is enumerable, so $X \times Y$ is an enumerable union of enumerable sets. Hence we apply (ii) and conclude that $X \times Y$ is enumerable.

Thus the product of two enumerable sets is enumerable. To show that, for all $n \in \mathbb{N}$, the product of n enumerable sets is enumerable, use induction on n.

2.1.6. Example.

(a) The set $\mathbb{Q}$ of rational numbers is enumerable.

(b) The set $\mathbb{R}$ of real numbers is not enumerable.

Proof of (a): We have shown that $\mathbb{Z}$ is enumerable, so $\mathbb{Z} \times \mathbb{Z}$ is enumerable by (iii) of lemma 2.1.5. So $\mathbb{Z} \times (\mathbb{Z} - \{0\})$ is also enumerable, by (i) of lemma 2.1.5.

Now define $f : \mathbb{Z} \times (\mathbb{Z} - \{0\}) \to \mathbb{Q}$ by $f(m, n) = \dfrac{m}{n}$. Clearly f is a function **onto** $\mathbb{Q}$, so by 2.1.2, $\mathbb{Q}$ is enumerable.

Proof of (b): By (i) of the lemma, it suffices to show that $\mathbb{R} \cap [0, 1)$ is not enumerable.

To get a contradiction, suppose we have an enumeration $\{r_0, r_1, \ldots\}$ of $\mathbb{R} \cap [0, 1)$. Write each r_j in decimal form:

$$r_0 = 0.d_{0,0}d_{1,0}d_{2,0} \cdots$$
$$r_1 = 0.d_{0,1}d_{1,1}d_{2,1} \cdots$$
$$\vdots$$

(Here, $d_{i,0}$ is the $i+1$th digit after the decimal point of the real number r_0. Formally we can obtain $d_{i,0}$ as

$$d_{i,0} = \lfloor 10^{i+1} r_0 \rfloor \bmod 10$$

where $\lfloor x \rfloor$ is the integer part of x, and mod 10 means taking the remainder ofter division by 10. The other $d_{i,j}$'s are obtained from the other r_j's similarly.)

We show that there is an element of $\mathbb{R} \cap [0,1)$ that is different from all of the i_j. For each $n \in \mathbb{N}$, define

$$e_n = \begin{cases} 7 & \text{if } d_{n,n} \neq 7 \\ 5 & \text{if } d_{n,n} = 7. \end{cases}$$

Then, let $r = 0.e_0e_1e_2\cdots$ (or formally, $r = \sum_{i=0}^{\infty} \frac{e_i}{10^{i+1}}$). Then, given $n \in \mathbb{N}$, $r \neq r_n$, since $e_n \neq d_{n,n}$.

(Note: It is possible to have two different decimal expansion that represent the same number, for example $0.4999\ldots = 0.5000\ldots$. However, this can only happen if one expansion ends in an string of 9's and the other in an infinite string of 0's. But our number i uses only the digits 5 and 7.)

We have the following very useful corollary to Lemma 2.1.5:

2.1.7. Corollary. Let $\mathcal{L}$ be a language of first order logic with enumerably many symbols. Then

(i) The set of terms of $\mathcal{L}$ is enumerable.

(ii) The set of formulas of $\mathcal{L}$ is enumerable.

Proof: Let $\mathcal{E}$ be the set of expressions of $\mathcal{L}$, where an expression is a finite string of symbols from the alphabet of $\mathcal{L}$. We will show that $\mathcal{E}$ is enumerable, and the corollary will then follow, using Lemma 5(i).

For each $k \in \mathbb{N}$, let $\mathcal{E}_k$ be the subset of $\mathcal{E}$ consisting of the expressions of length k. But $\mathcal{L}$ is enumerable, and so $\mathcal{E}_k$ is the finite product of enumerable sets. Hence, $\mathcal{E}_k$ is enumerable.

But $\mathcal{E} = \bigcup\{\mathcal{E}_k : k \in \mathbb{N}\}$, so we see that $\mathcal{E}$ is an enumerable union of enumerable sets. Therefore, $\mathcal{E}$ is enumerable.

We now give an important application of enumerability to **sentential logic**:

Recall that a **truth assignment** is a mapping S from a set $\{A_1, \ldots\}$ of sentential symbols to $\{T, F\}$ and that $\overline{S}$ is defined inductively, as follows:

$$\begin{aligned} \alpha \text{ is a block: } & \overline{S}(\alpha) = S(\alpha) \\ \alpha \text{ is } (\beta @ \gamma), \ & \overline{S}(\alpha) = G_@(\overline{S}(\beta), \overline{S}(\gamma)) \\ & (\text{where } @ \in \{\wedge, \vee, \rightarrow, \leftrightarrow, |\}, \text{ see 1.2.3}) \\ \alpha \text{ is } (\neg\beta) : & \overline{S}(\alpha) = G_\neg(\overline{S}(\beta)) \end{aligned}$$

We say S **satisfies** α if $\overline{S}(\alpha) = T$, and we say S **satisfies** $\{\alpha_0, \ldots, \alpha_k\}$ if $\overline{S}(\alpha_i) = T$ for $i = 0, \ldots, k$. Also, we say S satisfies Φ if $\overline{S}(\alpha) = T$, for all $\alpha \in \Phi$. Finally, we say that a set of sentential formulas Φ is **satisfiable** if there is a truth assignment S which satisfies Φ.

2.1.8. Theorem. Compactness Theorem for Sentential Logic. Let Φ be an enumerable set of sentential formulas. Then Φ is satisfiable iff every finite subset of Φ is satisfiable.

Proof: Let $\Phi = \{\alpha_1, \alpha_2, \ldots\}$. We omit the trivial direction and assume that every finite subset of Φ is satisfiable. To show that Φ is satisfiable, we will construct a truth assignment S mapping the sentential symbols of Φ to $\{T, F\}$ such that $\overline{S}(\alpha) = T$, for all $\alpha \in \Phi$.

First, let $B_0 = \emptyset$, and for each $n > 0$ let B_n be the set of sentential symbols appearing in α_n, and let $B = B_0 \cup B_1 \cup B_2 \cdots$.

Our construction will be done in stages, so we need to introduce the concept of **extension.** If S and S' are truth assignments, we say that S' **extends** S if for all sentential symbols $\mathtt{A}$:

$$\text{If } S(\mathtt{A}) \text{ is defined, then } S'(\mathtt{A}) \text{ is also defined, and } S'(\mathtt{A}) = S(\mathtt{A}).$$

We will begin our construction with the "empty" assignment S_0 (i.e., $S_0(\mathtt{A})$ is undefined for all sentential symbols $\mathtt{A}$).

We will then show that, given a particular assignment $S_n : B_0 \cup \cdots \cup B_n \to \{T, F\}$ which satisfies $\{\alpha_1, \ldots, \alpha_n\}$, we can extend S_n to a particular assignment $S_{n+1} : B_0 \cup \cdots \cup B_{n+1} \to \{T, F\}$ which satisfies $\{\alpha_1, \ldots, \alpha_{n+1}\}$. We will then take the union of these assignments, thus obtaining an assignment S satisfying Φ.

Before proceeding to the construction, we need one more definition. Let $S_n : B_0 \cup \cdots \cup B_n \to \{T, F\}$ be a truth assignment which satisfies $\{\alpha_1, \ldots, \alpha_n\}$. We say that S_n is **k-good** if there is a truth assignment $S : B_0 \cup \cdots \cup B_{n+k} \to \{T, F\}$ such that

(i) S is an extension of S_n

(ii) S satisfies $\{\alpha_1, \ldots, \alpha_{n+k}\}$.

If S_n is k-good for all $k \in \mathbb{N}$, we say that S_n is **very good.**

Claim 1: For each $k \in \mathbb{N}$, S_0 is k-good, i.e., S_0 is very good.

Proof: Fix a natural number k. By assumption, $\{\alpha_1, \ldots, \alpha_k\}$ is satisfiable, so let $S : B_k \to \{T, F\}$ be a truth assignment which satisfies $\{\alpha_1, \ldots, \alpha_k\}$. As S extends S_0, S_0 is k-good.

The crucial point of this proof is the next claim:

Claim 2: If $S_n : B_0 \cup \cdots \cup B_n \to \{T, F\}$ is very good, then there is an assignment $S_{n+1} : B_0 \cup \cdots \cup B_{n+1} \to \{T, F\}$ such that

(i) S_{n+1} is an extension of S_n

(ii) S_{n+1} satisfies $\{\alpha_1, \ldots, \alpha_{n+1}\}$
(iii) S_{n+1} is very good.

Proof: Since S_n is very good, it satisfies $\{\alpha_1, \ldots, \alpha_n\}$ and is, in particular, 1-good. So there is an assignment $S_{n+1} : B_0 \cup \cdots \cup B_{n+1} \to \{T, F\}$ which extends S_n and satisfies $\{\alpha_1, \ldots, \alpha_{n+1}\}$. In fact, there may be more than one such assignment. Thus, (i) and (ii) are easily satisfied. The only problem is that we must make sure that one of these assignments is very good.
So, let $\mathcal{S}_{n+1}$ be the set of assignments with domain $B_0 \cup \cdots \cup B_{n+1}$ which extend S_n and satisfy $\{\alpha_1, \ldots, \alpha_{n+1}\}$. Since S_n is very good, we have that for each $k \in \mathbb{N}$, S_n is $k+1$-good, and this implies that there is an element of $\mathcal{S}_{n+1}$ which is k-good. Hence, for each $k \in \mathbb{N}$, there is an element of $\mathcal{S}_{n+1}$ which is k-good. But $\mathcal{S}_{n+1}$ is finite, so there must be an element of $\mathcal{S}_{n+1}$ which is very good.

We are now ready to construct inductively the assignment $S : B \to \{T, F\}$ which satisfies Φ. Start with S_0. Now, given a very good $S_n : B_0 \cup \cdots \cup B_n \to \{T, F\}$, use Claim 2 to pick an extension S_{n+1} of S_n which satisfies $\{\alpha_1, \ldots, \alpha_{n+1}\}$ and is very good.
Claim 3: $S = \bigcup\{S_n : n \in \mathbb{N}\}$ satisfies Φ.
Proof: S is well-defined because, by construction, each S_{n+1} is an extension of S_n. To see that S satisfies Φ, pick $\alpha \in \Phi$. Then α is α_n, for some $n \in \mathbb{N}$. But S_n satisfies $\{\alpha_1, \ldots, \alpha_n\}$, hence S_n satisfies α_n. But S is an extension of S_n, hence S satisfies φ_n.

We have the following example of sentential compactness.

A **map** is a pair $\langle C, N\rangle$, where C is interpreted as a set of countries and N is a set of unordered pairs $\{c, c'\}$ from C. If c and d are in C, and pair $\{c, d\}$ is in N then we interpret this as "c and d have a common border". In this case, we say that c and d are **neighbors.** If C is infinite, then we call $\langle C, N\rangle$ an **infinite map.** A **submap** of $\langle C, N\rangle$ is a map $\langle C', n'\rangle$, where $C' \subseteq C$, and $N' \subseteq N$.

A map $\langle C, N\rangle$ can be **colored with four colors** if

> each country can be assigned exactly one of four colors in such a way that no two neighbors are assigned the same color

or in other words:

> if there is a function $F : C \to \{1, 2, 3, 4\}$ such that $F(c) \neq F(d)$ whenever $\{c, d\} \in N$.

We call F a **coloring** of $\langle C, N\rangle$.

2.1.9. Example. An infinite map can be colored with four colors iff every finite submap can be colored with four colors.

Proof: Let $\langle C, N\rangle$ be an infinite map, with set of countries $= \{c_1, c_2, \dots\}$. We want to color C with the four colors $\{1,2,3,4\}$.

We will use an infinite set of sentential symbols:

$$\{\mathtt{A}_{1,1}, \mathtt{A}_{2,1}, \mathtt{A}_{3,1}, \mathtt{A}_{4,1}, \mathtt{A}_{5,1} \dots \\ \mathtt{A}_{1,2}, \mathtt{A}_{2,2}, \mathtt{A}_{3,2}, \mathtt{A}_{4,2}, \mathtt{A}_{5,2} \dots \\ \mathtt{A}_{1,3}, \mathtt{A}_{2,3}, \mathtt{A}_{3,3}, \mathtt{A}_{4,3}, \mathtt{A}_{5,3} \dots \\ \mathtt{A}_{1,4}, \mathtt{A}_{2,4}, \mathtt{A}_{3,4}, \mathtt{A}_{4,4}, \mathtt{A}_{5,4} \dots\}$$

(where we assume all these symbols are distinct). The interpretation of $\mathtt{A}_{i,j}$ is : Country c_i will be colored with color j.

Let Φ consist of the following three types of formulas

(i) $\mathtt{A}_{i,j} \to \neg\mathtt{A}_{i',j}$ and $\mathtt{A}_{i',j} \to \neg\mathtt{A}_{i,j}$, whenever $j \in \{1,2,3,4\}$ and $\{c_i, c_{i'}\} \in N$ are adjacent countries.

(ii) $\mathtt{A}_{i,1} \vee \mathtt{A}_{i,2} \vee \mathtt{A}_{i,3} \vee \mathtt{A}_{i,4}$, for any i.

(iii) $\mathtt{A}_{i,j} \to \neg\mathtt{A}_{i,j'}$, whenever $j \neq j'$.

Type (i) formulas say that if c_i and $c_{i'}$ are neighbors and c_i has color j, then $c_{i'}$ does not have color j. For each pair $\{c_i, c_{i'}\} \in N$ of neighbors, Φ contains eight type (i) formulas, and the conjunction of these formulas say that i and i' do not have the same color.

Type (ii) formulas say that a country has at least one of the four colors. For each $c_i \in C$, Φ contains one type (ii) formula.

Type (iii) formulas say that if a country has color j, then it does not have color j'. For each country $c_i \in C$, Φ contains twelve formulas of this type, and the conjunction of these twelve formulas say that c_i has at most one color.

Assume that every finite submap of $\langle C, N\rangle$ can be colored with four colors. We will show that every finite subset of Φ is satisfiable. Then, by the compactness theorem, Φ is satisfiable. Finally, we will show that this implies that $\langle C, N\rangle$ can be colored with four colors.

Let Σ be a finite subset of Φ. Let $C(\Sigma)$ be the set of countries $c_i \in C$ such that there is a $j \in \{1,2,3,4\}$ such that the sentential symbol $\mathtt{A}_{i,j}$ appears in some formula σ in Σ. Formally,

$$C(\Sigma) = \{c \in C : (\exists\sigma \in \Sigma)(\exists j \in \{1,2,3,4\})(\mathtt{A}_{i,j} \in BKS(\sigma))\},$$

where $BKS(\sigma)$ is the set of sentential symbols of σ. Since Σ is finite, $C(\Sigma)$ is finite. Let $N(\Sigma) = \{\{c,d\} \in N : c, d \in C(\Sigma)\}$. Then $\langle C(\Sigma), N(\Sigma)\rangle$ is a finite submap of $\langle C, N\rangle$. By hypothesis, $\langle C(\Sigma), N(\Sigma)\rangle$ can be colored with four colors. So let $F : C \to \{1,2,3,4\}$ be a coloring of $\langle C(\Sigma), N(\Sigma)\rangle$. Now define a truth assignment $S : B(\Sigma) \to \{T, F\}$ on the set $B(\Sigma)$ of sentential symbols of Σ by

$$S(\mathtt{A}_{i,j}) = T \text{ iff } F(c_i) = j.$$

Claim: S satisfies Σ.

Proof: Let $\sigma \in \Sigma$. If σ is $\mathtt{A}_{i,j} \to \neg\mathtt{A}_{i',j}$, then $\overline{S}(\mathtt{A}_{i,j} \to \neg\mathtt{A}_{i',j}) = F$ iff $S(\mathtt{A}_{i,j}) = T$ and $S(\mathtt{A}_{i',j}) = T$ iff $F(c_i) = j$ and $F(c_{i'}) = j$, which contradicts the fact that F is a coloring. Hence $\overline{S}(\sigma) = T$.

Suppose σ is $\mathtt{A}_{i,1} \vee \mathtt{A}_{i,2} \vee \mathtt{A}_{i,3} \vee \mathtt{A}_{i,4}$. Let $j = F(c_i)$. Then $S(\mathtt{A}_{i,j}) = T$. But j must be one of 1, 2, 3 and 4. Hence, $\overline{S}(\sigma) = T$.

Finally, if σ is $\mathtt{A}_{i,c} \to \neg\mathtt{A}_{i,d}$, then letting $F(c_i) = j$ at most one of $S(A_{i,c})$, $S(A_{i,d})$ can be $= T$ — depending on whether $c = j$ or $d = j$. So $\overline{S}(\sigma) = T$.

Thus every finite subset of Φ is satisfiable. By compactness, Φ is satisfiable. Let $S : B \to \{T, F\}$ be a truth assignment which satisfies Φ. We define a function $F : C \to \{1,2,3,4\}$. Fix a country c_i. Then $\overline{S}(\mathtt{A}_{i,1} \vee \mathtt{A}_{i,2} \vee \mathtt{A}_{i,3} \vee \mathtt{A}_{i,4}) = T$, so there is at least one element j of $\{1,2,3,4\}$ such that $S(\mathtt{A}_{i,j}) = T$. But also, $\overline{S}(\mathtt{A}_{i,j} \to \neg\mathtt{A}_{i,j'}) = T$, for each $j' \in \{1,2,3,4\} - \{j\}$. Hence, $S(\mathtt{A}_{i,j'}) = F$, for each $j' \neq j$. Thus, j is unique. Define $F(c_i) = j$.

Claim: F is a coloring of $\langle C, N\rangle$.

Proof: Exercise.

We give a second example of an application of sentential compactness.

2.1.10. Definition. A binary relation $\prec$ on a set A is a **partial order** if it is both transitive and asymmetric, that is, for every $a, b, c \in A$

(i) if $a \prec b$ and $b \prec c$ then $a \prec c$ ("$\prec$ is **transitive**").

(ii) if $a \prec b$ then not $b \prec a$ ("$\prec$ is **asymmetric**").

(In particular, this implies $a \not\prec a$ for all a.)

A binary relation $<$ is a **complete order** (sometimes called a "total" or "linear" order) if it is a partial order and for every $a, b \in A$

(iii) $a < b$ or $a = b$ or $b < a$ ("$\prec$ is **trichotomic**")

We say that a partial order $\prec$ can be extended to a complete order if there exists a complete order $<$ that preserves the partial order (*i.e.*, if $a \prec b$ then $a < b$).

We will show every countable partial order can be extended to a complete order by showing this first for finite sets and then using the compactness theorem.

2.1.11. Definition. If $\prec$ is a partial order on a set X, then an element $x \in X$ is called "minimal", if there is no $y \prec x$.
x is called "least", if for all $y \in X$ that are distinct from x we have $x \prec y$.

2.1.12. Remark. In any partial order there is at most one least element. If x is least, then x is minimal, but not necessarily conversely. In a complete order, if there is a minimal element then it is also the least element. A partial order may not have a minimal element, but every finite partial order has a minimal element.

We leave the proof as an exercise.

2.1.13. Lemma. For every natural number $n > 0$: If $(X, \prec)$ is a partial order with n elements, then there is a complete order $<$ on X that extends $\prec$.

Proof: By induction on n. If $n = 1$, then the partial order is a linear order.

Now let $n = k+1$. Let $\prec$ be a partial order on X, where X is a set with $k+1$ elements. Let $x_0 \in X$ be a minimal element. $X' := X - \{x_0\}$ is a set with k elements, and it is partially ordered by $\prec' := \{(x, y) : x \prec y,\ x, y \in X - \{x_0\}\}$. So by induction assumption there is a complete order $<'$ on X'. We now define a complete order on X as follows: For all $x, y \in X$,

$$x < y \text{ iff either } x <' y, \text{ or } x = x_0 \neq y.$$

We have to check that this defines indeed a complete order on X that extends $\prec$:

Transitivity: If $x < y < z$, then neither y nor z can be equal to x_0, as x was minimal. Hence $y <' z$. Now either $x = x_0$, in which case $x < z$, or $x <' y <' z$, in which case $x <' z$, so again $x < z$.

Asymmetry: $x < x$ is equivalent to "$x <' x$ or $x = x_0 \neq x$," which is impossible.

Trichotomy: If $x \neq y$ we have to show $x < y$ or $y < x$. If either x or y is equal to x_0, then this is clear. Otherwise we have that either $x <' y$ or $y <' x$, so again we are done.

Finally, if $x \prec y$, then $y \neq 0$. Either $x = x_0$ (so $x < y$), or $x \neq x_0$, so $x \prec' y$, hence $x <' y$, so $x < y$.

2.1.14. Lemma. If $\prec$ is a partial order on a countable set X, then $\prec$ can be extended to a linear order on X.

Proof: Let $X = \{x_1, x_2, x_3, \dots\}$ (all x_i distinct). We will use an infinite set of sentential symbols:

$$\{\mathsf{A}_{1,1}, \mathsf{A}_{2,1}, \mathsf{A}_{3,1}, \mathsf{A}_{4,1}, \mathsf{A}_{5,1} \dots$$
$$\mathsf{A}_{1,2}, \mathsf{A}_{2,2}, \mathsf{A}_{3,2}, \mathsf{A}_{4,2}, \mathsf{A}_{5,2} \dots$$
$$\mathsf{A}_{1,3}, \mathsf{A}_{2,3}, \mathsf{A}_{3,3}, \mathsf{A}_{4,3}, \mathsf{A}_{5,3} \dots$$
$$\mathsf{A}_{1,4}, \mathsf{A}_{2,4}, \mathsf{A}_{3,4}, \mathsf{A}_{4,4}, \mathsf{A}_{5,4} \dots$$
$$\dots$$
$$\dots\}.$$

(again we assume all these symbols are distinct.) The interpretation of $A_{i,j}$ is : x_i will be $< x_j$.

Let Φ consist of the following four types of formulas

(i) $\mathsf{A}_{i,j} \rightarrow \neg\mathsf{A}_{j,i}$ for all i, j.
(ii) $\mathsf{A}_{i,j}$, whenever $x_i \prec x_j$.
(iii) $\mathsf{A}_{i,j} \wedge \mathsf{A}_{j,k} \rightarrow \mathsf{A}_{i,k}$ for all i, j, k.
(iv) $\mathsf{A}_{i,j} \vee \mathsf{A}_{j,i}$ whenever $i \neq j$.

Now we can prove the following claims:
Claim 1: Every finite subset of Φ is satisfiable.
Claim 2: Φ is satisfiable.
Claim 3: $\prec$ can be extended to a linear order on X.

We leave the proof of claim 1 to the reader. Claim 2 follows from claim 1 by the compactness theorem.

Proof of claim 3: Assume that S is a truth assignment satisfying Φ. The we define $<$ by

$$x_i < x_j \quad \text{iff} \quad S(\mathsf{A}_{i,j}) = T$$

We have to show that $<$ is a complete order extending $\prec$. $<$ extends $\prec$, because if $x_i \prec x_j$, then $\mathsf{A}_{i,j} \in \Phi$, by (ii), so $S(\mathsf{A}_{i,j}) = T$, so $x_i < x_j$.

Similarly, $<$ satisfies the axioms of a partial order because of clauses (i) and (iii) above.

Finally, $<$ is a complete order because of clause (iv) above.

We now leave sentential calculus and again turn our attention to first order logic.

2.1.15. Definition. A **theory** Γ is any set of sentences (= closed formulas) in a first order language $\mathcal{L}$. Often we don't give a special name to the language we are using, and just call it $\mathcal{L}(\Gamma)$. This also makes for very convenient notation if we compare two related theories which are formulated in two different languages.

The intention is that the formulas in Γ are true in certain models that we want to investigate. For example, if we want to investigate groups, we could choose as our theory the set of group axioms.

If we want to work in number theory, we would choose as our theory the set of all sentences that are valid in $\mathbb{N}$. This set is called $Th(\mathbb{N})$. However, since this set is a very complicated set, we will (in a subsequent chapter) look at a subset of $Th(\mathbb{N})$, which is given by a simple set of "axioms."

2.1.16. Definition. Let Γ be a theory. A theory Γ' is an **extension** of Γ if $\mathcal{L}(\Gamma) \subseteq \mathcal{L}(\Gamma')$ and for every formula φ in $\mathcal{L}(\Gamma)$ such that $\Gamma \vdash \varphi$, we also have $\Gamma' \vdash \varphi$. If, in addition, $\mathcal{L}(\Gamma') = \mathcal{L}(\Gamma)$, we say that Γ' is a **simple extension** of Γ. If Γ' is an extension of Γ, we also say that Γ is a **subtheory** of Γ', and we write $\Gamma \subseteq \Gamma'$.

2.1.17. Lemma. If Γ' is an extension of Γ, and Γ' is consistent, then Γ is consistent.

2.1.18. Theorem. Compactness Theorem for First-Order Logic. Let Γ be a theory. Then Γ is consistent iff every finite subtheory of Γ is consistent.

Proof: ($\Rightarrow$) Suppose Γ is consistent. Let Γ_0 be a finite subtheory of Γ. Then $\Gamma_0 \subseteq \Gamma$, so by the previous lemma, Γ_0 is consistent.

($\Leftarrow$) Suppose Γ is inconsistent. Let φ be such that $\Gamma \vdash \varphi \wedge \neg\varphi$. That is, there is a derivation in Γ whose last step is $\varphi \wedge \neg\varphi$. But derivations are finite, so only finitely many members of Γ appear in the derivation. It follows that there is a finite subtheory Γ_0 of Γ such that $\Gamma_0 \vdash \varphi \wedge \neg\varphi$. Hence Γ_0 is inconsistent.

2.1.19. Remark. This proof was much simpler then the proof of 2.1.8. The reason is that the notion of consistency that we have for first order theories is **syntactical** (defined purely in terms of the formulas themselves, whereas satisfiability for sentential logic is **semantical** (referring to an intended truth value of the formulas). To show that a first order theory is inconsistent, we just have to find a derivation, which is a syntactic object. To show that a sentential theory is not satisfiable, we have to check all possible interpretations, i.e., truth assignments.

However, later we will see that consistency of a first order theory is equivalent to a "semantic" property, namely, of having a model.

2.1.20. Definition. A consistent first order theory Γ is said to be **complete** if for every sentence φ of $\mathcal{L}(\Gamma)$, either $\Gamma \vdash \varphi$ or $\Gamma \vdash \neg\varphi$. [Note that the consistency of Γ precludes the possibility that $\Gamma \vdash \varphi$ **and** $\Gamma \vdash \neg\varphi$.] If Γ is not complete, we say that Γ is **incomplete.**

(If Γ is inconsistent, then for every formula φ we have $\Gamma \vdash \varphi$ and $\Gamma \vdash \neg\varphi$. So inconsistent theories could be called complete. However, since inconsistent theories are not interesting, we include the assumption of consistency into the definition of "complete.")

Note that not every consistent theory is complete. We will see in Chapter 3 that the Peano theory of arithmetic is incomplete. We also have the following example.

2.1.21. Example. The theory of groups is incomplete.

Formally, a group $\mathcal{G} = \langle G, +, 0\rangle$ is given by a set G, a binary operation, $+^{\mathcal{G}} : G \times G \to G$ which maps pairs of elements from G into the set G, and a distinguished element $0^{\mathcal{G}}$, which satisfies the following conditions:

(1) $+^{\mathcal{G}}$ is associative.
(2) for all $g \in G$, $g +^{\mathcal{G}} 0^{\mathcal{G}} = 0^{\mathcal{G}} +^{\mathcal{G}} g = g$.
(3) for each element a of G, there is an element b of G such that $a +^{\mathcal{G}} b = 0^{\mathcal{G}}$.

Hence any group $\langle G, +, 0\rangle$ is a model for the theory GROUPS, whose only nonlogical axioms are

(1) $\forall x \forall y \forall z((x + y) + z = x + (y + z))$.
(2) $\forall x(x + 0 = x \wedge 0 + x = x)$.
(3) $\forall x \exists y(x + y = 0 \wedge y + x = 0)$.

2.1.22. Claim. The theory GROUPS is incomplete.

Proof: Let φ be $\forall x \forall y(x+y = y+x)$. Then GROUPS $\nvdash \varphi$ and GROUPS $\nvdash \neg\varphi$. Why is this so? Well, suppose GROUPS $\vdash \neg\varphi$. Then, by the validity theorem, GROUPS $\models \neg\varphi$. But $\langle \mathbb{Z}, +\rangle$, the group of integers under addition, clearly satisfies the three nonlogical axioms of GROUPS, and is thus a model of GROUPS. Hence $\neg\varphi$ must be true in $\langle \mathbb{Z}, +\rangle$, *i.e.*, φ must be false. But for any two integers m and n, we know that $m + n = n + m$. So φ is true in $\langle \mathbb{Z}, +\rangle$. Contradiction.

Similarly, the group S_3 of permutations on three elements contains elements x and y such that $x + y \neq y + x$. Hence, if GROUPS $\vdash \varphi$, then GROUPS $\models \varphi$, so φ must be true in S_3. But φ is not true in S_3. Contradiction.

In short, we showed that the commutative law cannot be derived from the group axiom by exhibiting a model of all group axioms in which the commutative law does not hold, and we showed that the negation of the commutative law cannot be derived from the group axioms either, because there is a model in which commutativity does hold.

We are now ready to prove the main result of the section.

2.1.23. Theorem. Every consistent theory has a complete simple extension.

Proof: Let Γ be a consistent theory. By corollary 2.1.7, the set of sentences of $\mathcal{L}(\Gamma)$ is enumerable, so let $\{\varphi_1, \varphi_2, \ldots\}$ be an enumeration of this set. We will construct a complete theory Γ^* from Γ inductively, as follows:

Step One: Set $\Gamma_0 = \Gamma$.
Step Two: Having constructed $\Gamma_0, \Gamma_1, \ldots, \Gamma_n$, define Γ_{n+1}:
If $\Gamma_n \cup \{\varphi_n\}$ is consistent, let $\Gamma_{n+1} = \Gamma_n \cup \{\varphi_n\}$,
otherwise, let $\Gamma_{n+1} = \Gamma_n$.
Step Three: Let $\Gamma^* = \bigcup\{\Gamma_n : n \in \mathbb{N}\}$.

So we get a sequence $\Gamma = \Gamma_0 \subset \Gamma_1 \subset \cdots \subset \Gamma_n \subset \cdots \subset \Gamma^*$.

We claim that Γ^* is consistent and that Γ^* is complete.

First we show that each Γ_n is consistent. For this it is enough that for all n:

If Γ_n is consistent, then Γ_{n+1} is consistent.

This is clear since either $\Gamma_{n+1} = \Gamma_n$, or $\Gamma_{n+1} = \Gamma_n \cup \{\varphi_n\}$ which has to be consistent by construction.

To see that Γ^* is consistent, let Γ' be a finite subtheory of Γ^*. Then, there is an $n \in \mathbb{N}$ such that $\Gamma' \subseteq \Gamma_n$, for otherwise Γ' would not be finite. Γ_n is consistent, so by Lemma 11, Γ' is consistent. Hence every finite subtheory of Γ^* is consistent. By first-order compactness, Γ^* is consistent.

To see that Γ^* is complete, let φ be a sentence of $\mathcal{L}(\Gamma)$. Then φ is φ_n, for some $n \in \mathbb{N}$. If $\varphi_n \in \Gamma^*$, then $\Gamma^* \vdash \varphi_n$. If φ_n is not in Γ^*, then it must be the case that $\Gamma_n \cup \{\varphi_n\}$ is inconsistent. Hence, $\Gamma_n \vdash \neg\varphi_n$. So $\Gamma^* \vdash \neg\varphi_n$.

2.1.24. Definition. For any theory Γ, we define the **deductive closure** of Γ as

$$cl_{\vdash}(\Gamma) := \{\varphi : \varphi \text{ is closed, and } \Gamma \vdash \varphi\}$$

If $\Gamma_0 \subseteq \Gamma \subseteq cl_{\vdash}(\Gamma_0)$, we call Γ_0 "axioms" for Γ.

2.1.25. Lemma. For any theories Γ, Γ', we have

(1) $\Gamma \subseteq cl_{\vdash}(\Gamma)$.

(2) $cl_{\vdash}(cl_{\vdash}(\Gamma)) = cl_{\vdash}(\Gamma)$.
(3) If $\Gamma \subseteq \Gamma'$, then $cl_{\vdash}(\Gamma) \subseteq cl_{\vdash}(\Gamma')$.

Proof: Exercise.

2.1.26. Definition. We say that Γ_1, Γ_2 are equivalent, if $cl_{\vdash}(\Gamma_1) = cl_{\vdash}(\Gamma_2)$, i.e., if the same theorems can be derived from both theories.

2.1.27. Remark. We often do not distinguish between equivalent theories. For example, any theory Γ with $cl_{\vdash}(\Gamma) = cl_{\vdash}(\text{GROUPS})$ may be called "the theory of groups."

Exercises

Enumerability

1. Show that every subset of an enumerable set is enumerable.

2. Find a set S which is not enumerable, but which is the union of an infinite number of enumerable sets.

3. For each one of the following sets, prove or disprove that the set is enumerable.

(a) The set $S = \{a + b\sqrt{2} : a, b \in \mathbb{Z}\}$.
(b) The set of all possible board configurations in chess.
(c) The set of all subsets of $\mathbb{Z}$.
(d) The set of all functions $f : \mathbb{Z} \to \mathbb{Z}$.
(e) The set of all constant functions $f : \mathbb{R} \to \mathbb{Z}$.
(f) The set of all unary relations on $\mathbb{Z}$.
(g) The set of all subsets of people in the world.
(h) The set of all truth assignments for a given enumerable set of sentential formulas.

4. Prove or disprove the following claim: If C is an inductive structure with a finite number of blocks and a finite number of operations, then the number of elements in C is enumerable. What happens if the number of blocks is enumerable?

An Application of the Compactness Theorem

5. We can graphically represent a partial order on a finite set A by its **Hasse diagram**. Elements of the set A are represented by points, and if an element b is a direct successor of a (i.e., $a < b$ but there is no c with $a < c < b$), then we write b above a and connect a and b.

For example, the following are Hasse diagrams of partial orders on the 4 element set $\{a, b, c, d\}$:

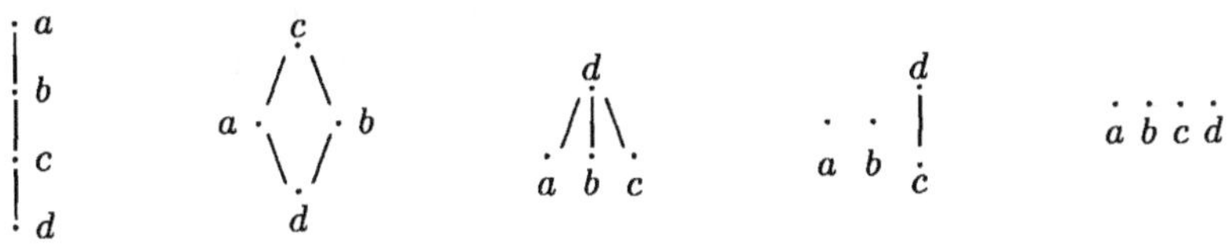

For each of these partial orderings, list **all** pairs (x, y) with the property $x < y$, and identify all minimal and least elements.

6. Prove 2.1.12.

7. Complete the proof of 2.1.14.

$\vdash$ and $cl_{\vdash}$

8. Prove lemma 2.1.25: For any theories Γ, Γ', we have
(1) $\Gamma \subseteq cl_{\vdash}(\Gamma)$.
(2) $cl_{\vdash}(cl_{\vdash}(\Gamma)) = cl_{\vdash}(\Gamma)$.
(3) If $\Gamma \subseteq \Gamma'$, then $cl_{\vdash}(\Gamma) \subseteq cl_{\vdash}(\Gamma')$.

9. We say that two sets of sentences, Σ_1 and Σ_2, are **equivalent** if for every sentence φ, if $\varphi \in \Sigma_1$ then $\Sigma_2 \vdash \varphi$, and if $\varphi \in \Sigma_2$ then $\Sigma_1 \vdash \varphi$.
We say that a set of sentences Σ is **independent** if for every sentence $\varphi \in \Sigma$, $\Sigma - \{\varphi\} \nvdash \varphi$.

(a) Prove that if Σ is a finite set of sentences, then it has an independent equivalent subset.
(b) Give an example of a set of sentences Σ that does not have an independent equivalent subset.

2.2. The Completeness Theorem

The main goal of this section is to prove:

2.2.1. Theorem. The Completeness Theorem. A theory Γ is consistent iff it has a model.

The $\Leftarrow$ direction is easy. Suppose a theory Γ has a model $\mathcal{M}$. If Γ is inconsistent, let φ be a sentence in $\mathcal{L}(\Gamma)$ such that $\Gamma \vdash \varphi \wedge \neg\varphi$. Now by the validity theorem, we get $\mathcal{M} \models \varphi \wedge \neg\varphi$. So $\mathcal{M} \models \varphi$ and $\mathcal{M} \models \neg\varphi$, which is impossible.

Therefore, Γ is consistent.

Before proving the $\Rightarrow$ direction, we present a statement which is equivalent to the completeness theorem:

(*) If φ is a formula, then: $\Gamma \vdash \varphi$ iff $\Gamma \models \varphi$.

To see that 2.2.1 implies (*), first note that $\Gamma \vdash \varphi \rightarrow \Gamma \models \varphi$ is just the validity theorem. Now, if $\Gamma \models \varphi$, then φ is valid in every model of Γ. Hence $\Gamma \cup \{\neg\varphi\}$ does not have a model. By 2.2.1, $\Gamma \cup \{\neg\varphi\}$ is inconsistent. So $\Gamma \vdash \varphi$.

To see that (*) implies 2.2.1, Γ has no model iff $\Gamma \models \varphi \wedge \neg\varphi$ for any φ. ($\Gamma \models \varphi$ means that every model of Γ satisfies φ, or equivalently, that there is no model of Γ that does not satisfy φ. Since there are no models of Γ at all, this is certainly true. The same reasoning also shows $\Gamma \models \neg\varphi$).

By (*), $\Gamma \models \varphi \wedge \neg\varphi$ implies $\Gamma \vdash \varphi \wedge \neg\varphi$ so Γ is inconsistent. Conversely, if Γ is inconsistent, then Γ has no model, so we get 2.2.1 from (*).

Hence (2.2.1) and (*) are equivalent.

To prove that every consistent theory has a model (without assuming (*)) is not as trivial as proving that every theory which has a model is consistent. We need to construct a model for Γ, but what can we use for the universe? All we have to work with is a theory Γ and a language $\mathcal{L}(\Gamma)$. The solution is to build the universe out of the language. Specifically, the universe will be obtained from the closed terms of $\mathcal{L}(\Gamma)$. We will define a relation on the set of closed terms by declaring two terms to be related iff the theory Γ proves that they are equal. We will show that this relation is in fact an equivalence relation, and then take the set of equivalence classes as the universe.

The closed terms of a language consist of the constants and the functions applied to the constants. So we have an immediate problem. What if $\mathcal{L}(\Gamma)$ contains no constants? The solution is to add a set C of constants to $\mathcal{L}$, forming $\mathcal{L}'$. In order to handle this, we need some results involving constants.

2.2.2. Fact. If φ_1, φ_2 are formulas not involving the constants $c_0, \ldots, c_n$, then

$$\varphi_1(x_0/c_0, \ldots, x_n/c_n) = \varphi_2(x_0/c_0, \ldots, x_n/c_n) \quad \text{implies} \quad \varphi_1 = \varphi_2$$

Proof: Easy exercise.

In informal proofs in mathematics we often use a phrase like "Fix a natural number n" or "let n be an arbitrary (but fixed) natural number." In a formal proof, this corresponds to adding a constant symbol n to the language. Whatever we prove about n will be true about any natural number, because n was arbitrary. This line of reasoning is justified by the following theorem:

2.2.3. Theorem on Constants. Let Γ be a theory, and let $\mathcal{L} = \mathcal{L}(\Gamma)$. Let $C = \{c_0, c_1, \ldots\}$ be an enumerable set of constant symbols not in $\mathcal{L}$. Let Γ' be the theory with the same nonlogical axioms as Γ but with language $\mathcal{L}'$, which is the language $\mathcal{L}$ extended by all constants of C. (So $\mathcal{L}'$ also contains terms and formulas that can be formed using the new constants.) Then, if φ is a formula in $\mathcal{L}$, we have: $\Gamma \vdash \varphi$ iff $\Gamma' \vdash \varphi(x_0/c_0, \ldots, x_k/c_k)$.

Proof: ($\Rightarrow$) This is the easy direction. Since $\Gamma \vdash \varphi$, also $\Gamma' \vdash \varphi$. By applying the generalization theorem we get

$$\Gamma' \vdash \forall x_o \varphi$$

Now using a substitution axiom and modus ponens we ge t

$$\Gamma' \vdash \varphi(x_o/c_o)$$

Applying the last two steps repeatedly, we get

$$\Gamma' \vdash \varphi(x_o/c_o, \ldots, x_n/c_n)$$

($\Leftarrow$) Let φ be a formula of $\mathcal{L}$, and suppose $\Gamma' \vdash \varphi(x_o/c_o, \ldots, x_k/c_k)$.

By induction on the derivation of the formula $\varphi(x_o/c_o, \ldots, x_k/c_k)$ from Γ' we will show that $\Gamma \vdash \varphi$.

Case 1: $\varphi(x_o/c_o, \ldots, x_k/c_k)$ is an axiom.

If $\varphi(x_o/c_o, \ldots, x_k/c_k)$ is a nonlogical axiom of Γ', then it is clear that $\Gamma \vdash \varphi$, since Γ and Γ' have the same nonlogical axioms.

If $\varphi(x_o/c_o, \ldots, x_k/c_k)$ is a tautology, then φ is also a tautology, so $\Gamma \vdash \varphi$.

Now assume that $\varphi(x_o/c_o, \ldots, x_k/c_k)$ is a pure generalization axiom,

i.e., of the form $\psi \to \forall y\psi$.
So φ was of the form $\psi_1 \to \forall y\psi_2$ where $\psi = \psi_1(x_0/c_0, \ldots, x_k/c_k)$ and $\forall y\psi = (\forall y\psi_2)(x_0/c_0, \ldots, x_k/c_k)$.
We may assume that y is different from all the x_i. If not, then y is equal to one of the x_i, say to x_0. Then c_0 does not appear in $\forall y\psi$ (since y was not substituted) and hence also not in ψ. So x_0 does not occur freely in φ_1, so $\psi \to \forall y\psi$ is $\varphi(x_1/c_1, \ldots, x_k/c_k)$.
Now we investigate the relationship between ψ_1 and ψ_2. First we notice that ψ_1 must be equal to $\psi(c_0/x_0, \ldots, c_k/x_k)$. Since y is different from all the x_i, then we must also have $\psi_2 = \psi(c_0/x_0, \ldots, c_k/x_k)$. So φ was also of the form $\psi_1 \to \forall y\psi_1$, with y not free in φ_1, i.e., an axiom.
This deals with the pure form of the axiom. We leave the general case of a generalization axiom as well as the other groups to the reader.

Case 2: $\varphi(x_0/c_0, \ldots, x_k/c_k)$ is derived from the formulas $\psi(x_0/c_0, \ldots, x_k/c_k)$ and $\psi(x_0/c_0, \ldots, x_k/c_k) \to \varphi(x_0/c_0, \ldots, x_k/c_k)$ by modus ponens. Then, by induction hypothesis, $\Gamma \vdash \psi \to \varphi$ and $\Gamma \vdash \psi$. So, by modus ponens, we have $\Gamma \vdash \varphi$.

Why would we want to add constants to the language? Suppose the following. We are working in a consistent theory Γ which proves $\exists x\varphi$, but there is no constant symbol c in $\mathcal{L}(\Gamma)$ such that $\Gamma \vdash \varphi(x/c)$. In this situation, we say that φ has no **witness** in $\mathcal{L}$. To correct this problem, we can add a new constant, c_φ, to $\mathcal{L}$, and a new axiom, $\exists x\varphi \to \varphi(x/c_\varphi)$, to Γ, called the special axiom for c_φ. Then, clearly, φ will have a witness in $\mathcal{L}$. Moreover, our new theory will be consistent. Before proving this, we give a definition and a construction.

2.2.4. Definition. A theory Γ is **Henkin** if for every sentence of the form $\exists x\theta$ in $\mathcal{L}(\Gamma)$, there is a constant c in $\mathcal{L}(\Gamma)$ such that $\Gamma \vdash \exists x\theta \to \theta(x/c)$.

Given a theory Γ, we now show how to construct an extension Γ_H of Γ such that Γ_H is Henkin.

First we prove the following lemma:

2.2.5. Lemma. If Γ is a consistent theory in the language $\mathcal{L}$, then there is a language $\mathcal{L}' \supseteq \mathcal{L}$ (with the same function and relation symbols, but new constants) and a **consistent** theory $\Gamma' \supseteq \Gamma$ such that:
For all sentences of the form $\exists y\,\varphi$ in the language $\mathcal{L}$
there is a constant symbol $c \in \mathcal{L}'$ such that the formula
$\exists y\,\varphi \to \varphi(y/c)$ is in Γ'.

Proof: Let $\{\varphi_0, \varphi_1, \ldots\}$ be an enumeration of the sentences of $\mathcal{L}$ which are of the form $\exists x\theta$. So for each natural number i there is a formula θ_i and a variable y_i such that $\varphi_i = \exists y_i\,\theta_i$. (Note that for $i \neq j$ it is possible that

y_i and y_j are the same variable.) For each i we will add a special constant c_i to $\mathcal{L}$, and we will add a special axiom, $\exists x\theta_i \to \theta_i(y_i/c_i)$ to Γ. Formally, let $\{c_0, c_1, \dots\}$ be an enumerable set of distinct constant symbols not in $\mathcal{L}$, and let

$$\Gamma' = \Gamma \cup \{\exists x\theta_i \to \theta_i(y_i/c_i) : i \in \mathbb{N}\}.$$

We have to show that Γ' is consistent. So suppose Γ' is inconsistent. Then some finite subtheory of Γ' must be inconsistent. So there is some k such that

$$\Gamma \cup \{\exists x\theta_i \to \theta_i(y_i/c_i) : i < k\}$$

is consistent but

$$\Gamma \cup \{\exists x\theta_i \to \theta_i(y_i/c_i) : i < k\} \cup \{\exists x\theta_k \to \theta_k(c_k)\}$$

is inconsistent. Let Γ^* be

$$\Gamma \cup \{\exists x\theta_i \to \theta_i(y_i/c_i) : i < k\}.$$

Then $\Gamma^* \vdash \neg(\exists x\theta_k \to \theta_k(c_k))$. So $\Gamma^* \vdash \exists x\theta_k \wedge \neg\theta_k(c_k)$. So $\Gamma^* \vdash \exists x\theta_k$. Also, since $\Gamma^* \vdash \neg\theta_k(c_k)$, the theorem on constants and the generalization rule imply that $\Gamma^* \vdash \forall x \neg\theta_k(x)$, which is a contradiction. Therefore, Γ' is consistent.

Now each φ_i from $\mathcal{L}$ in Γ has a witness in Γ', but there may be formulas φ in $\mathcal{L}'$ which do not have a witness. So we repeat the process.

2.2.6. Theorem. Assume that Γ is a consistent theory in the language $\mathcal{L}$. Then there exists a language $\mathcal{L}_H \supseteq \mathcal{L}$ and a consistent theory $\Gamma_H \supseteq \Gamma$ in the language $\mathcal{L}_H$ such that Γ_H is Henkin.

Proof: We let $\Gamma_0 := \Gamma$, $\mathcal{L}_0 := \mathcal{L}$. For each n, we construct Γ_{n+1} and $\mathcal{L}_{n+1}$ from Γ_n and $\mathcal{L}_n$ as in lemma 2.2.5, i.e., we let $\Gamma_{n+1} := \Gamma'_n$. We then take

$$\Gamma_H = \bigcup\{\Gamma_n : n \in \mathbf{N}\}.$$

Clearly, Γ_H is Henkin, and $\mathcal{L}(\Gamma_H)$ is enumerable. We call Γ_H the **Henkinization of Γ**.

We have to show that Γ_H is consistent. If Γ_H is inconsistent, then there must be a finite inconsistent subset, so for some n, Γ_n must be inconsistent. But by the previous lemma (using induction on n) we see that each Γ_n must be consistent.

Our strategy for the proof of the completeness theorem will be the following: Given a consistent theory Γ in a language $\mathcal{L}$, we first find a complete consistent Henkin theory Γ'. Then we construct a model $\mathcal{M}'$ for Γ'. Γ' is a theory in a bigger language $\mathcal{L}'$, so $\mathcal{M}'$ is not literally a model of Γ. However, if we "forget" the interpretations of the new constant/function/relation symbols, we get a model $\mathcal{M}$ for the original language $\mathcal{L}$, which will be a model of Γ. The relationship between the models $\mathcal{M}$ and $\mathcal{M}'$ is formalized in the definition below:

2.2.7. Definition. Let $\mathcal{L}$ be a language, and let $\mathcal{L}'$ be an extension of $\mathcal{L}$ (i.e. $\mathcal{L}$ is a sublanguage of $\mathcal{L}'$, with possibly fewer constant symbols or function symbols or relation symbols). Let $\mathcal{M}$ be a model for $\mathcal{L}$, and let $\mathcal{M}'$ be a model for $\mathcal{L}'$. We say that $\mathcal{M}'$ is an **expansion** of $\mathcal{M}$ if

(i) $M' = M$.

(ii) every predicate $\mathtt{R}$ in $\mathcal{L}$ is also a predicate in $\mathcal{L}'$, and $\mathtt{R}^{\mathcal{M}} = \mathtt{R}^{\mathcal{M}'}$.

(iii) every function $\mathtt{F}$ in $\mathcal{L}$ is also a function in $\mathcal{L}'$, and $\mathtt{F}^{\mathcal{M}} = \mathtt{F}^{\mathcal{M}'}$.

(iv) every constant $\mathtt{c}$ in $\mathcal{L}$ is also a constant in $\mathcal{L}'$, and $\mathtt{c}^{\mathcal{M}} = \mathtt{c}^{\mathcal{M}'}$.

If $\mathcal{M}'$ is an expansion of $\mathcal{M}$, then we call $\mathcal{M}$ the **restriction of** $\mathcal{M}'$ to $\mathcal{L}$, and we write $\mathcal{M} = \mathcal{M}' \restriction \mathcal{L}$.

2.2.8. Example. $\mathcal{M} := (\mathbb{Z}, +, 0)$ is a group. $\mathcal{M}' := (\mathbb{Z}, +, \cdot, 0)$ is a ring. $\mathcal{M} = \mathcal{M}' \restriction \mathcal{L}$, where $\mathcal{L}$ is the language of groups.
More generally, any ring is an expansion of its additive group.

2.2.9. Restriction Lemma. Let Γ and Γ' be theories in languages $\mathcal{L}$ and $\mathcal{L}'$, respectively, with Γ' an extension of Γ and $\mathcal{L}'$ an extension of $\mathcal{L}$. Let $\mathcal{M}'$ be a model of Γ'. Then $\mathcal{M}' \restriction \mathcal{L}$ is a model of Γ.

Proof: Let $\mathcal{M} = \mathcal{M}' \restriction \mathcal{L}$.
We claim the following

(1) For all closed $\mathcal{M}$-terms τ (in the language $\mathcal{L}$), $\tau^{\mathcal{M}} = \tau^{\mathcal{M}'}$.

(2) For all closed $\mathcal{M}$-formulas φ (in the language $\mathcal{L}$), $\mathcal{M} \models \varphi$ iff $\mathcal{M}' \models \varphi$.

(3) $\mathcal{M} \models \Gamma$.

Proof of (1): By induction on τ, using assumption 2.2.7(iv)–(v).

Proof of (2): By induction on φ. For atomic formulas, we use 2.2.7(iii). We leave the details to the reader, noting only that the universes M and M' are the same sets, so there are no problems in the inductive step $\varphi = \forall \mathbf{x} \psi$.

Proof of (3): Every formula $\varphi \in \Gamma$ is also in Γ'. Hence $\mathcal{M}' \models \varphi$, so by (2), $\mathcal{M} \models \varphi$.

Before we start the actual construction of our model from the constants in the language, we make a few general remarks about equivalence relations:

Assume that A and B are sets, and $f : A \to B$ is a function. We define the relation $\sim_f$ on A as follows:

$$a_1 \sim_f a_2 \quad \text{iff } f(a_1) = f(a_2).$$

Then it is easy to see that the relation $\sim_f$ is reflexive, symmetric and transitive, i.e.,

For all $a \in A$, $a \sim_f a$.
For all $a_1, a_2 \in A$, if $a_1 \sim_f a_2$ then $a_2 \sim_f a_1$.
For all $a_1, a_2, a_3 \in A$, if $a_1 \sim_f a_2$ and $a_2 \sim_f a_3$, then $a_1 \sim_f a_3$.

Conversely, we have the following:

2.2.10. Fact. If $R \subseteq A \times A$ is any reflexive, symmetric and transitive relation (we call such a relation an **equivalence relation**) then there is a set B and a function $f : A \to B$ such that $(a_1, a_2) \in R$ iff $f(a_1) = f(a_2)$.

Proof: We let the set B be the set of "equivalence classes," i.e. for each $a \in A$ we let the equivalence class of a be the set

$$a/R := \{a' \in A : (a, a') \in R\}.$$

The function $f(a) := a/R$ will then satisfy the requirements, as can be easily checked by the reader.

Now we are ready to construct a model from the constants. Let Γ be a consistent theory. Let Γ_H be the Henkinization of Γ. By theorem 2.2.6, Γ_H is consistent. Hence, by theorem 2.1.23, Γ_H has a complete simple extension Γ^*. Since Γ^* is a **simple** extension of Γ_H, it follows that Γ^* is also a Henkin theory. We will show that Γ^* has a model $\mathcal{M}^*$, and this will suffice. By the restriction lemma, $\mathcal{M}^* \restriction \mathcal{L}(\Gamma)$ will be a model of Γ. Thus, the heart of the Henkin proof is the construction of the model $\mathcal{M}^*$ of Γ^*.

As noted previously, we build the universe of $\mathcal{M}^*$ using the closed terms of $\mathcal{L}(\Gamma^*)$. Define a relation $\sim$ on the set of closed terms by

$$\tau \sim \mu \text{ iff } \Gamma^* \vdash \tau = \mu.$$

It is not hard to show that $\sim$ is an equivalence relation. By the identity axioms, $\Gamma^* \vdash \mathtt{x} = \mathtt{x}$, so $\Gamma^* \vdash \tau = \tau$ by substitution. Hence $\tau \sim \tau$, so $\sim$ is reflexive. We leave symmetry and transitivity as an exercise.

The universe, M^*, is the set of equivalence classes under $\sim$. To complete the definition of $\mathcal{M}^*$, we also need to define $F^{\mathcal{M}^*}$ for each function symbol F in $\mathcal{L}(\Gamma^*)$, $\mathtt{c}^{\mathcal{M}^*}$ for each constant symbol in $\mathcal{L}(\Gamma^*)$ and $R^{\mathcal{M}^*}$ for each relation symbol R.

If $\mathtt{c}$ is a constant symbol in $\mathcal{L}(\Gamma^*)$, we let

$$\mathtt{c}^{\mathcal{M}^*} = [\mathtt{c}].$$

If τ is a closed term of $\mathcal{L}(\Gamma^*)$, let $[\tau]$ denote the equivalence class of τ. Let $\mathtt{F}$ be an n-ary function symbol. We define

$$\mathtt{F}^{\mathcal{M}^*}([\tau_1]\cdots[\tau_n]) = [\mathtt{F}\tau_1\cdots\tau_n].$$

For $\mathtt{R}$, an n-ary predicate symbol, we define

$$([\tau_1]\cdots[\tau_n]) \in \mathtt{R}^{\mathcal{M}^*} \text{ iff } \Gamma^* \vdash \mathtt{R}\tau_1\cdots\tau_n.$$

In order to show that these definitions are sound, we must show that they do not depend on the particular representative chosen from the equivalence class. In other words, suppose $\tau \in [\mu]$, so that $[\tau] = [\mu]$. Then we certainly want $\mathtt{F}^{\mathcal{M}^*}([\mu])$ to equal $F^{\mathcal{M}^*}([\tau])$. But this is not a problem since the theory proves that τ and μ are equal. Formally, let $\mathtt{F}$ be a unary function symbol. Then

$$\begin{aligned}
[\tau] = [\mu] &\rightarrow \tau \sim \mu \\
&\rightarrow \Gamma^* \vdash \tau = \mu \\
&\rightarrow \Gamma^* \vdash \mathtt{F}\tau = \mathtt{F}\mu \\
&\rightarrow \mathtt{F}\tau \sim \mathtt{F}\mu \\
&\rightarrow [\mathtt{F}\tau] = [\mathtt{F}\mu] \\
&\rightarrow \mathtt{F}^{\mathcal{M}^*}([\tau]) = \mathtt{F}^{\mathcal{M}^*}([\mu]).
\end{aligned}$$

Similarly, for any unary predicate symbol $\mathtt{R}$,

$$\begin{aligned}
[\tau] = [\mu] &\rightarrow \tau \sim \mu \\
&\rightarrow \Gamma^* \vdash \tau = \mu \\
&\rightarrow \Gamma^* \vdash \mathtt{R}\tau \leftrightarrow \mathtt{R}\mu \\
&\rightarrow \Gamma^* \vdash \mathtt{R}\tau \text{ iff } \Gamma^* \vdash \mathtt{R}\mu \\
&\rightarrow [\tau] \in \mathtt{R}^{\mathcal{M}^*} \text{ iff } [\mu] \in \mathtt{R}^{\mathcal{M}^*}.
\end{aligned}$$

Using the equality axiom, it is straightforward to generalize these proofs to n-ary functions and n-ary predicates. Hence we see that our definitions are sound.

We claim that $\mathcal{M}^* \models \Gamma^*$. Before we prove this, we need a preliminary result which shows that each closed term in $\mathcal{L}(\Gamma^*)$ has the required interpretation in $\mathcal{M}^*$.

2.2.11. Lemma. If τ is a closed term in $\mathcal{L}(\Gamma^*)$, then $\tau^{\mathcal{M}^*} = [\tau]$.

Proof: by induction on the complexity of τ. Since τ is a closed term, it does not contain any variables. Hence it must either be a constant symbol, or be of the form $\mathrm{F}\tau_1 \cdots \tau_n$.
If τ is a constant symbol c, then by definition we have $\tau^{\mathcal{M}^*} = \mathrm{c}^{\mathcal{M}^*} = [\mathrm{c}] = [\tau]$. Otherwise, we have:

$$\begin{aligned}\tau^{\mathcal{M}^*} &= (F\tau_1 \cdots \tau_n)^{\mathcal{M}^*} \\ &= F^{\mathcal{M}^*}((\tau_1)^{\mathcal{M}^*} \cdots (\tau_n)^{\mathcal{M}^*}) \quad \text{by definition of interpretation} \\ &= F^{\mathcal{M}^*}([\tau_1] \cdots [\tau_n]) \quad \text{by induction hypothesis} \\ &= [F\tau_1 \cdots \tau_n] \quad \text{by definition of } F^{\mathcal{M}^*} \\ &= [\tau].\end{aligned}$$

We are now ready to show the following

2.2.12. Theorem. If φ is a sentence of $\mathcal{L}(\Gamma^*)$, then $\Gamma^* \vdash \varphi$ iff $\mathcal{M}^* \models \varphi$.

Proof: by induction on φ. If φ is atomic, then φ is $R\tau_1 \cdots \tau_n$, or φ is $\tau_1 = \tau_2$. Hence either

$$\begin{aligned}\Gamma^* \vdash \varphi &\text{ iff } \Gamma^* \vdash \mathrm{R}\tau_1 \cdots \tau_n \\ &\text{ iff } ([\tau_1] \cdots [\tau_n]) \in \mathrm{R}^{\mathcal{M}^*} \\ &\text{ iff } ((\tau_1)^{\mathcal{M}^*} \cdots (\tau_n)^{\mathcal{M}^*}) \in \mathrm{R}^{\mathcal{M}^*} \\ &\text{ iff } \mathcal{M}^* \models \mathrm{R}(\tau_1, \ldots, \tau_n) \\ &\text{ iff } \mathcal{M}^* \models \varphi\end{aligned}$$

or

$$\begin{aligned}\Gamma^* \vdash \varphi &\text{ iff } \Gamma^* \vdash \tau = \mu \\ &\text{ iff } \tau \sim \mu \\ &\text{ iff } [\tau] = [\mu] \\ &\text{ iff } \tau^{\mathcal{M}^*} = \mu^{\mathcal{M}^*} \\ &\text{ iff } \mathcal{M}^* \models \tau = \mu \\ &\text{ iff } \mathcal{M}^* \models \varphi.\end{aligned}$$

This covers the cases where φ is atomic.

For the case where φ is $(\neg\psi)$, we need to use the completeness of our theory.

$$\begin{aligned}\Gamma^* \vdash \varphi \text{ iff } & \Gamma^* \vdash \neg\psi \\ \text{iff } & \Gamma^* \nvdash \psi \qquad \text{since } \Gamma^* \text{ is complete !} \\ \text{iff } & \mathcal{M}^* \not\models (\psi) \qquad \text{by induction hypothesis} \\ \text{iff } & \mathcal{M}^* \models (\neg\psi) \\ \text{iff } & \mathcal{M}^* \models (\varphi).\end{aligned}$$

If φ is $\psi \wedge \chi$, then we have

$$\begin{aligned}\Gamma^* \vdash \varphi \text{ iff } & \Gamma^* \vdash \psi \wedge \chi \\ \text{iff } & \Gamma^* \vdash \psi \text{ and } \Gamma^* \models \chi \\ \text{iff } & \mathcal{M}^* \models \psi \text{ and } \mathcal{M}^* \models \chi \\ \text{iff } & \mathcal{M}^* \models \psi \wedge \chi \\ \text{iff } & \mathcal{M}^* \models \varphi.\end{aligned}$$

Finally, we have to deal with the case $\varphi = \forall \mathtt{x}\psi$. Here we will use the fact that our theory is a Henkin theory. First note that for every closed term τ in $\mathcal{L}(\Gamma^*)$ there is a constant $\mathtt{c}$ in $\mathcal{L}^*$ such that $\Gamma^* \models \mathtt{c} = \tau$. This constant can be found by applying the fact that Γ^* is a Henkin theory to the formula $\exists \mathtt{x}\, \mathtt{x} = \tau$.

Now assume that $\Gamma \vdash \forall \mathtt{x}\psi$. Then for any element $m \in \mathcal{M}^*$ we can find a term τ such that $m = [\tau]$ is the equivalence class of τ.

We have $\Gamma \vdash \psi(\mathtt{x}/\tau)$. So by induction hypothesis, $\mathcal{M}^* \models \psi(\mathtt{x}/\tau)$. As $\tau^{\mathcal{M}^*} = [\tau] = m$, we have $\mathcal{M}^* \models \psi(\mathtt{x}/m)$. This can be done for any m, so $\mathcal{M}^* \models \forall \mathtt{x}\psi$.

For the converse direction, assume $\mathcal{M} \models \forall \mathtt{x}\psi$. Let $\mathtt{c}$ be a constant such that the formula

$$\exists \mathtt{x} \neg\psi \;\rightarrow\; \neg\psi(\mathtt{x}/\mathtt{c})$$

is in Γ^*. So

$$\Gamma^* \vdash \psi(\mathtt{x}/\mathtt{c}) \rightarrow \neg\exists \mathtt{x} \neg\psi$$

or equivalently,

$$\Gamma^* \vdash \psi(\mathtt{x}/\mathtt{c}) \rightarrow \forall \mathtt{x}\, \psi. \qquad (*)$$

Since $\mathcal{M}^* \models \forall \mathtt{x}\, \psi$, also $\mathcal{M}^* \models \psi(\mathtt{x}/\mathtt{c})$. So by induction hypothesis, $\Gamma^* \vdash \psi(\mathtt{x}/\mathtt{c})$, so by $(*)$, $\Gamma \vdash \forall \mathtt{x}\, \psi$.

Now the obvious

2.2.13. Corollary. $\mathcal{M}^*$ is a model of Γ^*.

Proof: Let φ be a nonlogical axiom of Γ^*. Since $\Gamma^* \vdash \varphi$, $\mathcal{M}^* \models \varphi$, by the theorem.

As mentioned previously, $\mathcal{M}^* \restriction \mathcal{L}(\Gamma)$ is a model of Γ. This completes the proof of the completeness theorem.

Exercises

The Compactness Theorem for First Order Logic

1. Let Γ be a set of sentences such that every finite subset of Γ has a model. Show that Γ has a model.

2. Let Σ_1 and Σ_2 be sets of sentences (not necessarily finite) such that there is no model $\mathcal{M}$ such that both $\mathcal{M} \models \Sigma_1$ and $\mathcal{M} \models \Sigma_2$.
Prove that there exists a sentence φ such that:
Every model of Σ_1 satisfies φ and every model of Σ_2 satisfies $\neg\varphi$.

3. We use the language of groups, i.e., we have a binary relation symbol $+$ and a constant symbol 0.
(a) Find a sentence φ such that for every model $\mathcal{M}$ of our language, $\mathcal{M} \models \varphi$ iff $\mathcal{M}$ is a group.
(b) Find a theory Γ such that for every model $\mathcal{M}$ of our language, $\mathcal{M} \models \Gamma$ iff $\mathcal{M}$ is an infinite group.
(c) Show that if Γ is a theory such that every finite group is a model of Γ, then there is an infinite group that is a model of Γ.
(d) Show that if φ is a sentence such that every infinite group is a model of φ, then there is a finite group that is a model of φ.

Henkin Theories

4. Assume that Γ is a theory satisfying the following:
(i) Γ is a Henkin theory.
(ii) For any two constants c, d, either $\Gamma \vdash \mathsf{c} = \mathsf{d}$ or $\Gamma \vdash \mathsf{c} \neq \mathsf{d}$ (i.e., $\Gamma \vdash \neg\, \mathsf{c} = \mathsf{d}$).
(iii) There are two constants a, b such that $\Gamma \vdash \mathsf{a} \neq \mathsf{b}$.
Show that Γ is a complete theory.
(Hint: For any sentence φ, consider the sentence

$$\exists \mathsf{x}\big[(\varphi \wedge \mathsf{x} = \mathsf{a}) \vee (\neg\varphi \wedge \mathsf{x} = \mathsf{b})\big]$$

and apply (i).)

5. Show that (i) and (ii) from the previous exercise do not imply that Γ is complete. (Hint: As a counterexample, use a theory Γ that has only one element models.)

6. Consider the following condition:
(iii)′ $\Gamma \not\vdash \forall x \forall y\, x = y$. Does (i)∧(ii)∧(iii)′ imply that Γ is complete? (Hint: It would be enough to show that it implies (iii).)

2.3. Nonstandard Models of Arithmetic

2.3.1. Definition. Let $\mathcal{L}$ be the language with nonlogical symbols $+, \cdot, <, 0, S$, where $+$ and $\cdot$ are binary function symbols, $<$ is a binary relation symbol, 0 is a constant symbol, and S is a unary function symbol.

2.3.2. Definition.
(1) Let $\mathbb{N} = \langle \mathbb{N}, +, \cdot, <, 0, S \rangle$ be the natural numbers with the usual operations and relations (i.e. $0^{\mathbb{N}} = 0$, $+^{\mathbb{N}}$ is the addition of natural numbers, $S^{\mathbb{N}}$ is the successor operation, etc.).
(2) Let Th($\mathbb{N}$) (the "theory of $\mathbb{N}$") be the set of all sentences (closed formulas) that are valid in $\mathbb{N}$.
(3) Every model $\mathcal{M}$ of Th($\mathbb{N}$) is called a **model of arithmetic.** The model $\mathbb{N}$ is called the natural model of Th($\mathbb{N}$).

As usual, we write M for the universe of any model $\mathcal{M}$. But we write $\mathbb{N}$ for the set of natural numbers as well as for the model.

2.3.3. Definition. For every term τ and for every natural number n there is a term $S^n\tau$ ("the n-th successor of τ"), that is defined (by induction) as follows:

$S^0\tau$ is the term τ itself.
Given $S^n\tau$, let $S^{n+1}\tau$ be the term $S(S^n\tau)$.

Notice that this is a definition by "external" induction, i.e., induction on the natural numbers. This is related to, but quite different from, the induction axiom of Peano Arithmetic (see Chapter 4), in which the inter-

pretation of the variables can range over all elements of any given model of PA (that may be different from the natural numbers).
Note also that the term $S^n\tau$ is defined only for natural numbers n. For arbitrary elements a of some model $\mathcal{M}$ of Th($\mathbb{N}$), $S^a\tau$ is not defined.

The terms S^n0 play a special role: They correspond to the natural numbers, as clearly $(S^n0)^{\mathbb{N}} = n$, for all $n \in \mathbb{N}$.

2.3.4. Definition.

(1) Let $\mathcal{M}$ be a model of arithmetic. The **finite** or **standard** elements of $\mathcal{M}$ are the elements that are of the form $a = (S^n0)^{\mathcal{M}}$, for some natural number n, the other elements are called **nonstandard** or **infinite**.

(2) For any model $\mathcal{M}$ of Th($\mathbb{N}$), we let

$$M_{\mathsf{fin}} = \{a \in M : a \text{ is finite}\} = \{(S^n0)^{\mathcal{M}} : n \in \mathbb{N}\}.$$

(3) A model $\mathcal{M}$ of Th($\mathbb{N}$) is called **standard**, if it has no nonstandard elements, i.e. if $\mathcal{M} = \mathcal{M}_{\mathsf{fin}}$. Otherwise, it is called **nonstandard**.

2.3.5. Theorem.

(a) There are nonstandard models of arithmetic.

(b) For every standard model $\mathcal{M}$ of Th($\mathbb{N}$) there is a unique isomorphism $f : \mathbb{N} \to \mathcal{M}$.

(c) No nonstandard model is isomorphic to $\mathbb{N}$.

Proof of (a): We will consider the language $\mathcal{L}' = \mathcal{L}$ augmented by a constant symbol c. Let

$$\Gamma = \text{Th}(\mathbb{N}) \cup \{0 < c, S0 < c, SS0 < c, \ldots\}.$$

Claim: Γ is consistent.

Proof of the claim: It is enough to show that every finite subset of Γ is consistent. Let $\Gamma_0 \subseteq \text{Th}(\mathbb{N}) \cup \{0 < c, S0 < c, SS0 < c, \ldots\}$ be finite, then $\Gamma_0 \subseteq \text{Th}(\mathbb{N}) \cup \{0 < c, S0 < c, SS0 < c, \ldots, S^n0 < c\}$ for some $n \in \mathbb{N}$. Let

$$\mathbb{N}' = \langle \mathbb{N}, +, \cdot, <, 0, S, n+1 \rangle,$$

so $\mathbb{N} = \mathbb{N}' \restriction \mathcal{L}$, $\mathtt{c}^{\mathbb{N}'} = n+1$, i.e. we consider the model $\mathbb{N}$ where c is interpreted as $n+1$. Then $\mathbb{N}' \models \Gamma_0$, so Γ_0 is consistent.

As Γ is consistent, it has a model. Now let

$$\mathcal{M} = \langle M, +^{\mathcal{M}}, \cdot^{\mathcal{M}}, <^{\mathcal{M}}, \mathtt{0}^{\mathcal{M}}, \mathtt{S}^{\mathcal{M}}, \mathtt{c}^{\mathcal{M}} \rangle$$

be a model of Γ, and let $a = \mathtt{c}^{\mathcal{M}}$. Then for every natural number n, $\mathcal{M} \models (\mathtt{c} > \mathtt{S}^n\mathtt{0})$. Since $\forall \mathtt{x}(\mathtt{x} > \mathtt{y} \rightarrow \mathtt{x} \neq \mathtt{y}) \in \mathrm{Th}(\mathbb{N})$, $\mathcal{M} \models \mathtt{c} > \mathtt{S}^n\mathtt{0}$ implies $\mathcal{M} \models \mathtt{c} \neq \mathtt{S}^n\mathtt{0}$, and therefore $\mathcal{M} \models (a \neq \mathtt{S}^n\mathtt{0})$.
Let $\mathcal{M}' = \mathcal{M} \restriction \mathcal{L}$, then $\mathcal{M}'$ is a nonstandard model of arithmetic, because $\mathcal{M}' \models a \neq \mathtt{S}^n\mathtt{0}$ for all natural number n.

This finishes the proof of (a).

Proof of (b): Let $\mathcal{M}$ be a standard model of arithmetic. There can be at most one isomorphism from $\mathbb{N}$ to $\mathcal{M}$: For any isomorphism $f : \mathbb{N} \rightarrow \mathcal{M}$, we must have $f(0) = f(\mathtt{0}^{\mathbb{N}}) = \mathtt{0}^{\mathcal{M}}$, $f(1) = f(\mathtt{S0})^{\mathbb{N}} = (\mathtt{S0})^{\mathcal{M}}$, etc. Hence, define the function i from the natural numbers to the universe of $\mathcal{M}$ by

$$i(n) = (\mathtt{S}^n\mathtt{0})^{\mathcal{M}} \qquad \text{for all natural numbers } n.$$

i is a bijection: Since $\mathcal{M}$ is a standard model, i is onto. If n and m are two distinct natural numbers, then $(\mathtt{S}^n\mathtt{0})^{\mathbb{N}} = n \neq m = (\mathtt{S}^m\mathtt{0})^{\mathbb{N}}$, hence $\mathbb{N} \models \mathtt{S}^n\mathtt{0} \neq \mathtt{S}^m\mathtt{0}$, hence $\mathcal{M} \models \mathtt{S}^n\mathtt{0} \neq \mathtt{S}^m\mathtt{0}$, so $i(n) \neq i(m)$. Therefore i is 1-1.

i is an isomorphism: We will only show $i(n +^{\mathbb{N}} m) = i(n) +^{\mathcal{M}} i(m)$, leaving the formulation and proofs of the analogous statements for $\cdot$, $<$ and 0 to the reader.
Let $k = n + m$. Then $\mathbb{N} \models \mathtt{S}^k\mathtt{0} = \mathtt{S}^n\mathtt{0} + \mathtt{S}^m\mathtt{0}$, so $\mathcal{M} \models \mathtt{S}^k\mathtt{0} = \mathtt{S}^n\mathtt{0} + \mathtt{S}^m\mathtt{0}$, so $i(k) = i(m) +^{\mathcal{M}} i(n)$.
This completes the proof of (b).

(c) Assume that $f : \mathcal{M} \rightarrow \mathbb{N}$ is an isomorphism. As $\mathcal{M}$ is nonstandard, there is an $a \in M$ such that for all natural numbers n,

$$\mathcal{M} \models a \neq \mathtt{S}^n\mathtt{0}.$$

Let $f(a) = k$. Then $\mathcal{M} \models (a \neq \mathtt{S}^k(\mathtt{0}))$, so $\mathbb{N} \models (f(a) \neq \mathtt{S}^k\mathtt{0})$, or $\mathbb{N} \models k \neq \mathtt{S}^k\mathtt{0}$. This is a contradiction.

The Structure of Nonstandard Models

What do we know about a nonstandard model? We will see that there are many different nonstandard models of $\mathrm{Th}(\mathbb{N})$. However, they all share some basic structure.

To analyze $\mathcal{M}$, we have to keep in mind that every closed formula in our language that is valid in $\mathbb{N}$ is also valid in $\mathcal{M}$. Conversely, if φ is not valid in $\mathbb{N}$, then $\neg\varphi$ is valid in $\mathbb{N}$ and hence in $\mathcal{M}$, so φ is not valid in $\mathcal{M}$. Hence

$$\text{For all closed formulas } \varphi,\ \mathcal{M} \models \varphi \text{ iff } \mathbb{N} \models \varphi$$

The difference between $\mathcal{M}$ and $\mathbb{N}$ stems from the fact that some properties of $\mathbb{N}$ are **not expressible** in our language. For example, the statement

$$\forall x(x = 0 \lor x = S0 \lor x = SS0 \lor \cdots) \qquad (*)$$

is valid in $\mathbb{N}$, but it is not in Th($\mathbb{N}$), because it is not even in our language.

We know a lot about the standard elements of $\mathcal{M}$, because they are represented by closed terms, so there are closed formulas "talking" about them. (E.g. if $a = (SS0)^{\mathcal{M}}$, $b = (SSSS0)^{\mathcal{M}}$, we know that $a +^{\mathcal{M}} a = b$.) But we know little about the nonstandard elements, because there are no closed formulas that refer to them, except via quantifiers. (E.g. for every (standard or nonstandard) element a we know $a \geq 0^{\mathcal{M}}$.)

We leave it as an exercise to show:

For every closed term τ there exists a number n such that $\tau^{\mathcal{M}} = (S^n0)^{\mathcal{M}}$ (so closed terms always represent standard elements).

For the following, fix a nonstandard model $\mathcal{M}$, and let $\mathbb{N}$ be the standard model. Define $i : \mathbb{N} \to \mathcal{M}$ by $i(n) = (S^n0)^{\mathcal{M}}$, i.e. the interpretation of the term S^n0 in the model $\mathcal{M}$.

We leave as an exercise the proof that i is an isomorphism between $\mathbb{N}$ and $\mathcal{M}_{\text{fin}} = \langle M_{\text{fin}}, +, \cdot, 0, S\rangle$ (where S is really the function $S^{\mathcal{M}}$ restricted to $\mathcal{M}_{\text{fin}}$, etc.).

The proof is similar to 2.3.5(b).

2.3.6. Fact. If $a \in M_{\text{fin}}$, $b \in M - M_{\text{fin}}$, then $a <^{\mathcal{M}} b$.

Before we prove this fact, we have to introduce some notation:

Let x be a variable. For every natural number n we will define a formula ψ_n with free variable x.

We let ψ_0 be the formula $x \neq 0$.

If ψ_n is given, let $\psi_{n+1} = (\psi_n \land x \neq S^{n+1}0)$.

E.g. ψ_3 is the formula

$$x \neq 0 \land x \neq S0 \land x \neq SS0 \land x \neq SSS0$$

Instead of ψ_n we will write

$$(x \neq 0 \land \cdots \land x \neq S^n0).$$

(It seems pedantic to insist that such an obvious concept has to be defined by induction. However, we want to make it clear that n is NOT a variable and does NOT appear in the formula $\psi_n(\mathrm{x})$ – for example there is no variable n in the formula $\psi_3(\mathrm{x})$, nor does the number 3 explicitly appear in it.)

2.3.7. Fact. For any natural number n,

$$\mathbb{N} \models \forall \mathrm{x}((\mathrm{x} \neq 0 \wedge \cdots \wedge \mathrm{x} \neq \mathrm{S}^n 0) \leftrightarrow \mathrm{x} > \mathrm{S}^n 0)$$

We leave the proof (by induction on n) to the reader.

Now we can prove fact 2.3.6: Let $a = (\mathrm{S}^n 0)^{\mathcal{M}}$, where $n \in \mathbb{N}$ is a natural number. We have

$$\mathbb{N} \models \forall \mathrm{x}(\mathrm{x} \neq 0 \wedge \mathrm{x} \neq \mathrm{S}0 \wedge \cdots \wedge \mathrm{x} \neq \mathrm{S}^n 0 \rightarrow \mathrm{x} > \mathrm{S}^n 0)$$

hence

$$\mathcal{M} \models \forall \mathrm{x}(\mathrm{x} \neq 0 \wedge \mathrm{x} \neq \mathrm{S}0 \wedge \cdots \wedge \mathrm{x} \neq \mathrm{S}^n 0 \rightarrow \mathrm{x} > \mathrm{S}^n 0)$$

so in particular

$$\mathcal{M} \models (b \neq 0 \wedge b \neq \mathrm{S}0 \wedge \cdots \wedge b \neq \mathrm{S}^n 0 \rightarrow b > \mathrm{S}^n 0).$$

Since $\mathcal{M} \models (b \neq 0 \wedge b \neq \mathrm{S}0 \wedge \cdots \wedge b \neq \mathrm{S}^n 0)$, we get

$$\mathcal{M} \models (b > \mathrm{S}^n 0) \quad \text{i.e., } \mathcal{M} \models b > a.$$

2.3.8. Fact. Let $\varphi(\mathrm{x})$ be a formula (of $\mathcal{L}$) with only one free variable, x. Then

(a) $\mathcal{M} \models \varphi(a)$ for all $a \in M_{\text{fin}}$ iff $\mathcal{M} \models \varphi(b)$ for all $b \in M$.

(b) $\mathcal{M} \models \varphi(b)$ for some $b \in M - M_{\text{fin}}$ iff $\mathcal{M} \models \varphi(a)$ for infinitely many $a \in M_{\text{fin}}$.

(c) $\mathcal{M} \models \varphi(b)$ for all $b \in M - M_{\text{fin}}$ iff $\mathcal{M} \models \varphi(a)$ for all but finitely many $a \in M_{\text{fin}}$.

Proof of (a):

$$\begin{aligned} \mathcal{M} \models \varphi(a) \quad \text{for all } a \in M_{\text{fin}} &\leftrightarrow \quad \text{for all } n \in \mathbb{N},\ \mathcal{M} \models \varphi(\mathrm{S}^n 0) \\ &\leftrightarrow \quad \text{for all } n \in \mathbb{N},\ \mathbb{N} \models \varphi(\mathrm{S}^n 0) \\ &\leftrightarrow \quad \text{for all } n \in \mathbb{N},\ \mathbb{N} \models \varphi(n) \\ &\leftrightarrow \quad \mathbb{N} \models \forall \mathrm{x} \varphi(\mathrm{x}) \\ &\leftrightarrow \quad \mathcal{M} \models \forall \mathrm{x} \varphi(\mathrm{x}). \end{aligned}$$

(b) follows easily from (c).

Proof of (c):

$$\begin{aligned}
\forall a \in M - M_{\text{fin}},\ \mathcal{M} \models \varphi(a) &\rightarrow \forall b \in M - M_{\text{fin}},\ \mathcal{M} \models (\forall \mathrm{x}\, \mathrm{x} > b \rightarrow \varphi(\mathrm{x}))\\
&\rightarrow \exists b \in M,\ \mathcal{M} \models (\forall \mathrm{x}\, \mathrm{x} > b \rightarrow \varphi(\mathrm{x}))\\
&\rightarrow \mathcal{M} \models \exists \mathrm{y} \forall \mathrm{x}(\mathrm{x} > \mathrm{y} \rightarrow \varphi(\mathrm{x}))\\
&\rightarrow \mathbb{N} \models \exists \mathrm{y} \forall \mathrm{x}(\mathrm{x} > \mathrm{y} \rightarrow \varphi(\mathrm{x}))\\
&\rightarrow \text{there is a } k \in \mathbb{N} \text{ such that}\\
&\qquad\qquad \text{for all } n > k \text{ in } \mathbb{N},\ \mathbb{N} \models \varphi(k).
\end{aligned}$$

Conversely, if for all $n > k$, $\mathbb{N} \models \varphi(n)$, then

$$\mathbb{N} \models \forall \mathrm{x}(\mathrm{x} > \mathrm{S}^k 0 \rightarrow \varphi(\mathrm{x})) \quad \rightarrow \quad \mathcal{M} \models \forall \mathrm{x}(\mathrm{x} > \mathrm{S}^k 0 \rightarrow \varphi(\mathrm{x}))$$

so for all $a \in M - M_{\text{fin}}$, $\mathcal{M} \models \varphi(a)$.

2.3.9. Corollary. "x is finite" cannot be expressed by a formula.

Proof: Assume that $\varphi(\mathrm{x})$ is a formula such that for all $a \in M$, $\mathcal{M} \models \varphi(a)$ iff $a \in M_{\text{fin}}$. This is a contradiction to 2.3.8(a).

2.3.10. Definition. Let $s = \mathrm{S}^{\mathcal{M}} : \mathcal{M} \rightarrow \mathcal{M}$ be the successor function of $\mathcal{M}$, and define the predecessor function $p : \mathcal{M} \rightarrow \mathcal{M}$ by

$$p(a) = \begin{cases} b & \text{iff } \mathcal{M} \models a > 0 \text{ and } s(b) = a \\ 0 & \text{if } a = 0^{\mathcal{M}}. \end{cases}$$

This definition makes sense, because

$$\mathbb{N} \models \forall \mathrm{x}\, (\mathrm{x} \neq 0 \rightarrow \exists! \mathrm{y}\, \mathrm{S}\mathrm{y} = \mathrm{x})$$

hence

$$\mathcal{M} \models \forall \mathrm{x}\, (\mathrm{x} \neq 0 \rightarrow \exists! \mathrm{y}\, \mathrm{S}\mathrm{y} = \mathrm{x}).$$

(We write $\exists! \mathrm{z} \varphi(\mathrm{z})$ as an abbreviation for $\exists \mathrm{z}\, \varphi(\mathrm{z}) \wedge \forall \mathrm{x} \forall \mathrm{y}(\varphi(\mathrm{x}) \wedge \varphi(\mathrm{y}) \rightarrow \mathrm{x} = \mathrm{y})$, i.e., "there exists a unique z satisfying φ.")

2.3.11. Remark. $\mathcal{M}$ is not only equipped with a predecessor function, we can also define subtraction: The formula $\forall \mathrm{x} \forall \mathrm{y}(\mathrm{x} \leq \mathrm{y} \rightarrow \exists! \mathrm{z}\, \mathrm{x} + \mathrm{z} = \mathrm{y})$ is in Th($\mathbb{N}$), so it is valid in $\mathcal{M}$.
Hence, for every $a \leq^{\mathcal{M}} b$ in M,

$$\mathcal{M} \models \exists! \mathrm{z}(a + \mathrm{z} = b),$$

so we let $b - a =$ the unique $c \in M$ satisfying $\mathcal{M} \models a + c = b$.

We can similarly extend all other definable functions from $\mathbb{N}$ to $\mathcal{M}$. Also relations that are definable by formulas can be naturally extended from $\mathbb{N}$ to $\mathcal{M}$.

On $\mathcal{M}$ we can define the following relation:

2.3.12. Definition.

$$x \sim y \leftrightarrow \text{ for some natural number } n,\ s^n(x) = y \text{ or } s^n(y) = x.$$

2.3.13. Fact.

(1) $\sim$ is an equivalence relation.

(2) $0 \sim x$ iff $x \in M_{\text{fin}}$. (Hence the relation $\sim$ is not definable by any formula in our language: If there were a formula $\varphi(x, y)$ such that $(\mathcal{M} \models \varphi(a, b)) \leftrightarrow (a \sim b)$, then we would have $a \in M_{\text{fin}} \leftrightarrow \mathcal{M} \models a \sim 0$, contradicting 2.3.9.)

(3) All equivalence classes are convex, i.e.

$$\text{If } a < b < c \text{ and } a \sim c, \text{ then } a \sim b$$

Proof of (3): Assume $\mathcal{M} \models \mathrm{S}^n a = c$. ($\mathrm{S}^n c = a$ is impossible, because $\mathcal{M} \models \mathrm{S}^n c = a \rightarrow c \leq a$.) As

$$\mathbb{N} \models \forall \mathrm{x} \forall \mathrm{y} \forall \mathrm{z}[\mathrm{x} < \mathrm{y} < \mathrm{z} \wedge \mathrm{S}^n \mathrm{x} = \mathrm{z} \rightarrow (\mathrm{y} = \mathrm{Sx} \vee \mathrm{y} = \mathrm{SSx} \vee \cdots \vee \mathrm{y} = \mathrm{S}^{n-1}\mathrm{x})]$$

we must have

$$\mathcal{M} \models a < b < c \wedge \mathrm{S}^n a = c \rightarrow (b = \mathrm{S}a \vee b = \mathrm{SS}a \vee \cdots \vee b = \mathrm{S}^{n-1}a).$$

(Note: As in 2.3.6, the formula in the previous line should be defined by induction on the natural number n.)

2.3.14. Definition.

(1) Given any $a \in M$, let $a/\sim$ be the equivalence class of a:

$$a/\sim\, = \{b \in M : a \sim b\}.$$

(2) Let $M/\sim$ be the set of all equivalence classes:

$$M/\sim\, = \{a/\sim\, : a \in M\}.$$

As the equivalence classes are convex, we can define a linear order on the equivalence classes.

2.3.15. Definition. Let A, B be two distinct equivalence classes. We let $A < B$ iff for some element $a \in A$ and for some element b in B we have $a < b$.
For two not necessarily distinct equivalence classes A and B we let $A \leq B$ iff $A < B$ or $A = B$.

2.3.16. Remark. If $A < B$ are two equivalence classes as above, then we have in fact that for all $a \in A$ and for all $b \in B$: $a < b$. So we could have equivalently defined:

$$a/{\sim} < b/{\sim} \quad \text{iff} \quad a \not\sim b \text{ and } a < b.$$

Proof: Assume not, so we have $b < a$ and $a_0 < b_0$, where $a, a_0 \in A$, $b, b_0 \in B$. We have two cases: Either $b < a_0$, then $b < a_0 < b_0$, so by 2.3.13(3) we have $a \sim b$ hence $A = B$, a contradiction. The other case is that $b > a_0$, then $a_0 < b < a$ so again $A = B$.

We leave as an exercise the proof that this relation indeed defines a linear order on the set $M/{\sim}$ of equivalence classes, with least element $0/{\sim} = M_{\text{fin}}$.

(Similarly, $\sim$ respects addition:

$$\text{If } a \sim a' \text{ and } b \sim b', \text{ then } a + b \sim a' + b'$$

so we could define addition on the equivalence classes by $a/{\sim} + b/{\sim} = (a + b)/{\sim}$. (See exercise 3.))

2.3.17. Fact.

(1) If a is in M_{fin}, then $a/{\sim} = M_{\text{fin}} \simeq \mathbb{N}$.
(2) If a is in $M - M_{\text{fin}}$, $\langle a/{\sim}, <, s\rangle$ is isomorphic to $\langle \mathbf{Z}, <, \sigma\rangle$, where $\mathbf{Z}$ are the integers, s is the restriction of $S^{\mathcal{M}}$ to $a/{\sim}$, and σ is the successor function in the integers.

Proof of (2): $a/{\sim}$ contains exactly the elements

$$\ldots, p^2(a), p(a), a, s(a), s^2(a), \ldots,$$

and the map sending positive elements $n \in \mathbf{Z}$ to $s^n(a)$, negative elements $-n \in \mathbf{Z}$ to $p^n(a)$ and 0 to a respects the $<$-relation and the successor function.

Note however that this isomorphism is not unique, as you could choose any other element $b \sim a$ as the image of 0. Also, $\mathbf{Z}$ has additional structure that cannot be found in $a/{\sim}$: There is no distinguished element in $a/{\sim}$ that would naturally correspond to 0, and there are no operations corresponding to the addition and multiplication of $\mathbf{Z}$.

2.3.18. Theorem.

(1) For any $a < b \in M$, if $a/\sim < b/\sim$, then there exists a $c \in M$ such that $a/\sim < c/\sim < b/\sim$.

(2) $\mathcal{M} = \langle M/\sim, < \rangle$ has a first element, but no second element and no last element.

Proof of (1): Let $a < b$. Since $\mathcal{M} \models \forall x \forall y \exists z (z+z = x+y \vee z+z = x+Sy)$, there exists a $c \in M$ such that $\mathcal{M} \models c + c = a + b$ or $\mathcal{M} \models c + c = a + Sb$. Without loss of generality we may assume that $c + c = a + b$ (otherwise replace b by $S^{\mathcal{M}}b$ in the rest of the proof). So c is the arithmetic mean of a and b. We will show that

$$a/\sim < c/\sim$$

(the proof of $c/\sim < b/\sim$ being similar).

As $\mathcal{M} \models \forall x \forall y \forall z (x \leq y \wedge z + z = x + y \rightarrow x \leq z)$, we have $a \leq c$ and hence $a/\sim \leq c/\sim$. Assume, by way of contradiction, that $a/\sim = c/\sim$. Then for some natural number n, $\mathcal{M} \models S^n a = c$. But for every n,

$$\mathcal{M} \models \forall x \forall y \forall z (x + y = z + z \wedge z = S^n x \rightarrow y = S^{2n} x)$$

so $\mathcal{M} \models b = S^{2n} a$, $a \sim b$, a contradiction.

Proof of (2): M_{fin} is the first element of $\mathcal{M}/\sim$, since $\mathcal{M} \models \forall x \neg (x < 0)$. There can be no second element, because for any infinite $b \in M$, by (1) there exists a $c \in M$ such that $M_{\text{fin}} < c/\sim < a/\sim$.

Finally, given any a in $M - M_{\text{fin}}$, $a + a$ cannot be in the same class as a, because for every natural number n, $\mathcal{M} \models \forall x (x + x = S^n x \rightarrow x = S^n 0)$. As $\mathcal{M} \models a + a > a$, there is no last equivalence class.

So $\mathcal{M}$ looks somewhat like the picture in figure 1. (In the picture, 0 means $0^{\mathcal{M}}$, and $a > b$ are two arbitrary infinite elements in different equivalence classes.)

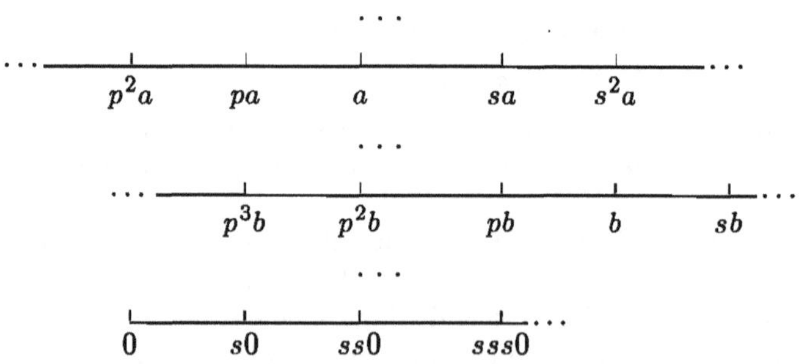

Figure 1

We have shown that there are nonstandard models, and that any two standard models are isomorphic. Is it also the case that any two nonstandard models are isomorphic? The answer to this question is "No." In fact, there are very many nonisomorphic countable models of Th($\mathbb{N}$):

2.3.19. Theorem. There are uncountably many nonisomorphic countable models of Th($\mathbb{N}$).

Proof: Given any list $\mathcal{M}_0, \mathcal{M}_1, \ldots$ of countably many models of Th($\mathbb{N}$), we will construct a model that is not isomorphic to any M_n.

We will use the fact that every infinite set has uncountably many subsets.

Let $\mathbb{P} \subseteq \mathbb{N}$ be the set of prime numbers. For any natural number n and any $a \in M_n$ let

$$A_{n,a} = \{p \in \mathbb{P} : \mathcal{M}_n \models \mathsf{S}^p\mathsf{0}|a\}.$$

where $\tau_1|\tau_2$ is an abbreviation of $\exists \mathbf{z}\ \tau_1 \cdot \mathbf{z} = \tau_2$.

Each $A_{n,a}$ is a subset of $\mathbb{P}$. $A_{n,a}$ is the set of those (finite) prime numbers which divide a in the model $\mathcal{M}_n$.

(Remark: $A_{n,a}$ may be finite or infinite.)

For each n there are only countably many sets $A_{n,a}$, and as there are only countably many n, the set

$$\{A_{n,a} : a \in M_n, n \in \mathbb{N}\}$$

is countable. But we know that there are uncountably many subsets of $\mathbb{P}$. Hence there exists a set $A \subseteq \mathbb{P}$ that is different from any of the $A_{n,a}$'s.

Fix such a set A. Then

$$A \subseteq \mathbb{P} \quad \text{and} \quad \forall n \in \mathbb{N}\ \forall a \in M_n\ A \neq A_{n,a}.$$

Given this set A we will construct a model of Th($\mathbb{N}$) that is not isomorphic to any $\mathcal{M}_n$.

Let $\mathcal{L}'$ be the language $\mathcal{L}$ expanded by a constant symbol c. Let

$$\Gamma = \mathrm{Th}(\mathbb{N}) \cup \{\mathsf{S}^p\mathsf{0}|\mathsf{c} : p \in A\} \cup \{\neg\mathsf{S}^p\mathsf{0}|\mathsf{c} : p \in \mathbb{P} - A\}.$$

Claim: Γ is consistent.

Proof: By the compactness theorem it is enough to show that every finite subtheory is consistent. So let $\Gamma_0 \subseteq \Gamma$ be finite. We will show that Γ_0 is consistent.

There are only finitely many primes $p \in A$ such that the formula $\mathtt{S}^p0|\mathtt{c}$ is in Γ_0. Let n be the product of these primes. Define a model $\mathcal{M}$ (for the language $\mathcal{L}'$) by letting $\mathcal{M}\lceil\mathcal{L} = \mathbb{N}$, and $\mathtt{c}^{\mathcal{M}} = n$. It is is clear that $\mathcal{M} \models \mathrm{Th}(\mathbb{N})$, so in order to show $\mathcal{M} \models \Gamma_0$ we have to show

$$M \models \Gamma_0 \cap (\{\mathtt{S}^p|\mathtt{c} : p \in A\} \cup \{\neg\mathtt{S}^p|\mathtt{c} : p \notin A\}).$$

For $p \notin A$, we know $p \not| n$, so $\mathcal{M} \models p \not| n$.
For $p \in A$, if also $(\mathtt{S}^p0|\mathtt{c}) \in \Gamma_0$, then by definition $p|n$, so $\mathcal{M} \models p|n$.

Hence Γ is consistent.

Now let $\mathcal{M}$ be a model of all of Γ, and let $b = \mathtt{c}^{\mathcal{M}}$. To finish the proof of our theorem, we will show that $\mathcal{M}\lceil\mathcal{L}$ is not isomorphic to any $\mathcal{M}_n$.

Assume that $f : \mathcal{M} \to \mathcal{M}_n$ is an isomorphism. Let $a = f(b)$. Then for every $p \in \mathbb{P}$,

$$p \in A \leftrightarrow \mathcal{M} \models (\mathtt{S}^p0)|\mathtt{c} \leftrightarrow \mathcal{M} \models (\mathtt{S}^p0)|b \leftrightarrow \mathcal{M}_n \models (\mathtt{S}^p0)|a \leftrightarrow p \in A_{n,a}$$

Hence $A = A_{n,a}$, which contradicts our choice of A.

Exercises

1. Show
For every closed term τ there exists a number n such that $\tau^{\mathcal{M}} = (\mathtt{S}^n\mathtt{0})^{\mathcal{M}}$.

2. Let $\mathcal{M}$ be a nonstandard model of arithmetic. Show that
(1) There is an element of $a \in M$ such that for all prime numbers p we have $\mathcal{M} \models \mathtt{S}^n\mathtt{0}|a$.
(2) There is an element of $a \in M$, $a > 1$ such that for no prime number p we have $\mathcal{M} \models \mathtt{S}^n\mathtt{0}|a$.

3. Assume that $\mathcal{M}$ is a nonstandard model of arithmetic, and let a, a', b, b' be elements of M. Assume $a \sim a'$ and $b \sim b'$. Show $a + b \sim a' + b'$.

Define an "addition" operation $\oplus$ on $M/{\sim}$ such that for all $a, b \in \mathcal{M}$ the following holds:

$$(a + b)/{\sim} = (a/{\sim}) \oplus (b/{\sim}).$$

Show that if $\mathcal{M}$ is a nonstandard model of arithmetic, then there is no binary operation $\odot$ on $M/{\sim}$ satisfying

$$(a \cdot b)/{\sim} = (a/{\sim}) \odot (b/{\sim}).$$

(Hint: Consider $a \in M - M_0$, $b_1 = 0$, $b_2 = 1$.)

Chapter 3

Model Theory

3.0. Introduction

Model theory is the branch of mathematics that studies models of a mathematical theory from the most general point of view. The goal is to obtain for a given theory Γ, a classification (modulo isomorphism) of all models for this theory by only looking at its logical properties.

The most basic problem in model theory is to find the number of nonisomorphic models that a theory has in a fixed cardinality. To answer this question it is necessary to find techniques for building models of a given theory.

We have seen an example of such a technique in Chapter 2, where we built a model from the constant symbols (or closed terms). We will refine this technique in section 3.4, where we will study the number of nonisomorphic **countable** models of a given theory. We will give examples of a theory with only one countable model, with 3 countable models, with countably many countable models, and with uncountably many countable models. We will also prove the theorem that no complete theory has exactly 2 countable models.

In section 3.1 we will study the relationships between two models of the same theory, and we will consider increasing chains of models, another technique of building models with prescribed properties. As an application of this technique we will prove Robinson's joint consistency theorem and Craig's interpolation theorem.

In section 3.2 we will introduce another technique for building models, namely, the method of ultraproducts. This technique played an important role in the development of model theory. We will use this framework to give another proof of the compactness theorem.

We think that this chapter will give the reader a solid background for understanding the more recent developments in model theory, the "classification theory of models."

In this section it will be more convenient to use a first order language which has only the connectives $\neg$ and $\wedge$, and only the quantifier $\exists$. Formulas of the form $\forall \mathtt{x} \cdots$ are to be interpreted as abbreviations for $\neg\exists\mathtt{x}\neg\cdots$. Our first order language will have an enumerable (= countable or finite) set of nonlogical symbols (so the language itself will be countable).

3.0.1. Reminder. Recall the definition of expansion and restriction: If $\mathcal{L}_1 \subseteq \mathcal{L}_2$ are languages, $\mathcal{M}_1$ a model for the language $\mathcal{L}_1$, and $\mathcal{M}_2$ model for the language $\mathcal{L}_2$, then we say that $\mathcal{M}_2$ is an expansion of the model $\mathcal{M}_1$, or $\mathcal{M}_1$ is the restriction of $\mathcal{M}_2$ to the language $\mathcal{L}_1$, iff:

$M_1 = M_2$.

For all constant symbols $\mathtt{c}$, all relation symbols $\mathtt{R}$, and all function symbols $\mathtt{f}$ in $\mathcal{L}_1$ we have $\mathtt{c}^{\mathcal{M}_1} = \mathtt{c}^{\mathcal{M}_2}$, $\mathtt{R}^{\mathcal{M}_1} = \mathtt{R}^{\mathcal{M}_2}$, $\mathtt{f}^{\mathcal{M}_1} = \mathtt{f}^{\mathcal{M}_2}$.

3.0.2. Notation. In this chapter, we fix a list $\mathtt{x}_1$, $\mathtt{x}_2$, ... of distinct variables. We call a formula an n-formula if its free variables are among $\mathtt{x}_1$, ..., $\mathtt{x}_n$. In particular, a 0-formula will be a closed formula. Similarly, an $\mathcal{M}$-formula is called $\mathcal{M}$-n-formula, if its free variables are among $\mathtt{x}_1$, ..., $\mathtt{x}_n$. (For the definition of an $\mathcal{M}$-formula, see 1.3.18 in chapter 1, or 3.1.6 below.)

If φ is an n-formula, and τ_1, ..., τ_n are terms (or $\mathcal{M}$-terms for some model $\mathcal{M}$), then we write $\varphi(\tau_1, \ldots, \tau_n)$ as an abbreviation for the formula $\varphi(\mathtt{x}_1/\tau_1, \ldots, \mathtt{x}_n/\tau_n)$. In particular, $\varphi(\mathtt{x}_1, \ldots, \mathtt{x}_n)$ is just the formula φ. We use similar notation for $\mathcal{M}$-terms and $\mathcal{M}$-formulas.

Similarly, if t is a set of n-formulas, then we write $t(\tau_1, \ldots, \tau_n)$ for the set

$$\{\varphi(\mathtt{x}_1/\tau_1, \ldots, \mathtt{x}_n/\tau_n) \ : \ \varphi \in t\}.$$

We defined a **theory** to be a set of closed formulas, and we defined for any theory Γ: $\Gamma \models \varphi$ iff all models $\mathcal{M}$ satisfying Γ also satisfy φ, or equivalently, if there is no model $\mathcal{M}$ satisfying $\Gamma \cup \{\neg\varphi\}$. If we consider two languages $\mathcal{L}_1 \subseteq \mathcal{L}_2$, and $\Gamma \cup \{\varphi\}$ is in the language $\mathcal{L}_1$, then this definition could be ambiguous, since it is not clear if only models $\mathcal{M}$ of the language $\mathcal{L}_1$ are considered, or also models of $\mathcal{L}_2$. So we will use the notation $\Gamma \models_{\mathcal{L}_1} \varphi$ and $\Gamma \models_{\mathcal{L}_2} \varphi$, where $\Gamma \models_{\mathcal{L}} \varphi$ means

there is no model $\mathcal{M}$ (for the language $\mathcal{L}$) such that $\mathcal{M} \models \Gamma \cup \{\neg\varphi\}$

Similarly, we will write $\mathcal{M} \models_{\mathcal{L}} \varphi$ when we want to emphasize that $\mathcal{M}$ is a model for the language $\mathcal{L}$, and φ is a formula in that language. If $\mathcal{M}$ is a model for a bigger language, we write $\mathcal{M} \models_{\mathcal{L}} \varphi$ to abbreviate $\mathcal{M} \upharpoonright \mathcal{L} \models \varphi$.

However, in the restriction lemma (see 2.2.9 in Chapter 2) we saw that this distinction turns out to be immaterial:

3.0.3. Fact. If $\mathcal{L}_1 \subseteq \mathcal{L}_2$, $\mathcal{M}_2$ is a model for the language $\mathcal{L}_2$, and $\mathcal{M}_1 = \mathcal{M}_2 \upharpoonright \mathcal{L}_1$, then for any φ in $\mathcal{L}_1$: $\mathcal{M}_1 \models \varphi$ iff $\mathcal{M}_2 \models \varphi$.

3.0.4. Lemma. If $\mathcal{L}_1 \subseteq \mathcal{L}_2$, $\Gamma \cup \{\varphi\}$ a set of formulas in $\mathcal{L}_1$, then $\Gamma \models_{\mathcal{L}_1} \varphi$ iff $\Gamma \models_{\mathcal{L}_2} \varphi$.

Proof: If $\mathcal{M}_1 \models_{\mathcal{L}_1} \Gamma \cup \{\neg\varphi\}$, then any expansion $\mathcal{M}_2$ of $\mathcal{M}_1$ for the language $\mathcal{L}_2$ will model $\Gamma \cup \{\neg\varphi\}$. Conversely, if $\mathcal{M}_2 \models_{\mathcal{L}_2} \Gamma \cup \{\neg\varphi\}$, then $\mathcal{M}_1 := \mathcal{M}_2 \upharpoonright \mathcal{L}_1$ will be a model of $\Gamma \cup \{\neg\varphi\}$.

3.0.5. Reminder. In section 2.2, we proved the "theorem on constants:" If φ is an n-formula in the language $\mathcal{L}$, $\mathtt{c}_1, \ldots, \mathtt{c}_n$ are distinct constant symbols not in $\mathcal{L}$, and Γ is a theory in $\mathcal{L}$, then:

$$\Gamma \vdash \varphi(\mathtt{x}_1, \ldots, \mathtt{x}_n) \text{ iff } \Gamma \vdash \varphi(\mathtt{c}_1, \ldots, \mathtt{c}_n).$$

3.0.6. Reminder. Recall that for any set t of formulas, we let $cl_{\vdash}(t) := \{\varphi : t \vdash \varphi\}$ = the deductive closure of t.

3.0.7. Definition. Fix a language $\mathcal{L}$. Let $\mathcal{M}, \mathcal{N}$ be models for the language $\mathcal{L}$. We say that $\mathcal{M}$ is a **submodel** of $\mathcal{N}$ ($\mathcal{M} \subseteq \mathcal{N}$) iff

(a) $M \subseteq N$
(b) For all constant symbols $\mathtt{c}$ in $\mathcal{L}$, $\mathtt{c}^{\mathcal{M}} = \mathtt{c}^{\mathcal{N}}$.
(c) For all n-ary function symbols $\mathtt{f}$ in $\mathcal{L}$, $\mathtt{f}^{\mathcal{M}} = \mathtt{f}^{\mathcal{N}} \upharpoonright M^n$.
(d) For all n-ary relation symbols $\mathtt{R}$ in $\mathcal{L}$, $\mathtt{R}^{\mathcal{M}} = \mathtt{R}^{\mathcal{N}} \cap M^n$.

Recall that whenever $\mathcal{M}$, $\mathcal{M}'$, $\mathcal{M}_1$, ... are models, we write M, M', M_1, ... for the corresponding universes.

Definition 3.0.7(c) means that for example the unary functions $\mathtt{f}^{\mathcal{M}}$ (mapping M into M) and $\mathtt{f}^{\mathcal{N}}$ (mapping N into N) agree on M. Similarly, if for example $\mathtt{R}$ is a binary relation symbol, then we demand that for all $a, b \in M$, $\mathcal{M} \models \mathtt{R}(a, b)$ iff $\mathcal{N} \models \mathtt{R}(a, b)$.

(See exercises 1 and 2.)

3.0.8. Fact. Assume that $\mathcal{N}$ is a model (for the language $\mathcal{L}$), $M \subseteq N$, and

(1) For all constant symbols $\mathtt{c}$ in $\mathcal{L}$, $\mathtt{c}^{\mathcal{N}} \in M$.

(2) For all n-ary function symbols $\mathtt{f}$ in $\mathcal{L}$, and for all elements $a_1, \ldots, a_n$ in M, also $\mathtt{f}^{\mathcal{N}}(a_1, \ldots, a_n) \in M$.

Then there is a unique model $\mathcal{M} \subseteq \mathcal{N}$ with universe M.

Proof: We define $\mathtt{c}^{\mathcal{M}}$, $\mathtt{f}^{\mathcal{M}}$, $\mathtt{R}^{\mathcal{M}}$ such that the conditions in 3.0.7 are satisfied. That is, we let $\mathtt{c}^{\mathcal{M}} := \mathtt{c}^{\mathcal{N}}$, and if $\mathtt{f}$ is an n-ary function, then for any elements $m_1, \ldots, m_n$ we let

$$\mathtt{f}^{\mathcal{M}}(m_1, \ldots, m_n) := \mathtt{f}^{\mathcal{N}}(m_1, \ldots, m_n).$$

This will be an element of M, by our assumption on M. Finally, we let, for any n-ary relation symbol $\mathtt{R}$ and any $m_1, \ldots, m_n \in M$:

$$\mathcal{M} \models \mathtt{R}(m_1, \ldots, m_n) \text{ iff } \mathcal{N} \models \mathtt{R}(m_1, \ldots, m_n).$$

Exercises

1. Find all submodels of $(\mathbb{N}, +, \cdot, 0)$ that contain 2 and 5.

2. Let $\mathcal{L}$ be the language of group theory, i.e., with a binary operation symbol $+$ and a constant symbol 0.

(a) If $\mathcal{G} = (G, +, 0)$ is a group, what are the submodels of G?

(b) Find a language $\mathcal{L}' \supseteq \mathcal{L}$ to describe groups such that every group $\mathcal{G}$ can be expanded to a model $\mathcal{G}'$ for the language $\mathcal{L}'$ such that the subgroups of $\mathcal{G}$ are exactly the (restrictions of) submodels of $\mathcal{G}'$. (Hint: Consider subtraction, or even unary $-$.)

3.1. Elementary Substructures and Chains

3.1.1. Definition. For any model $\mathcal{M}$, we let $\mathrm{Th}(\mathcal{M})$ (the "theory" of $\mathcal{M}$) be the set of all closed formulas that are valid in $\mathcal{M}$:

$$\mathrm{Th}(\mathcal{M}) := \{\varphi : \varphi \text{ closed}, \mathcal{M} \models \varphi\}.$$

We may write $\mathrm{Th}_{\mathcal{L}}(\mathcal{M})$ to emphasize that $\mathcal{M}$ is a model for the language $\mathcal{L}$. If $\mathcal{M}$ is a model for a language bigger than $\mathcal{L}$, we write $\mathrm{Th}_{\mathcal{L}}(\mathcal{M})$ for $\mathrm{Th}(\mathcal{M}\restriction\mathcal{L})$.

3.1.2. Fact. $\mathrm{Th}(\mathcal{M})$ is a complete and consistent theory.

Proof: Exercise.

3.1.3. Definition. We say that that two models $\mathcal{M}_1$, $\mathcal{M}_2$ are elementarily equivalent ($\mathcal{M}_1 \equiv \mathcal{M}_2$) if their theories are the same:

$$\mathcal{M}_1 \equiv \mathcal{M}_2 \text{ iff } \mathrm{Th}(\mathcal{M}_1) = \mathrm{Th}(\mathcal{M}_2).$$

3.1.4. Fact. If $\mathcal{M}_1$ and $\mathcal{M}_2$ are isomorphic, then $\mathcal{M}_1 \equiv \mathcal{M}_2$.

Proof: Exercise.

3.1.5. Example. It is possible to find two countable models which are elementarily equivalent but not isomorphic.

Proof: We will use a language $\mathcal{L}$ which has no relation symbols and no function symbols, and countably many constant symbols c_1, c_2, ... Let $\mathcal{M}_1$ be defined by $M_1 = \{1, 2, \ldots\}$, $\mathsf{c}_i^{\mathcal{M}_1} = i$ for all $i > 0$. We let $\Gamma := \mathrm{Th}(\mathcal{M}_1)$.

Now consider a language $\mathcal{L}^*$ which has an additional constant symbol c. We let $\Gamma^* := \mathrm{Th}_{\mathcal{L}}(\mathcal{M}_1) \cup \{\mathsf{c} \neq \mathsf{c}_1, \mathsf{c} \neq \mathsf{c}_2, \mathsf{c} \neq \mathsf{c}_3, \ldots\}$. We claim that Γ^* is consistent.

It is enough to see that every finite subset is consistent. So we will show that every finite subset has a model.

Indeed, if Γ' is a finite subset of Γ^*, then we can find a model $\mathcal{M}'$ of Γ' by expanding $\mathcal{M}_1$: If c_n is not mentioned in Γ', then we can let $\mathsf{c}^{\mathcal{M}'} := n$.

Now let $\mathcal{M}_0^*$ be a model of Γ^*, and let $\mathcal{M}_0$ be $\mathcal{M}_0^*\restriction\mathcal{L}$. So $\mathcal{M}_0$ is a model of Γ. Let $a := \mathsf{c}^{\mathcal{M}_0^*}$. There can be no isomorphism $f : \mathcal{M}_0 \to \mathcal{M}_1$, because if $f(a) = n$, then $f(a) = n = \mathsf{c}_n^{\mathcal{M}_1} = f(\mathsf{c}_n^{\mathcal{M}_0})$ would imply $a = \mathsf{c}_n^{\mathcal{M}_0}$, so $\mathcal{M}_0^* \models \mathsf{c} = \mathsf{c}_n$, a contradiction.

Since $\mathcal{M}_0$ is a model of Γ, we have $\mathcal{M}_0 \equiv \mathcal{M}_1$.

(In fact a more careful analysis shows that the submodel of $\mathcal{M}_0$ with universe $\{c^{\mathcal{M}_0'}, c_1^{\mathcal{M}_0}, \ldots\}$ is also a model of Γ, i.e., we can find nonisomorphic but elementarily equivalent models that differ only by a single element. See exercise 8 in section 3.3.)

3.1.6. Definition. Let $\mathcal{M}$ be a model for a language $\mathcal{L}$, and let $A \subseteq M$. The language $\mathcal{L}(A)$ is defined as follows:

$\mathcal{L}(A)$ has the same relation symbols and the same function symbols as $\mathcal{L}$.

$\mathcal{L}(A)$ has all constant symbols from $\mathcal{L}$, but in addition for each element a of A there is a constant symbol $\underline{a}$. (As usual, we assume that no element of any model that we consider is also in the original language $\mathcal{L}$. Sometimes we will not distinguish between a and $\underline{a}$.)

So, e.g., closed formulas in the language $\mathcal{L}(A)$ are exactly the formulas of the form

$$\varphi(\underline{a}_1, \ldots, \underline{a}_n),$$

where φ is an n-formula in the original language $\mathcal{L}$, and $a_1, \ldots, a_n$ are elements of A.

In particular, formulas in the language $\mathcal{L}(M)$ are exactly the $\mathcal{M}$-formulas of 1.3.18.

$\mathcal{M}$ has a natural expansion $\mathcal{M}_A$ to a model of $\mathcal{L}(A)$ by interpreting the constant symbol $\underline{a}$ as the element a, i.e., $\underline{a}^{\mathcal{M}_A} = a$ for all $a \in A$. If θ is an $\mathcal{L}(A)$-formula, then we may write $\mathcal{M} \models \theta$ instead of $\mathcal{M}_A \models \theta$.

3.1.7. Definition. Whenever $\mathcal{M}_0 \subseteq \mathcal{M}_1 \subseteq \cdots$ (we call such a sequence a "chain" of models), we define the "limit" $\mathcal{N}$ of this chain, written $\bigcup_n \mathcal{M}_n$ or $\lim_n \mathcal{M}_n$ as follows:

N, the universe of $\mathcal{N}$, is the set $\bigcup_{n=0}^{\infty} M_n$.

For any k-ary relation symbol $\mathtt{R}$ in our language, we let

$$\mathtt{R}^{\mathcal{N}} := \{(a_1, \ldots, a_k) : \text{For some } n,\ \mathcal{M}_n \models \mathtt{R}(a_1, \ldots, a_k)\}.$$

Note that if $\mathcal{M}_n \models \mathtt{R}(a_1, \ldots, a_k)$ for SOME n, then $\mathcal{M}_n \models \mathtt{R}(a_1, \ldots, a_k)$ for ALL n for which this is well-defined, i.e., for all n with $a_1, \ldots, a_k \in M_n$.

For every constant symbol $\mathtt{c}$, $\mathtt{c}^{\mathcal{N}} := \mathtt{c}^{\mathcal{M}_0}$ (which is also equal to $\mathtt{c}^{\mathcal{M}_n}$ for any n).

For every function symbol $\mathtt{f}$, we let $\mathtt{f}^{\mathcal{N}}$ be the unique function that agrees with all $\mathtt{f}^{\mathcal{M}_n}$, that is: If $\mathtt{f}$ is k-ary, and $a_1, \ldots, a_k$ are in $\mathcal{M}_n$, then $\mathbf{f}^{\mathcal{N}}(a_1, \ldots, a_k)$ is the common value of all $\mathbf{f}^{\mathcal{M}_m}(a_1, \ldots, a_k)$, for all $m \geq n$.

This limit operation will be very useful in constructing models. However, often we want to construct a model satisfying a particular theory. In general it is impossible to predict $\text{Th}(\lim_n \mathcal{M}_n)$ from $\text{Th}(\mathcal{M}_0)$, $\text{Th}(\mathcal{M}_1)$, $\ldots$, as the following example shows:

3.1.8. Example. There is a chain of models $\mathcal{M}_0 \subseteq \mathcal{M}_1 \subseteq \cdots$ all satisfying the same (complete) theory, such that $\lim_n \mathcal{M}_n$ does not satisfy this theory.

Proof: We use the language with one binary relation symbol $<$. Let $\mathcal{M}_0 := \{0, 1, 2, \ldots\}$, and in general let

$$\mathcal{M}_n := \{-n, \ldots, -1, 0, 1, 2, \ldots\}$$

with the usual interpretation of $<$. Then

(1) For all n, $\mathcal{M}_n \simeq \mathcal{M}_0$, using the isomorphism that sends x to $x + n$.
(2) Hence, $\text{Th}(\mathcal{M}_0) = \text{Th}(\mathcal{M}_1) = \cdots$
(3) $\mathcal{M} = \lim_n \mathcal{M}_n = \{\ldots, -2, -1, -0, 1, 2, \ldots\}$ is a linearly ordered model satisfying a different theory than all the $\mathcal{M}_n$. In particular, $\mathcal{M} \models \forall \mathrm{x} \exists \mathrm{y}\, \mathrm{y} < \mathrm{x}$.

Thus, though each $\mathcal{M}_n$ has a minimal element, $\mathcal{M}$ has no minimal element, since the $\mathcal{M}_n$ do not agree **which** element is minimal. This motivates the definition of "elementary submodel" in 3.1.10, below.

3.1.9. Definition. The "complete diagram" of $\mathcal{M}$ is the set

$$\mathit{Diag}(\mathcal{M}) := \{\theta : \theta \text{ in } \mathcal{L}(M) \text{ is closed, and } \mathcal{M}_M \models \theta\}.$$

Note that if $\mathcal{M} \subseteq \mathcal{N}$, then $\mathcal{L}(M) \subseteq \mathcal{L}(N)$. Thus, the following definition makes sense:

3.1.10. Definition. Let $\mathcal{M} \subseteq \mathcal{N}$. We say that $\mathcal{M}$ is an **elementary submodel** of $\mathcal{N}$, abbreviated $\mathcal{M} \prec \mathcal{N}$ if

$$\mathit{Diag}(\mathcal{M}) \subseteq \mathit{Diag}(\mathcal{N})$$

or equivalently, if for every n-formula φ in $\mathcal{L}$ and for all $a_1, \ldots, a_n$ in M:

$$\mathcal{M} \models \varphi(a_1, \ldots, a_n) \quad \text{iff} \quad \mathcal{N} \models \varphi(a_1, \ldots, a_n). \qquad (!)$$

So if $\mathcal{M} \prec \mathcal{N}$, then $\mathcal{M} \subseteq \mathcal{N}$ and $\mathcal{M} \equiv \mathcal{N}$. The converse is in general not true. Note that $\mathcal{M} \equiv \mathcal{N}$ only says that for all (closed) formulas φ in the language $\mathcal{L}$ we have $\mathcal{M} \models \varphi$ iff $\mathcal{N} \models \varphi$, whereas in $\mathcal{M} \prec \mathcal{N}$ we demand that this is true for all formulas θ in the much bigger language $\mathcal{L}(M)$.

3.1.11. Example. We use the language of group theory. Let $\mathcal{N}$ be the set of integers (with usual addition), and $\mathcal{M}$ the set of even integers. Then

(1) $\mathcal{M} \subseteq \mathcal{N}$.
(2) $\mathcal{M}$ is isomorphic to $\mathcal{N}$.
(3) $\mathcal{M} \equiv \mathcal{N}$.
(4) $\mathcal{M} \not\prec \mathcal{N}$.

Proof: (1) is clear. For (2), note that the function $f(x) = 2 \cdot x$ is an isomorphism from $\mathcal{N}$ onto $\mathcal{M}$. (3) follows from (2).

Proof of (4): In $\mathcal{N}$, 2 is "divisible by 2," i.e., can be written as $x + x$ for some x. This is not true in $\mathcal{M}$. More formally, the formula $\exists \mathrm{x}\ \mathrm{x}+\mathrm{x} = 2$ is in $Diag(\mathcal{N})$ but not in $Diag(\mathcal{M})$.

Often, when we want to check if $\mathcal{M} \prec \mathcal{N}$, we can easily check validity in $\mathcal{N}$, but not so easily validity in $\mathcal{M}$. In these cases, the following lemma (called the Tarski-Vaught criterion) is very helpful.

3.1.12. Lemma. Assume $\mathcal{M} \subseteq \mathcal{N}$. Then $\mathcal{M} \prec \mathcal{N}$ iff for every closed $\mathcal{M}$-formula of the form $\exists \mathrm{x}\, \theta$

If there is $b \in N$ such that $\mathcal{N} \models \theta(\mathrm{x}/b)$, (i.e., if $\mathcal{N} \models \exists \mathrm{x}\, \theta$) $(*)$
then there is $a \in M$ such that $\mathcal{N} \models \theta(\mathrm{x}/a)$.

(Note that to check this condition we only need to know about the satisfaction relation in $\mathcal{N}$, not in $\mathcal{M}$. This will be very convenient when we want to build a model $\mathcal{M}$ inside a given $\mathcal{N}$, because while we are still building the model $\mathcal{M}$ we will not know whether $\mathcal{M} \models \varphi$ will hold or not.)

Proof: First the easy direction: Assuming $\mathcal{M} \prec \mathcal{N}$, we have to prove $(*)$.

If there is $b \in N$ such that $\mathcal{N} \models \theta(\mathrm{x}/b)$, then $\mathcal{N} \models \exists \mathrm{x}\theta$. So since $\mathcal{M} \prec \mathcal{N}$, $\mathcal{M} \models \exists \mathrm{x}\theta$. So there is $a \in M$, $\mathcal{M} \models \theta(\mathrm{x}/a)$. Again by $\mathcal{M} \prec \mathcal{N}$, we obtain $\mathcal{N} \models \theta(\mathrm{x}/a)$.

Now we prove the "hard" direction:
Assume that $(*)$ is satisfied. We want to show $\mathcal{M} \prec \mathcal{N}$, i.e., 3.1.10(!). We will use induction on the formula θ.

If θ is atomic, then we can conclude

$$\mathcal{M} \models \theta \text{ iff } \mathcal{N} \models \theta$$

from the fact that $\mathcal{M}$ is a submodel of $\mathcal{N}$.

If θ is $\neg\theta_1$, then we have $\mathcal{M} \models \theta$ iff $\mathcal{M} \not\models \theta_1$ iff $\mathcal{N} \not\models \theta_1$ iff $\mathcal{N} \models \theta$.

The induction step for $\theta = \theta_1 \wedge \theta_2$ is similar.

Finally, assume that θ is of the form $\exists \mathtt{x}\theta_1$.

Assume that $\mathcal{M} \models \theta$. Then there exists $a \in M$ such that $\mathcal{M} \models \theta_1(\mathtt{x}/a)$. By induction hypothesis, this implies that $\mathcal{N} \models \theta_1(\mathtt{x}/a)$. So $\mathcal{N} \models \exists \mathtt{x}\theta_1$.

Conversely, if $\mathcal{N} \models \exists \mathtt{x}\theta_1$, then there is $b \in N$ such that $\mathcal{N} \models \theta_1(\mathtt{x}/b)$. So by our assumption (*), there is $a \in M$ such that $\mathcal{N} \models \theta_1(\mathtt{x}/a)$. Now we can use our inductive assumption to obtain $\mathcal{M} \models \theta_1(\mathtt{x}/a)$, so $\mathcal{M} \models \theta$.

3.1.13. Example. If Γ is a Henkin theory, $\mathcal{N}$ a model of Γ, and the model $\mathcal{M}$ is defined by

$$M := \{\mathtt{c}^{\mathcal{N}} : \mathtt{c} \text{ a constant symbol in } \mathcal{L}\}$$

then $\mathcal{M} \prec \mathcal{N}$.

Proof: First we have to show that there really is a model $\mathcal{M} \subseteq \mathcal{N}$ with universe M. By definition, M contains all the interpretations of constant symbols, so we have to show that M is closed under all functions $\mathtt{f}^{\mathcal{N}}$.

So let $\mathtt{f}$ be an n-ary function symbol in the language $\mathcal{L}$, and let $\mathtt{c}_1^{\mathcal{N}}$, ..., $\mathtt{c}_n^{\mathcal{N}}$ be elements of M, where $\mathtt{c}_1$, ..., $\mathtt{c}_n$ are constant symbols in the language $\mathcal{L}$. Since Γ is a Henkin theory, we can find a constant symbol $\mathtt{c}$ such that the formula

$$\exists \mathtt{x}\, \mathtt{f}(\mathtt{c}_1, \ldots, \mathtt{c}_n) = \mathtt{x} \to \mathtt{f}(\mathtt{c}_1, \ldots, \mathtt{c}_n) = \mathtt{c}$$

is in Γ. Letting $a := (\mathtt{f}(\mathtt{c}_1, \ldots, \mathtt{c}_n))^{\mathcal{N}}$, we have

$$\mathcal{N} \models \mathtt{f}(\mathtt{c}_1, \ldots, \mathtt{c}_n) = a,$$

so $\mathcal{N} \models \exists \mathtt{x} \mathtt{f}(\mathtt{c}_1, \ldots, \mathtt{c}_n) = \mathtt{x}$. Therefore also

$$\mathcal{N} \models \mathtt{f}(\mathtt{c}_1, \ldots, \mathtt{c}_n) = \mathtt{c}$$

hence $a = \mathtt{c}^{\mathcal{N}} \in \mathcal{N}$.

This shows that we can find a model $\mathcal{M}$ with universe M which is a submodel of $\mathcal{N}$. Now to show that $\mathcal{M} \prec \mathcal{N}$, we use 3.1.12. So let θ be an $\mathcal{M}$-formula with free variable $\mathtt{x}$, and assume $\mathcal{N} \models \exists \mathtt{x}\theta$. θ is of the form $\varphi(a_1, \ldots, a_n, \mathrm{x})$, where φ is an $n+1$-formula in $\mathcal{L}$, and $a_i = \mathrm{c}_i^{\mathcal{N}}$ for some constant symbols c_i. We can find a constant symbol c such that the formula

$$\exists \mathrm{x}\, \varphi(\mathrm{c}_1, \ldots, \mathrm{c}_n, \mathrm{x}) \rightarrow \varphi(\mathrm{c}_1, \ldots, \mathrm{c}_n, \mathrm{c})$$

is in Γ. Let $a := \mathrm{c}^{\mathcal{N}}$. Since $\mathcal{N} \models \exists \mathrm{x} \theta$, we also have $\mathcal{N} \models \exists \mathrm{x} \varphi(\mathrm{c}_1, \ldots, \mathrm{c}_n, \mathrm{x})$, and hence $\mathcal{N} \models \varphi(\mathrm{c}_1, \ldots, \mathrm{c}_n, \mathrm{c})$, and thus $\mathcal{N} \models \theta(\mathrm{x}/a)$.

Another useful fact about elementary submodels is the following:

3.1.14. Lemma (The many automorphisms lemma). Assume $\mathcal{M} \subseteq \mathcal{N}$. If (**) holds, then $\mathcal{M} \prec \mathcal{N}$. Here (**) is the following statement:

(**) Whenever $A \subseteq M$ is finite, $b \in N$, and $\mathcal{L}_0 \subseteq \mathcal{L}$ a language with finitely many constant/function/relation symbols, then there is an automorphism $f : \mathcal{N} \restriction \mathcal{L}_0 \rightarrow \mathcal{N} \restriction \mathcal{L}_0$ such that $f(b) \in M$, and $f(a) = a$ for all $a \in A$.

Proof: We will show that 3.1.12 is satisfied. So assume $\mathcal{N} \models \theta(\mathrm{x}/b)$, where θ is a formula in $\mathcal{L}(M)$. So θ is of the form $\varphi(a_1, \ldots, a_n, \mathrm{x})$, where φ is an $n+1$-formula in a language $\mathcal{L}_0 \subseteq \mathcal{L}$, containing only finitely many relation, function and constant symbols. Let $A := \{a_1, \ldots, a_n\}$, and let f be an automorphism as in (**). Since $\mathcal{N} \restriction \mathcal{L}_0 \models \varphi(a_1, \ldots, a_n, b)$, we have $\mathcal{N} \restriction \mathcal{L}_0 \models \varphi(f(a_1), \ldots, f(a_n), f(b))$, hence $\mathcal{N} \models \theta(\mathrm{x}/f(b))$, with $f(b) \in M$.

Note that (**) is a sufficient condition for $\mathcal{M} \prec \mathcal{N}$, but not necessary. (See exercise 5.)

3.1.15. Definition. Assume $f : \mathcal{M} \rightarrow \mathcal{N}'$. We say that f is an **elementary embedding** if there is $\mathcal{M}' \prec \mathcal{N}'$ such that f maps $\mathcal{M}$ isomorphically onto $\mathcal{M}'$. Equivalently, f is an elementary embedding iff for all $a_1, \ldots, a_n$ in M and all n-formulas in $\mathcal{L}$ we have

$$\mathcal{M} \models \varphi(a_1, \ldots, a_n) \text{ iff } \mathcal{N}' \models \varphi(f(a_1), \ldots, f(a_n)).$$

(See figure 1.)

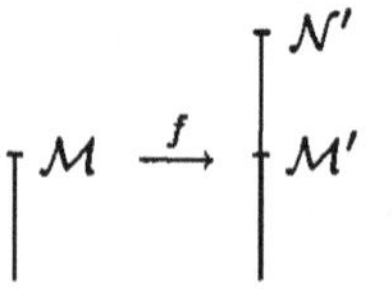

Figure 1.

3.1.16. Fact (Characterization of elementary embeddability). Assume $\mathcal{M}, \mathcal{N}$ are models for a language $\mathcal{L}$. Then $\mathcal{M}$ can be elementarily embedded into $\mathcal{N}$ iff there is an expansion of $\mathcal{N}$ to a model $\mathcal{N}^*$ for the language $\mathcal{L}(M)$ such that $\mathcal{N}^* \models Diag(\mathcal{M})$.

Proof: If $g : \mathcal{M} \to \mathcal{N}$ is an elementary embedding, then we can define a model $\mathcal{N}^*$ for the language $\mathcal{L}(M)$ as follows:

(1) $N^* = N$.

(2) $\mathtt{f}^{\mathcal{N}^*} = \mathtt{f}^{\mathcal{N}}$, $\mathtt{R}^{\mathcal{N}^*} = \mathtt{R}^{\mathcal{N}}$, $\mathtt{c}^{\mathcal{N}^*} = \mathtt{c}^{\mathcal{N}}$, for all function/relation/constant symbols $\mathtt{R}, \mathtt{f}, \mathtt{c}$ in the language $\mathcal{L}$.

(3) $\underline{a}^{\mathcal{N}^*} = g(a)$, for $a \in M$.

Conditions (1) and (2) just say that $\mathcal{N}^*$ is an expansion of $\mathcal{N}$. For any n-formula φ in L we have

$$\begin{aligned}\varphi(\underline{a}_1, \ldots, \underline{a}_n) \in Diag(\mathcal{M}) &\to \mathcal{M} \models \varphi(a_1, \ldots, a_n)\\ &\to \mathcal{N} \models \varphi(g(a_1), \ldots, g(a_n))\\ &\to \mathcal{N}^* \models \varphi(g(a_1), \ldots, g(a_n))\\ &\to \mathcal{N}^* \models \varphi(\underline{a}_1, \ldots, \underline{a}_n).\end{aligned}$$

So $\mathcal{N}^* \models Diag(\mathcal{M})$.

Conversely, if $\mathcal{N}^* \models Diag(\mathcal{M})$, then we can define $g : M \to N$ by $g(a) = \underline{a}^{\mathcal{N}^*}$, and a similar proof shows that g is an elementary embedding.

This fact, despite its simple proof, is very useful because it allows us to construct an elementary extension of a given model from its theory, using the results of Chapter 2.

3.1.17. Lemma (elementary embeddings vs. elementary substructures). If f is an elementary embedding of $\mathcal{M}$ into $\mathcal{N}'$, then there is a model $\mathcal{N}$ such that $\mathcal{M} \prec \mathcal{N}$, and an isomorphism g from $\mathcal{N}$ to $\mathcal{N}'$ such that g extends f. (See figure 2.) Consequently, if there is a model $\mathcal{N}'$ with some property X into which $\mathcal{M}$ can be embedded, then there is a model $\mathcal{N}$ with the property X such that $\mathcal{M} \prec \mathcal{N}$, assuming that the property X is preserved under isomorphism.

Proof: Let $\mathcal{M}' \subseteq \mathcal{N}'$ be the range of f. Let A be a set disjoint from M such that there is a bijection h between A and $N' - M'$ (i.e., a 1-1 function from A onto $N' - M'$). Let $N = M \cup A$. and let $g := f \cup h$, i.e. $g(x) = f(x)$ if $x \in M$, and $g(x) = h(x)$ if $x \in A$. Now define a model $\mathcal{N}$ with universe N such that g is an isomorphism. That is, for any constant symbol $\mathtt{c}$ we define $c^{\mathcal{N}} = c^{\mathcal{M}} = g^{-1}(c^{\mathcal{N}'})$, for any n-ary function symbol $\mathbf{f}$ and for all $a_1, \ldots, a_n$ in N we let $b_1 := g(a_1), \ldots, b_n := g(a_n)$, and we let

$$\mathbf{f}^{\mathcal{N}}(a_1, \ldots, a_n) = g^{-1}\left(\mathbf{f}^{\mathcal{N}'}(b_1, \ldots, b_n)\right).$$

Similarly, if $\mathtt{R}$ is an n-ary relation symbol $\mathbf{f}$ and for all $a_1, \ldots, a_n$ in N, we let $b_1 := g(a_1), \ldots, b_n := g(a_n)$, and we let

$$(a_1, \ldots, a_n) \in \mathtt{R}^{\mathcal{N}} \text{ iff } (b_1, \ldots, b_n) \in \mathtt{R}^{\mathcal{N}'}$$

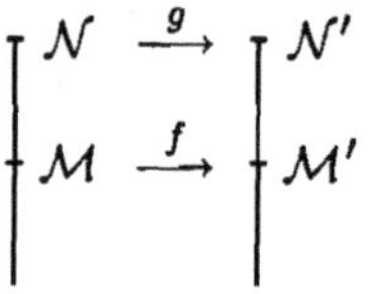

Figure 2.

3.1.18. Definition. An "elementary chain" is a sequence $\mathcal{M}_0, \mathcal{M}_1, \ldots$ of models with the property that for all n, $\mathcal{M}_n \prec \mathcal{M}_{n+1}$.

3.1.19. Fact. If $\mathcal{M}_0 \prec \mathcal{M}_1$ and $\mathcal{M}_1 \prec \mathcal{M}_2$, then $\mathcal{M}_0 \prec \mathcal{M}_2$.

Proof: $Diag(\mathcal{M}_0) \subseteq Diag(\mathcal{M}_1) \subseteq Diag(\mathcal{M}_2)$, so $Diag(\mathcal{M}_0) \subseteq Diag(\mathcal{M}_2)$.

3.1.20. Lemma (limits of elementary chains). If $\mathcal{M}_0 \prec \mathcal{M}_1 \prec \mathcal{M}_2 \prec \cdots$ is an elementary chain, and $\mathcal{M} = \bigcup_n \mathcal{M}_n$ is the limit of this chain, then for all n we have $\mathcal{M}_n \prec \mathcal{M}$. (In particular, $\mathcal{M} \equiv \mathcal{M}_0 \equiv \mathcal{M}_1 \equiv \cdots$)

Proof: First note that every $\mathcal{M}$-formula is in fact an $\mathcal{M}_n$-formula for some n. We will prove by induction on the closed $\mathcal{M}$-formula θ:

$$\text{for all } n, \text{ such that } \theta \text{ is a formula in } \mathcal{L}(M_n), \quad \mathcal{M} \models \theta \text{ iff } \mathcal{M}_n \models \theta.$$

In 3.1.7 we already saw that $\mathcal{M}_n \subseteq \mathcal{M}$. This deals with the case of atomic formulas θ.

The case where θ is of the form $\neg\theta_1$ or $\theta_1 \wedge \theta_2$ is easy and left to the reader.

Finally, assume that θ is of the form $\exists x\theta_1$. If $\mathcal{M}_n \models \exists x\theta_1$, then for some $a \in \mathcal{M}_n$ we have $\mathcal{M}_n \models \theta_1(x/a)$, so by induction hypothesis also $\mathcal{M} \models \theta_1(x/a)$, so $\mathcal{M} \models \theta$.

Conversely, assume that $\mathcal{M} \models \theta$, and let n be such that θ is a formula in $\mathcal{L}(M_n)$. We want to show $\mathcal{M}_n \models \theta$.

Since $\mathcal{M} \models \exists x\theta_1$, there must be an element a in $\mathcal{M}$ such that $\mathcal{M} \models \exists x\theta_1(x/a)$. Let $m \geq n$ be such that $a \in \mathcal{M}_m$. So $\mathcal{M}_m \models \theta_1(x/a)$, by induction hypothesis. So $\mathcal{M}_m \models \theta$. As $\mathcal{M}_n \prec \mathcal{M}_m$, also $\mathcal{M}_n \models \theta$.

Application: Joint Consistency and Interpolation

If Γ_1 and Γ_2 are consistent theories, then $\Gamma_1 \cup \Gamma_2$ is not necessarily consistent. Indeed, there may be a formula φ such that $\Gamma_1 \vdash \varphi$ and $\Gamma_2 \vdash \neg\varphi$. However, assume that Γ_1 and Γ_2 are theories in languages $\mathcal{L}_1$ and $\mathcal{L}_2$ and Γ_1 and Γ_2 agree on all formulas in the common language $\mathcal{L}_0 := \mathcal{L}_1 \cap \mathcal{L}_2$. Is then $\Gamma_1 \cup \Gamma_2$ consistent? The following theorem ("Robinson's joint consistency theorem") will show that this is indeed the case.

3.1.21. Theorem. Assume that $\mathcal{L}_1$ and $\mathcal{L}_2$ are two languages of first order logic, and let $\mathcal{L}_0 := \mathcal{L}_1 \cap \mathcal{L}_2$. (So $\mathcal{L}_0$ contains only those predicate/function/constant symbols that are in both $\mathcal{L}_1$ and $\mathcal{L}_2$.)
Let $\mathcal{L}$ be a language containing $\mathcal{L}_1$ and $\mathcal{L}_2$.
Let Γ_0 be a complete consistent theory in the language $\mathcal{L}_0$.
Assume that Γ_1 and Γ_2 are consistent theories in the languages $\mathcal{L}_1$ and $\mathcal{L}_2$, respectively, and assume that $\Gamma_0 = \Gamma_1 \cap \Gamma_2$.
Then $\Gamma_1 \cup \Gamma_2$ is consistent.

3.1.22. Corollary. If Γ_1 is in the language $\mathcal{L}_1$, Γ_2 in the language $\mathcal{L}_2$, $\mathcal{L}_0 = \mathcal{L}_1 \cap \mathcal{L}_2$, then $\Gamma_1 \cup \Gamma_2$ is consistent iff $(cl_\vdash(\Gamma_1) \cap \mathcal{L}_0) \cup (cl_\vdash(\Gamma_2) \cap \mathcal{L}_0)$ is consistent.

Proof: Exercise.

Before we prove the theorem, we need the following two lemmas:

3.1.23. Lemma. Assume that Γ_0, Γ_1, Γ_2 are as in 3.1.21. Let $\mathcal{M}_1$ be a model for Γ_1. Then there exists a model $\mathcal{M}_2$ for Γ_2 such that $\mathcal{M}_1 \restriction \mathcal{L}_0 \prec \mathcal{M}_2 \restriction \mathcal{L}_0$. (See figure 3.)

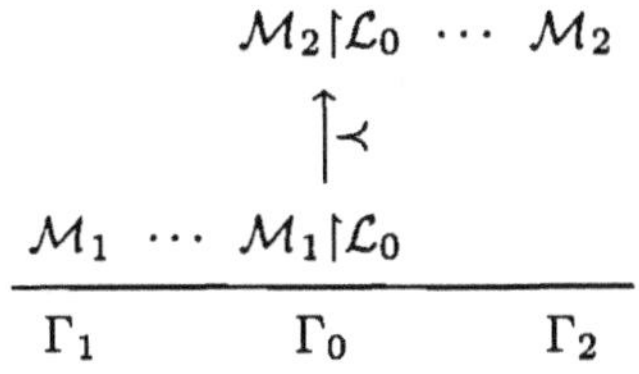

Figure 3.

Proof: Let $\Gamma := \Gamma_2 \cup Diag(\mathcal{M}_1 \restriction \mathcal{L}_0)$. We claim that Γ is consistent. If Γ is inconsistent, then there is a formula $\varphi_2 \in cl_\vdash(\Gamma_2)$ and a formula $\theta_1 \in Diag(\mathcal{M}_1 \restriction \mathcal{L}_0)$ such that $\{\varphi_2, \theta_1\}$ is inconsistent. Since we can derive φ_2 from Γ_2, we can also derive the universal closure of φ_2, so we will assume that φ_2 is closed.

θ_1 is of the form $\varphi_1(\underline{a}_1, \ldots, \underline{a}_n)$ (where $\varphi_1 \in \mathcal{L}_0$), which is a formula in $\mathcal{L}_0(\{\underline{a}_1, \ldots, \underline{a}_n\})$, where the a_i are elements of M_1.

So $\varphi_2 \vdash \neg\varphi_1(\underline{a}_1, \ldots, \underline{a}_n)$. Since the constant symbols $\underline{a}_1$, $\ldots$, $\underline{a}_n$ do not appear in φ_2, we can apply the theorem on constants (3.0.5) and get $\varphi_2 \vdash \neg\varphi_1$, and by the generalization theorem $\varphi_2 \vdash \forall x_1 \cdots \forall x_n \neg\varphi_1$. So $\Gamma_2 \vdash \neg\exists x_1 \cdots \exists x_n \varphi_1$. So it is impossible that $\exists x_1 \cdots \exists x_n \varphi_1 \in \Gamma_0$, and hence, by completeness of Γ_0,

$$\neg\exists x_1 \cdots \exists x_n \varphi_1 \in \Gamma_0$$

But since $\mathcal{M}_1 \restriction \mathcal{L}_0 \models \theta_1$, we have $\mathcal{M}_1 \restriction \mathcal{L}_0 \models \varphi_1(a_1, \ldots, a_n)$, so

$$\mathcal{M}_1 \restriction \mathcal{L}_0 \models \exists x_1 \cdots \exists x_n \varphi_1$$

we must have, by completeness of Γ_0,

$$\exists x_1 \cdots \exists x_n \varphi_1 \in \Gamma_0$$

which is a contradiction, since Γ_0 is consistent. Now since Γ is consistent we can find a model $\mathcal{M}_2'$ of Γ. $\mathcal{M}_1 \restriction \mathcal{L}_0$ can be elementarily embedded into $\mathcal{M}_2' \restriction \mathcal{L}_0$, by 3.1.16. So by 3.1.17, we can find $\mathcal{M}_2$ as required.

3.1.24. Lemma. Assume $\mathcal{M}_1$ is a model for Γ_1, $\mathcal{M}_2$ a model for Γ_2, and $\mathcal{M}_1 \restriction \mathcal{L}_0 \prec \mathcal{M}_2 \restriction \mathcal{L}_0$.
Then there is a model $\mathcal{M}_3$ of Γ_1 such that $\mathcal{M}_1 \prec \mathcal{M}_3$, and $\mathcal{M}_2 \restriction \mathcal{L}_2 \prec \mathcal{M}_3 \restriction \mathcal{L}_2$. (See figure 4.)

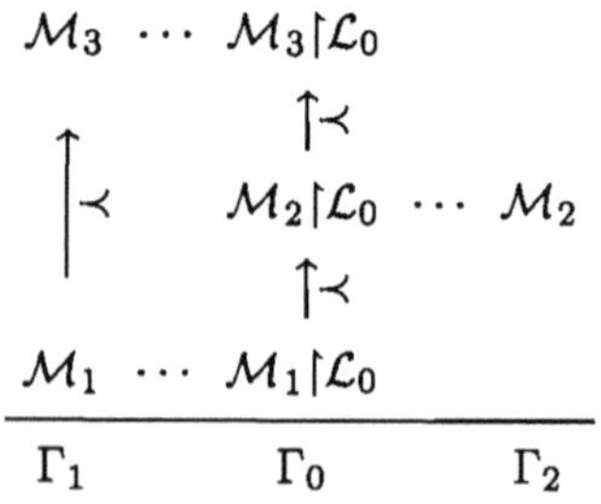

Figure 4.

Proof: Let $\Gamma := Diag(\mathcal{M}_1) \cup Diag(\mathcal{M}_2 \restriction \mathcal{L}_0)$. Γ is a theory in the language $\mathcal{L}_1(M_2)$, containing Γ_1. We claim that Γ is consistent. It is enough to prove that for all formula $\theta_1 \in Diag(\mathcal{M}_1)$ and all $\theta_0 \in Diag(\mathcal{M}_2 \restriction \mathcal{L}_0)$, we have $\theta_1 \wedge \theta_0$ consistent.

Assume that $\theta_1 \wedge \theta_0$ is inconsistent. θ_0 is of the form $\psi_0(\underline{b}_1, \ldots, \underline{b}_n)$, where $b_1, \ldots b_n \in M_2 - M_1$ and ψ_0 is in $\mathcal{L}_0(M_1)$. Since $\theta_1 \vdash \neg\theta_0$, we also have $\{\theta_1\} \vdash \neg\psi_0$ by the theorem on constants. So we get $\{\theta_1\} \vdash \forall x_1 \cdots \forall x_n \neg\psi_0$, so

$$\forall x_1 \cdots \forall x_n \neg\psi_0 \in Diag(\mathcal{M}_1).$$

But we also have $\mathcal{M}_2 \models \psi_0(b_1, \ldots, b_n)$, hence $\exists x_1 \cdots \exists x_n \psi_0 \in Diag(\mathcal{M}_2) \restriction \mathcal{L}_0$, so

$$\exists x_1 \cdots \exists x_n \psi_0 \in Diag(\mathcal{M}_1 \restriction \mathcal{L}_0) \subseteq Diag(\mathcal{M}_1)$$

which is a contradiction.
Since Γ is consistent, we can now find a model $\mathcal{M}_3'$ of Γ. Letting $M_1' := \{\underline{a}^{\mathcal{M}_3'} : a \in M_1\}$ and $M_2' := \{\underline{a}^{\mathcal{M}_3'} : a \in M_2\}$, we get submodels $\mathcal{M}_1'$ and $\mathcal{M}_2'$ which are isomorphic copies of $\mathcal{M}_1$ and $\mathcal{M}_2$. Moreover, $\mathcal{M}_1' \restriction \mathcal{L}_0 \prec \mathcal{M}_2' \restriction \mathcal{L}_0$. So using 3.1.17 we can find a model $\mathcal{M}_3$ as required.

3.1.25. Proof of 3.1.21. Now we can prove the theorem:
Let $\mathcal{M}_1$ be a model of Γ_1. By 3.1.23, we can find a model $\mathcal{M}_2$ of Γ_2 such that $\mathcal{M}_1 \restriction \mathcal{L}_0 \prec \mathcal{M}_2 \restriction \mathcal{L}_0$. By lemma 3.1.24, we can find a model $\mathcal{M}_3$ of Γ_1 with the properties $\mathcal{M}_1 \prec \mathcal{M}_3$, $\mathcal{M}_2 \restriction \mathcal{L}_0 \prec \mathcal{M}_3 \restriction \mathcal{L}_0$.
Again using 3.1.24, we can find $\mathcal{M}_4 \models \Gamma_2$, $\mathcal{M}_2 \prec \mathcal{M}_4$, $\mathcal{M}_3 \restriction \mathcal{L}_0 \prec \mathcal{M}_4 \restriction \mathcal{L}_0$. etc. (See figure 5.)
Finally, we let $N := \bigcup_n M_n$, and we define a model $\mathcal{N}$ with universe N such that for all n, $\mathcal{M}_n \restriction \mathcal{L}_0 \prec \mathcal{N} \restriction \mathcal{L}_0$, and for all k, $\mathcal{M}_{2k} \prec \mathcal{N} \restriction \mathcal{L}_2$, and for all k, $\mathcal{M}_{2k+1} \prec \mathcal{N} \restriction \mathcal{L}_1$. Clearly $\mathcal{N} \models \Gamma_1 \cup \Gamma_2$.

As a consequence, we get the following theorem ("Craig's interpolation theorem"):

3.1.26. Theorem. Assume that φ_1 is a formula in a language $\mathcal{L}_1$, φ_2 in the language $\mathcal{L}_2$, $\mathcal{L}_1$ and $\mathcal{L}_2$ are subsets of $\mathcal{L}$, and $\varphi_1 \vdash_{\mathcal{L}} \varphi_2$.
Then there exists a formula φ_0 in the language $\mathcal{L}_0 := \mathcal{L}_1 \cap \mathcal{L}_2$ such that

$$\varphi_1 \vdash_{\mathcal{L}_1} \varphi_0 \quad \text{and} \quad \varphi_0 \vdash_{\mathcal{L}_2} \varphi_2.$$

Proof: Let

$$\Gamma_0 := \{\psi \in \mathcal{L}_0 : \varphi_1 \vdash \psi \text{ or } \neg\varphi_2 \vdash \psi\}.$$

We claim that Γ_0 is inconsistent.

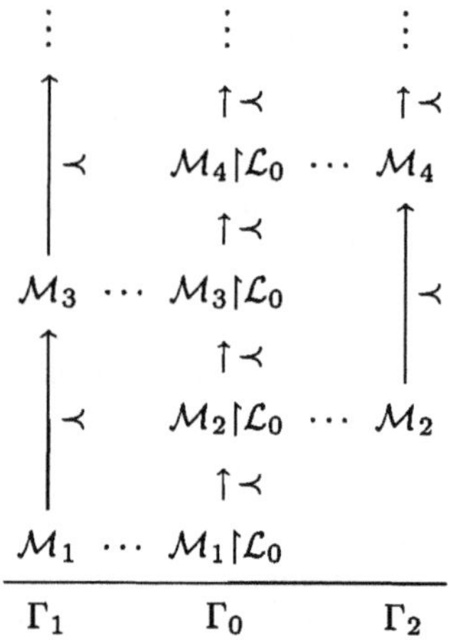

Figure 5.

Proof of the claim: So assume that Γ_0 is consistent, and let Γ_0' be a complete consistent extension (in the language $\mathcal{L}_0$ of Γ).

Let $\Gamma_1 := \Gamma_0' \cup \{\varphi_1\}$, Let $\Gamma_2 := \Gamma_0' \cup \{\neg\varphi_2\}$.

Since $\Gamma_1 \cup \Gamma_2$ is inconsistent (in the language $\mathcal{L}$), we can conclude from the joint consistency theorem that either Γ_1 or Γ_2 is inconsistent.

Case 1: Γ_1 is inconsistent. So there is a formula φ in Γ_0' such that $\{\varphi, \varphi_1\}$ is inconsistent. So $\varphi_1 \vdash \neg\varphi$, hence $\neg\varphi \in \Gamma_0 \subseteq \Gamma_0'$. Hence Γ_0' is inconsistent, because it contains both φ and $\neg\varphi$. But Γ_0' was chosen to be consistent.

Case 2: Similar.

This contradiction shows that Γ_0 is indeed inconsistent. So there is a finite inconsistent subset $\{\gamma_1, \ldots, \gamma_n, \delta_1, \ldots, \delta_m\}$, where $\varphi_1 \vdash \gamma_i$ for all i and $\neg\varphi_2 \vdash \delta_j$ for all j.

Let $\gamma := \gamma_1 \wedge \cdots \wedge \gamma_n$, $\delta := \delta_1 \wedge \cdots \wedge \delta_m$, then we have $\varphi_1 \vdash \gamma$, $\neg\varphi_2 \vdash \delta$, and $\{\gamma, \delta\}$ is inconsistent.

Thus $\varphi_1 \vdash \gamma$, $\gamma \vdash \neg\delta$, and $\neg\delta \vdash \varphi_2$. In particular, $\varphi \vdash_{\mathcal{L}_1} \gamma$ and $\gamma \vdash_{\mathcal{L}_2} \varphi_2$.

Exercises

1. Prove: For any model $\mathcal{M}$, Th($\mathcal{M}$) is a complete and consistent theory.

2. Show that if $\mathcal{M}_1$ and $\mathcal{M}_2$ are isomorphic, then $\mathcal{M}_1 \equiv \mathcal{M}_2$.

3. In this exercise we will show that if two models are elementarily equivalent and **finite**, then they are also isomorphic.

Fix a language $\mathcal{L}$ of first order logic, and write $\mathcal{L}(\mathtt{c})$ for the language which has all function, relation and constant symbols from $\mathcal{L}$, and an additional constant symbol c. If $m \in M$, then we write $(\mathcal{M}, m)$ for the model $\mathcal{M}'$ of $\mathcal{L}(\mathtt{c})$ which is an expansion of $\mathcal{M}$, and in which the constant symbol c is interpreted as m (similar to $\mathcal{M}_{\{m\}}$ in 3.1.6, but here we use a fixed constant symbol rather than a different one for each possible m).
(0) Assume that $\mathcal{M}$ is a finite model, and assume that $\mathcal{M}$ is elementarily equivalent to $\mathcal{N}$. Show that $\mathcal{N}$ is also finite, with the same number of elements as $\mathcal{M}$.
(1) Assume that $\mathcal{M}$ and $\mathcal{N}$ are two elementarily equivalent finite models. Show that for any m in M there is $n \in N$ such that $(\mathcal{M}, m)$ and $(\mathcal{N}, n)$ are elementarily equivalent.
(2) Assume that $\mathcal{M}$ is a finite model in which each element is the interpretation of some constant symbol. Assume that $\mathcal{N}$ is elementarily equivalent to $\mathcal{M}$. Show that $\mathcal{M}$ and $\mathcal{N}$ are isomorphic. (Hint: Any isomorphism f must satisfy $f(\mathtt{c}^{\mathcal{M}}) = \mathtt{c}^{\mathcal{N}}$ for any constant symbol c)
(3) Assume that $\mathcal{M}$ and $\mathcal{N}$ are as in (0). Show that $\mathcal{M}$ and $\mathcal{N}$ are isomorphic. (Hint: Use (1) inductively to arrive at a situation where (2) applies.)

4. Assume that $\mathcal{M}$ and $\mathcal{N}$ are two finite models, and $\mathcal{M} \equiv \mathcal{N}$. (So by the previous exercise, $\mathcal{M}$ and $\mathcal{N}$ are in fact isomorphic.) Show that there are natural numbers $n_1, \ldots, n_k$ such that the number of isomorphisms between $\mathcal{M}$ and $\mathcal{N}$ is $n_1! \cdot n_2! \cdots n_k!$.

(Hint: Define an equivalence relation $\sim$ on M by

$$a \sim b \text{ iff for all formulas } \varphi(\mathtt{x}),\ \mathcal{M} \models \varphi(a) \leftrightarrow \varphi(b)$$

Let $n_1, \ldots, n_k$ be the sizes of the equivalence classes.)

5. Find an example of $\mathcal{M} \prec \mathcal{N}$, where 3.1.14(**) is not satisfied. Hint: Consider a nonstandard model of arithmetic (see section 2.3).

6. (1) Let $\mathcal{M}_0 \subseteq \mathcal{M}_1 \subseteq \cdots$, and let $\mathcal{M}_\infty = \lim_n \mathcal{M}_n$, $\mathcal{M}_\infty \subseteq \mathcal{N}$. Assume that for all n:

Whenever $\exists x\,\theta$ is a closed formula in $\mathcal{L}(M_n)$ and $\mathcal{N} \models \exists x\theta$, then there is $a \in \mathcal{M}_{n+1}$, $\mathcal{N} \models \theta(x/a)$. Show that $\mathcal{M}_\infty \prec \mathcal{N}$.

(2) Prove the "Löwenheim-Skolem theorem:" For any model $\mathcal{N}$ (of a countable theory Γ) there is a **countable** elementary submodel $\mathcal{M} \prec \mathcal{N}$. (Hint: Use (1), starting with any countable submodel $\mathcal{M}_0$.)

7. Show that the two definitions in 3.1.15 are equivalent.

8. Complete the proof of 3.1.16.

9. Prove Corollary 3.1.22.

3.2. Ultraproducts and Compactness

An immediate consequence of the definition of derivation was the following fact (also called the "compactness theorem")

$$\text{If } \Gamma \vdash \varphi, \text{ then there is a finite } \Gamma_0 \subseteq \Gamma,\ \Gamma_0 \vdash \varphi$$

or equivalently

If every finite subset of Γ is consistent, then Γ is consistent.

Using the completeness theorem, we obtained (in section 2.2) a semantic version of the compactness theorem:

$$\text{If } \Gamma \models \varphi, \text{ then there is a finite } \Gamma_0 \subseteq \Gamma,\ \Gamma_0 \models \varphi$$

or equivalently

If every finite subset of Γ has a model, then Γ has a model.

The question arises whether this semantic version can be proved directly. Assume that for each finite subset $i \subseteq \Gamma$ we have a model $\mathcal{M}_i$. Is it possible somehow to "combine" these models into single model $\mathcal{M}$ satisfying Γ? In this section we will show how this can be achieved. We start with an example.

If $(A_0, +_0)$ and $(A_1, +_1)$ are groups, the the set of pairs

$$A_0 \times A_1 := \{(a_0, a_1) : a_0 \in A_0, a_1 \in A_1\}$$

can be naturally made into a group by defining

$$(a_0, a_1) + (a'_0, a'_1) := (a_0 + a'_0, a_1 + a'_1).$$

To generalize this concept to infinite products, first consider the following modification: We replace the pair (a_0, a_1) by the set $\{(0, a_0), (1, a_1)\}$, i.e., by the function that maps each $i \in \{0, 1\}$ to the corresponding a_i. Thus $A_0 \times A_1$ is replaced by the isomorphic structure

$$\{\vec{a} : \vec{a} \text{ is a function, } \mathrm{dom}(\vec{a}) = \{0,1\}, \text{ and } \forall i \in \{0,1\}\, \vec{a}(i) \in A_i\},$$

which we write as $\prod_{i\in\{0,1\}} A_i$. Addition is naturally defined by

$$\vec{a} + \vec{a}' = \vec{b}, \text{ where } \vec{b}(i) = \vec{a}(i) +_i \vec{a}'(i) \text{ for } i \in \{0,1\}.$$

In general, we have the following definition:

3.2.1. Definition. Assume I is a nonempty set, and for all $i \in I$, $(A_i, +_i)$ is a group. Then $\prod_{i\in I} A_i$, the product of the groups A_i, is the group defined as follows:

- The underlying set of $\prod_{i\in I} A_i$ is the set of all functions $\vec{a}$ with

  $\mathrm{dom}(\vec{a}) = I$
  For all $i \in I$, $\vec{a}(i) \in A_i$.
- The group operation $+$ is defined on $\prod_{i\in I} A_i$ by

$$\vec{a} + \vec{a}' = \vec{b}, \text{ where } \vec{b}(i) = \vec{a}(i) +_i \vec{a}'(i) \text{ for all } i \in I$$

We leave it to the reader to check that this is indeed a group.

If $(A_0, +_0, \cdot_0)$ and $(A_1, +_1, \cdot_1)$ are rings, then we similarly define the product structure $A_0 \times A_1$ or $\prod_{i\in\{0,1\}} A_i$, which turns out to be a ring. However, it is easy to see that the product of two fields is never a field. (See exercise 1.)

However, if A_0 and A_1 are fields, then there are homomorphic images of $A_0 \times A_1$ which are fields.

In this section we will show that there is always a homomorphic image of $\prod_{i\in I} A_i$ which shares many properties with the structures A_i.

First we introduce the concepts of **filters** and **ultrafilters**.

3.2.2. Definition. Let I be a nonempty set. $\mathcal{F}$ is called a filter on I if $\mathcal{F} \subseteq \mathcal{P}(I)$, (i.e., elements of $\mathcal{F}$ are subsets of I) and

(A) $\mathcal{F}$ is "closed under supersets:" If $X \in \mathcal{F}$, and $X \subseteq Y \subseteq I$, then $Y \in \mathcal{F}$.
(B) $\mathcal{F}$ is "closed under intersections:" If X_1 and X_2 in $\mathcal{F}$, then also $X_1 \cap X_2 \in \mathcal{F}$.
(C) $\mathcal{F}$ is nontrivial: $\emptyset \notin \mathcal{F}$, but $I \in \mathcal{F}$.

(See exercise 3 for a list of equivalent requirements.)

3.2.3. Example.
(1) The set $\{I\}$ is always a filter on the set I.
(2) The set $\{\{a,b\},\{a\}\}$ is a filter on $\{a,b\}$.
(3) The set $\{X \subseteq \mathbb{N} : X \text{ contains all even numbers}\}$ is a filter on $\mathbb{N}$.
(4) More generally, if $A \subseteq I$ is nonempty, then the set

$$\{X : A \subseteq X\}$$

is a filter on I. We call such filters **principal filters.**
(5) The set

$$\{X \subseteq \mathbb{N} : X \text{ contains all but finitely many natural numbers}\}$$

is a filter on $\mathbb{N}$.
(6) Combining examples (3) and (5), we have the set

$$\{X \subseteq \mathbb{N} : X \text{ contains all but finitely many even numbers}\}$$

which is also a filter on $\mathbb{N}$.

3.2.4. Definition. If $\mathcal{F} \subseteq \mathcal{P}(I)$ is a filter, then
(1) $\mathcal{F}$ is called **maximal** if there is no filter $\mathcal{F}'$ on I which properly extends $\mathcal{F}$. (I.e., whenever $\mathcal{F}'$ is a filter with the property that all sets in $\mathcal{F}$ are also in $\mathcal{F}'$, then we must have $\mathcal{F}' = \mathcal{F}$.)
(2) $\mathcal{F}$ is called **prime**, if for all X_1, $X_2 \subseteq I$:

$$X_1 \in \mathcal{F} \text{ or } X_2 \in \mathcal{F} \quad \text{iff} \quad X_1 \cup X_2 \in \mathcal{F}.$$

(Compare exercise 3(C').)
(3) F is called an **ultrafilter**, if

$$\forall X \subseteq I : X \in \mathcal{F} \text{ or } I - X \in \mathcal{F}.$$

We may think of sets in a family $\mathcal{F}$ as "large." Then "$\mathcal{F}$ is a filter" means that

If a set contains a "large" set, then the set itself is large.
The intersection of two "large" sets is large.
The whole set I is large, but the empty set is not large.

"$\mathcal{F}$ is an ultrafilter" means: For every set, either the set itself or its complement (with respect to I) is large.

3.2.5. Lemma. If $\mathcal{F}$ is a filter on I, then
$\mathcal{F}$ is an ultrafilter iff $\mathcal{F}$ is a maximal filter iff $\mathcal{F}$ is a prime filter.

Proof: We leave the proof of "prime implies ultra" and "ultra implies maximal" as an exercise, and prove only that every maximal filter is prime.

So let $\mathcal{F}$ be a maximal filter, and assume $Y_1 \cup Y_2 \in \mathcal{F}$. Define

$$\mathcal{F}' := \{Y \subseteq I : Y_1 \cup Y \in \mathcal{F}\}.$$

First we claim that $\mathcal{F}'$ satisfies 3.2.2(A)-(B).
If $X \in \mathcal{F}'$, and $X \subseteq Y$, then $Y_1 \cup X \in \mathcal{F}$, and $Y_1 \cup X \subseteq Y_1 \cup Y$, so since $\mathcal{F}$ satisfies (A), $Y_1 \cup Y \in \mathcal{F}$, so $Y \in \mathcal{F}$.
If X_1 and X_2 are in $\mathcal{F}'$, then

$$Y_1 \cup X_1 \in \mathcal{F} \qquad Y_1 \cup X_2 \in \mathcal{F}$$

and $Y_1 \cup (X_1 \cap X_2) = (Y_1 \cup X_1) \cap (Y_1 \cup X_2) \in \mathcal{F}$, so also $Y_1 \cup (X_1 \cap X_2) \in \mathcal{F}$, hence $X_1 \cap X_2 \in \mathcal{F}'$.

Our next claim is that $\mathcal{F} \subseteq \mathcal{F}'$. Indeed, let $Y \in \mathcal{F}$, then since $\mathcal{F}$ is closed under supersets, also $Y_1 \cup Y \in F$, so $Y \in \mathcal{F}'$. In particular, $I \in F'$.

Since $\mathcal{F}$ is maximal, $\mathcal{F}'$ cannot be a filter properly extending $\mathcal{F}$. So there are two cases:
Case 1: $\mathcal{F}'$ is not a filter. So we must have $\emptyset \in \mathcal{F}'$. The definition of $\mathcal{F}'$ now implies $Y_1 \in \mathcal{F}$.
Case 2: $\mathcal{F}' = \mathcal{F}$. Since $Y_2 \in \mathcal{F}'$, we also have $Y_2 \in \mathcal{F}$.

3.2.6. Definition. If $\mathcal{F}$ is a filter on I, $X \subseteq I$, define $\mathcal{F} + X$ to be the set

$$\{Y \subseteq I : \exists A \in \mathcal{F} : A \cap X \subseteq Y\}$$

(See exercise 7.)

3.2.7. Fact. Assume that $\mathcal{F}$ is a filter on I, and $X \subseteq I$.

(a) If $I - X \in \mathcal{F}$, then $\mathcal{F} + X = \mathcal{P}(I)$, i.e., every subset of I is in $\mathcal{F} + X$ (so $\mathcal{F} + X$ is not a filter).

(b) If $I - X \notin \mathcal{F}$, then $\mathcal{F} + X$ is a filter. In fact, $\mathcal{F} + X$ is the smallest filter extending $\mathcal{F}$ that contains the set X.

Proof: Exercise.

We have seen that not every filter is an ultrafilter. However, we have the following theorem:

3.2.8. Theorem. If $\mathcal{F}_0$ is a filter on the set I, then there exists an ultrafilter $\mathcal{F}$ on I with $\mathcal{F}_0 \subseteq \mathcal{F}$.

Hints for a formal proof can be found in the exercises. Very informally, the proof proceeds as follows:

If $\mathcal{F}_0$ is already an ultrafilter, then there is nothing to show. Otherwise there is a set $X_0 \subseteq I$ such that neither X_0 nor $I - X_0$ are in $\mathcal{F}_0$. So let $\mathcal{F}_1 := \mathcal{F}_0 + X_0$.

If $\mathcal{F}_1$ is already an ultrafilter, then we are done. Otherwise there is a set $X_1 \subseteq I$ such that neither X_1 nor $I - X_1$ are in $\mathcal{F}_1$. So let $\mathcal{F}_2 := \mathcal{F}_1 + X_1$.

If $\mathcal{F}_2$ is already ...

...

If this construction stops at some stage, we are done. Otherwise, let $F_\omega := \mathcal{F}_0 \cup \mathcal{F}_1 \cup \cdots$. If $\mathcal{F}_\omega$ is an ultrafilter, then we are done. Otherwise there is a set $X_\omega \subseteq I$ such that neither X_ω nor $I - X_\omega$ are in $\mathcal{F}_\omega$. So let $\mathcal{F}_{\omega+1} := \mathcal{F}_\omega + X_\omega$. This is a filter.

If $\mathcal{F}_{\omega+1}$ is already ...

...

After continuing this construction "forever," we obtain an ultrafilter. (For a formalization of "forever," see exercise 9.)

3.2.9. Definition. Assume that $I \neq \emptyset$, and for all $i \in I$ $A_i \neq \emptyset$, and $\mathcal{F}$ is an ultrafilter on I. Then we define the set $\prod_{i\in I} A_i$ as in 3.2.1 as the set of all functions $\vec{a}$ with

$\text{dom}(\vec{a}) = I$

For all $i \in I$, $\vec{a}(i) \in A_i$.

We now define an equivalence relation $\sim_{\mathcal{F}}$ on the set $\prod_{i\in I} A_i$ as follows:

If $\vec{a}$, $\vec{a}' \in \prod_{i\in I} A_i$, then $\vec{a} \sim_{\mathcal{F}} \vec{a}'$ iff

$$\{i : \vec{a}(i) = \vec{a}'(i)\} \in \mathcal{F}$$

i.e., if $\vec{a}$ and $\vec{a}'$ agree on a large set.

For $\vec{a}$ in $\prod_{i\in I} A_i$, we write $\vec{a}/\mathcal{F}$ for the equivalence class of $\vec{a}$ modulo the equivalence relation $\sim_{\mathcal{F}}$, i.e.,

$$\vec{a}/\mathcal{F} := \{\vec{b} : \vec{b} \sim_{\mathcal{F}} \vec{a}\}$$

and we write $\prod_{i\in I} A_i/\mathcal{F}$ for the set of all classes $\vec{a}/\mathcal{F}$. We call each element $\vec{b} \in \vec{a}/\mathcal{F}$ a **representative** of the class $\vec{a}/\mathcal{F}$.

Now assume that, for example, for all $i \in I$, $(\mathcal{A}_i, +_i)$ is a group. Then also $\prod_{i\in I} A_i$ carries a group structure, and $\sim_{\mathcal{F}}$ is a congruence relation, i.e., if $\vec{a} \sim_{\mathcal{F}} \vec{a}'$, and $\vec{b} \sim_{\mathcal{F}} \vec{b}'$, then $\vec{a} + \vec{b} \sim_{\mathcal{F}} \vec{a}' + \vec{b}'$.
[**Proof:** : Let

$$F_a := \{i \in I : \vec{a}(i) = \vec{a}'(i)\}$$
$$F_b := \{i \in I : \vec{b}(i) = \vec{b}'(i)\}$$
$$F := \{i \in I : \vec{a}(i) + \vec{b}(i) = \vec{a}'(i) + \vec{b}'(i)\}$$

Then $F_a \cap F_b \subseteq F$, so if $F_a, F_b \in \mathcal{F}$, then also $F \in \mathcal{F}$.]

So we can define a group operation on $\prod_{i\in I} A_i/\mathcal{F}$ by letting

$$\vec{a}/\mathcal{F} \ + \vec{b}/\mathcal{F} \ = (\vec{a} + \vec{b})/\mathcal{F}.$$

(Equivalently, we could define the zero function $\vec{0}$ by $\vec{0}(i) =$ the neutral element of $\mathcal{A}_i$, and show that the equivalence class of $N := \vec{0}/{\sim_{\mathcal{F}}}$ is a normal subgroup and get $\prod_{i\in I} A_i/\mathcal{F}$ as the quotient modulo N.)

More generally, we have the following definition:

3.2.10. Definition. Fix a language $\mathcal{L}$. Assume that I is a nonempty set, and that for all $i \in I$ $\mathcal{A}_i$ is a model for the language $\mathcal{L}$, and $\mathcal{F}$ is an ultrafilter on I. Then we define the structure $\mathcal{A} = \prod_{i\in I} \mathcal{A}_i/\mathcal{F}$ (called the ultraproduct of the structures $\mathcal{A}_i$) as follows:

The universe A of $\prod_{i\in I} \mathcal{A}_i/\mathcal{F}$ is $\prod_{i\in I} A_i/\mathcal{F}$. So, elements of $\prod_{i\in I} A_i/\mathcal{F}$ are equivalence classes of $\prod_{i\in I} A_i$.

For every constant symbol $\mathtt{c}$ in $\mathcal{L}$, let $\vec{c}$ be the function that maps each $i \in I$ to $\mathtt{c}^{\mathcal{A}_i}$, and let $\mathtt{c}^{\mathcal{A}} = \vec{c}/\mathcal{F}$.

If $\mathtt{f}$ is a k-ary function symbol in $\mathcal{L}$, we define $\mathtt{f}^{\mathcal{A}}$ as follows:

For any $m_1, \ldots, m_k \in A$, pick representatives $\vec{a}_1, \ldots,$ $\vec{a}_k$ in $\prod_{i\in I} A_i$ such that $m_1 = \vec{a}_1/\mathcal{F}, \ldots, m_k = \vec{a}_k/\mathcal{F}$, and define $\vec{b}$ by

$$\vec{b}(i) = \mathtt{f}^{\mathcal{A}_i}(\vec{a}_1(i), \ldots, \vec{a}_k(i))$$

and let $m = \vec{b}/\mathcal{F}$, or in other words

$$\mathtt{f}^{\mathcal{A}}(\vec{a}_1/\mathcal{F}, \ldots \vec{a}_k/\mathcal{F}) = \vec{b}/\mathcal{F}.$$

Note that $m = \vec{b}/\mathcal{F}$ does not depend on the choice of $\vec{a}_1, \ldots, \vec{a}_k$. Indeed, if we had chosen some other $\vec{a}'_1, \ldots, \vec{a}'_k$ (satisfying $\vec{a}_i \sim_{\mathcal{F}} \vec{a}'_i$) and then defined $\vec{b}'$ by

$$\vec{b}'(i) = \mathtt{f}^{\mathcal{A}_i}(\vec{a}'_1(i), \ldots, \vec{a}'_k(i))$$

we would have had $\vec{b}/\mathcal{F} = \vec{b}'/\mathcal{F}$ (i.e. $\vec{b} \sim_{\mathcal{F}} \vec{b}'$), as the following calculation shows:
Let

$$F_1 := \{i : \vec{a}_1(i) = \vec{a}'_1(i)\} \in \mathcal{F}$$

$$\ldots$$

$$F_k := \{i : \vec{a}_k(i) = \vec{a}'_k(i)\} \in \mathcal{F}$$

$$F := \{i : \vec{b}(i) = \vec{b}'(i)\}$$

then $F_1 \cap \cdots \cap F_k \subseteq F$, so $F \in \mathcal{F}$.

If $\mathtt{R}$ is a k-ary relation symbol, we let

$$(\vec{a}_1/\mathcal{F}, \ldots, \vec{a}_k/\mathcal{F}) \in \mathtt{R}^{\mathcal{A}} \text{ iff } \{i : (\vec{a}_1(i), \ldots, \vec{a}_k(i)) \in \mathtt{R}^{\mathcal{A}_i}\} \in \mathcal{F}$$

Again it is easy to show that this definition does not depend on the choice of the representatives $\vec{a}_i$.

The following theorem (called Łoś's theorem) is the reason for considering ultraproducts. It describes how the first order theory of an ultraproduct, and indeed its complete diagram, can be computed from the first order theories (or diagrams) of its factors.

3.2.11. Theorem. Assume that I is a nonempty set, and for each $i \in I$ we have a structure $\mathcal{A}_i$ for some language $\mathcal{L}$. Let $\mathcal{F}$ be an ultrafilter on I. Then:
For any closed formula σ, $\prod_{i\in I} \mathcal{A}_i/\mathcal{F} \models \sigma$ iff $\{i : \mathcal{A}_i \models \sigma\} \in \mathcal{F}$.
Moreover, for any n-formula φ

$$\prod_{i\in I} \mathcal{A}_i/\mathcal{F} \models \varphi(\vec{a}_1/\mathcal{F}, \ldots, \vec{a}_n/\mathcal{F}) \text{ iff } \{i : \mathcal{A}_i \models \varphi(\vec{a}_1(i), \ldots, \vec{a}_n(i))\} \in \mathcal{F}. \qquad (*)$$

Proof: By induction on φ. To lighten the notation, we will assume $n = 1$.
If φ is atomic, then $(*)$ is true by the definition of the structure $\prod_{i\in I} \mathcal{A}_i/\mathcal{F}$.
If φ is $\varphi_1 \wedge \varphi_2$, then let

$$F_1 := \{i : \mathcal{A}_i \models \varphi_1(\vec{a}(i))\}$$

$$F_2 := \{i : \mathcal{A}_i \models \varphi_2(\vec{a}(i))\}$$

$$F := \{i : \mathcal{A}_i \models \varphi(\vec{a}(i))\}.$$

Clearly, $F = F_1 \cap F_2$. Since $F_1 \cap F_2 \in \mathcal{F}$ iff $F_1 \in \mathcal{F}$ and $F_2 \in \mathcal{F}$, we have $\mathcal{A} \models \varphi$ iff $\mathcal{A} \models \varphi_1$ and $\mathcal{A} \models \varphi_2$ iff $F_1 \in \mathcal{F}$ and $F_2 \in F$ iff $F \in \mathcal{F}$.

Assume $\varphi = \neg\varphi_1$. Again let

$$F := \{i : \mathcal{A}_i \models \varphi(\vec{a}(i))\}$$
$$F_1 := \{i : \mathcal{A}_i \models \varphi_1(\vec{a}(i))\}$$

Then we have $F_1 = I - F$.

$$\begin{aligned} \mathcal{A} \models \varphi(\vec{a}) &\text{ iff } \mathcal{A} \not\models \varphi_1(\vec{a}) \\ &\text{ iff } F_1 \notin \mathcal{F} \\ &\text{ iff } I - F \notin \mathcal{F} \\ &\text{ iff } F \in \mathcal{F} \qquad \text{because } \mathcal{F} \text{ is an ultrafilter.} \end{aligned}$$

Now assume $\varphi(\mathrm{x}) = \exists \mathrm{y} \psi(\mathrm{x}, \mathrm{y})$. Let

$$F := \{i : \mathcal{A}_i \models \varphi(\vec{a}(i))\}.$$

If $\mathcal{A} \models \varphi(a)$, then there is a $\vec{b} \in \prod_{i \in I} A_i$ such that $\mathcal{A} \models \psi(\vec{a}, \vec{b})$. Let

$$F_1 := \{i : \mathcal{A}_i \models \psi(\vec{a}(i), \vec{b}(i))\}.$$

Then by induction hypothesis, $F_1 \in \mathcal{F}$, and clearly $F_1 \subseteq F$, so $F \in \mathcal{F}$.

Conversely, assume $F \in \mathcal{F}$. Define $\vec{b} \in \prod_{i \in I} A_i$ as follows:

If $i \in F$, let $\vec{b}(i) \in A_i$ be such that $\mathcal{A}_i \models \psi(\vec{a}(i), \vec{b}(i))$.

Otherwise, let $\vec{b}(i)$ be an arbitrary element of A_i.

So we have $\{i : \mathcal{A}_i \models \psi(\vec{a}(i), \vec{b}(i))\} = F \in \mathcal{F}$. So by induction assumption, $\mathcal{A} \models \psi(\vec{a}, \vec{b})$. Hence $\mathcal{A} \models \varphi(\vec{a})$.

Now we are ready to give an alternative (purely model theoretic) proof of the compactness theorem:

3.2.12. Theorem. Assume that Γ is a set of closed formulas such that every finite subset of Γ has a model. Then Γ has a model.

Proof: Let I be the set of finite subsets of Γ. Thus our indices will be finite sets of formulas. For each $i \in I$ let $\mathcal{A}_i$ be a model satisfying $\mathcal{A}_i \models i$ (i.e., for all $\varphi \in i$, $\mathcal{A}_i \models \varphi$).

For $i \in I$, let $X_i := \{j \in I : i \subseteq j\}$. Let $\mathcal{F}_0 := \{X : \exists i : X_i \subseteq X\}$. Note that $X_{i \cup j} = X_i \cap X_j$, so $\mathcal{F}_0$ will be a filter on I. Let $\mathcal{F} \supseteq \mathcal{F}_0$ be an ultrafilter. So for every i, $X_i \in \mathcal{F}$.

We claim that the model

$$\mathcal{A} := \prod_{i \in I} \mathcal{A}_i / \mathcal{F}$$

satisfies the theory Γ.
Proof: Let $\sigma \in \Gamma$. Then, since for all $i \in I$ we have $\mathcal{A}_i \models i$, we have

$$\text{If } \sigma \in j \text{ then } \mathcal{A}_j \models \sigma.$$

Hence

$$\{j : \mathcal{A}_j \models \sigma\} \supseteq \{j : \sigma \in j\} = \{j : \{\sigma\} \subseteq j\} = X_{\{\sigma\}} \in \mathcal{F}.$$

Thus $\{j : \mathcal{A}_j \models \sigma\} \in \mathcal{F}$, so we have $\mathcal{A} \models \sigma$.

Exercises

1. Assume that $(A_0, +_0, \cdot_0)$ and $(A_1, +_1, \cdot_1)$ are fields. Show that
(a) $A_0 \times A_1$ is a ring.
(b) If 0_0 and 0_1 are the zero elements of A_0 and A_1, then $(0_0, 0_1)$ is the zero element of $A_0 \times A_1$.
(c) $A_0 \times A_1$ has zero divisors, hence is not a field.

2. Show that A_0 and A_1 are homomorphic images of $A_0 \times A_1$.

3. Show that $\mathcal{F} \subseteq \mathcal{P}(I)$ is a filter iff
(AB') For all $X_1, X_2 \subseteq I$:

$$X_1 \in \mathcal{F} \text{ and } X_2 \in \mathcal{F} \text{ iff } X_1 \cap X_2 \in \mathcal{F}$$

(C') $\mathcal{F} \neq \mathcal{P}(I)$ and $\mathcal{F} \neq \emptyset$.

4. List all filters on the three element set $\{a, b, c\}$. (There are 7 of them.) Which of them are ultrafilters?

5. The filter $\mathcal{F}_0 := \{A \in \mathbb{N} : \mathbb{N} - A \text{ is finite}\}$ is called the **Fréchet filter** on $\mathbb{N}$. A filter is called principal if there is a nonempty set A such that for all $F \in \mathcal{F}$ we have $A \subseteq F$.
(1) Show that the Fréchet filter is not principal.

(2) Show that if $\mathcal{F}$ is any filter on $\mathbb{N}$ containing the Fréchet filter, then $\mathcal{F}$ is not principal.
(3) Show that if $\mathcal{F}$ is a nonprincipal filter on $\mathbb{N}$ then $\mathcal{F}$ contains the Fréchet filter.
(4) Describe all principal ultrafilters on $\mathbb{N}$.
(5) Let $I \neq \emptyset$. Show that I is infinite iff there is a nonprincipal filter on I.

6. Show that every prime filter is an ultrafilter, and that every ultrafilter is maximal.

7. Show that $\mathcal{F} + X = \{Y \subseteq I : Y \cup (I - X) \in \mathcal{F}\}$.

8. Prove 3.2.7.

9. Show that every filter can be extended to an ultrafilter.

(Hint: Let $\mathcal{F}_0$ be a filter on a set I. For each filter $\mathcal{F}$ that is not an ultrafilter, choose a set $X_{\mathcal{F}}$ such that neither $X_{\mathcal{F}}$ nor $I - X_{\mathcal{F}}$ are in $\mathcal{F}$.)

Now call a set $\mathfrak{A}$ of filters on I "good" if

(a) Every element $\mathcal{F} \in \mathfrak{A}$ is

- EITHER the filter $\mathcal{F}_0$
- OR of the form $\mathcal{F}' + X_{\mathcal{F}'}$ for some $\mathcal{F}' \in \mathfrak{A}$
- OR a union of filters in $\mathfrak{A}$.

(b) Every nonempty subset $\mathfrak{B} \subseteq \mathfrak{A}$ has a least element (i.e. a filter $\mathcal{F}^-$ such that for all $\mathcal{F} \in \mathfrak{B}$ we have $\mathcal{F}^- \subseteq \mathcal{F}$).

For example, the empty set, and the sets $\{\mathcal{F}_0\}$, $\{\mathcal{F}_0, \mathcal{F}_1\}$, ..., as well as the set $\{\mathcal{F}_0, \mathcal{F}_1, \ldots, \mathcal{F}_\omega, \mathcal{F}_{\omega+1}\}$ from 3.2.8 are "good."

First show that if $\mathfrak{A}$ is good, $\mathcal{F}^* \in \mathfrak{A}$ and $\mathfrak{B} \subseteq \mathfrak{A}$ is bounded by $\mathcal{F}^*$ (i.e., for every $\mathcal{F} \in \mathfrak{B}$, $\mathcal{F} \subseteq \mathcal{F}^*$), then the set $\bigcup \mathfrak{B}$, i.e. the union of all elements of $\mathfrak{B}$,

$$\bigcup \mathfrak{B} := \{X : \text{ for some } \mathcal{F} \in \mathfrak{B},\ X \in \mathcal{F}\}$$

is in $\mathfrak{A}$. (In fact, let $\mathcal{F}^{**}$ be a minimal upper bound of $\mathfrak{B}$, i.e., a minimal element $\mathfrak{A}$ having the property of $\mathcal{F}^*$ described above, and show that $\bigcup \mathfrak{B} = \mathcal{F}^{**}$.)

Next, show that whenever $\mathfrak{A}_1$ and $\mathfrak{A}_2$ are "good," then either $\mathfrak{A}_1 \subseteq \mathfrak{A}_2$ or $\mathfrak{A}_2 \subseteq \mathfrak{A}_1$. (If not, let $\mathcal{F}_1$ and $\mathcal{F}_2$ be the least elements of $\mathfrak{A}_1 - \mathfrak{A}_2$ and $\mathfrak{A}_2 - \mathfrak{A}_1$, respectively, and let $\mathfrak{B}_1 := \{\mathcal{F} \in \mathfrak{A}_1 : \mathcal{F} \subset \mathcal{F}_1\}$, $\mathfrak{B}_2 := \{\mathcal{F} \in \mathfrak{A}_2 : \mathcal{F} \subset \mathcal{F}_2\}$. Now note that $\mathfrak{B}_1$ is also a subset of $\mathfrak{A}_2$, so it is bounded by $\mathcal{F}_2$, and hence $\mathcal{F}_1 = \bigcup \mathfrak{B} \subseteq \mathcal{F}_2$, and conversely, so $\mathcal{F}_1 = \mathcal{F}_2$.)

Finally, let $\mathcal{F}^*$ be the union of all filters that appear in some "good" set:

$$\mathcal{F}^* := \{X \subseteq I : \text{There is some "good" } \mathfrak{A} \text{ and some } \mathcal{F} \in \mathfrak{A} \text{ with } X \in \mathcal{F}.\}$$

and show that $\mathcal{F}^*$ is an ultrafilter.

10. Show that the relation $\sim_{\mathcal{F}}$ defined in 3.2.9 is an equivalence relation.

11. Let $\mathcal{M}$ be a model, I a nonempty set, and for each i let $\mathcal{M}_i = \mathcal{M}$. Let $\mathcal{F}$ be an ultrafilter on I, and let $\mathcal{M}^* = \prod \mathcal{M}_i/\mathcal{F}$. ($\mathcal{M}^*$ is called an "ultrapower" of $\mathcal{M}$.)
(1) Show that $\mathcal{M}^* \equiv \mathcal{M}$.
(2) Show that $\mathcal{M}$ can be elementarily embedded into $\mathcal{M}^*$.
(3) Show that if I is finite, then $\mathcal{M} \simeq \mathcal{M}^*$.
(4) Show that if M is finite, then $\mathcal{M} \simeq \mathcal{M}^*$.

3.3. Types and Countable Models

Types are generalizations of theories. Whereas a theory considers a model as a whole, types describe the model from the point of view of a single element (or finitely many elements). The study of types is one of the most important tools in model theoretic research.

We fix a first order language $\mathcal{L}$.

3.3.1. Definition. Let $\mathcal{M}$ be a model for $\mathcal{L}$, and consider $a \in \mathcal{M}$. The **type of a in $\mathcal{M}$** is defined as

$$\text{type}_{\mathcal{M}}(a) := \{\varphi : \varphi \text{ a 1-formula of } \mathcal{L}, \mathcal{M} \models \varphi(a)\}.$$

Similarly, if $a_1, \ldots, a_n$ are elements of $\mathcal{M}$ (not necessarily distinct), we define

$$\text{type}_{\mathcal{M}}(a_1, \ldots, a_n) := \{\varphi : \varphi \text{ an } n\text{-formula of } \mathcal{L},\ \mathcal{M} \models \varphi(a_1, \ldots, a_n)\}.$$

3.3.2. Example. Let $\mathbb{N} = (\mathbb{N}, +, \cdot, <, 0)$ be the model of natural numbers. Then $\text{type}_{\mathbb{N}}(2)$ contains (among others) the following formulas:

$$\mathtt{x}_1 = \mathtt{SS0}$$
$$\mathtt{S0} + \mathtt{S0} = \mathtt{x}_1$$
$$\mathtt{x}_1 < \mathtt{Sx}_1$$
$$\mathtt{0} = \mathtt{0}$$
$$\mathtt{x}_1 \cdot \mathtt{x}_1 = \mathtt{x}_1 + \mathtt{x}_1.$$

(Note that the only free variable appearing in a formula in $\text{type}_{\mathbb{N}}(2)$ may be x_1, but there may also be formulas in $\text{type}_{\mathbb{N}}(2)$ that do not contain any free variables at all — in fact, $\text{Th}(\mathbb{N}) \subseteq \text{type}_{\mathbb{N}}(2)$.)

3.3.3. Definition. An **n-type** is a consistent set of n-formulas.

An n-type t is **complete** if for every n-formula φ we have either $\varphi \in t$ or $\neg\varphi \in t$. (We often do not distinguish between t and $cl_{\vdash}(t)$, so we may call a type already complete if only $cl_{\vdash}(t)$ is complete, i.e., if for every n-formula φ we have $t \vdash \varphi$ or $t \vdash \neg\varphi$.)

3.3.4. Fact. $\text{type}_{\mathcal{M}}(a)$ is always a complete type.

3.3.5. Fact. If t is an n-type, then there is a complete n-type t^* extending t.

Proof: The proof is similar to the corresponding proof for theories in Chapter 2. We let $\varphi_1, \varphi_2, \ldots$ be a list of all n-formulas. We define types $t_0 \subseteq t_1 \subseteq \cdots$ as follows:

$t_0 = t$. When t_k is given, we consider two cases:

Case 1: $t_k \vdash \varphi_k$. In particular, $t_k \cup \{\varphi_k\}$ is consistent. So we let $t_{k+1} := t_k \cup \{\varphi_k\}$.

Case 2: $t_k \nvdash \varphi_k$. So $t_k \cup \{\neg\varphi_k\}$ is consistent. We let $t_{k+1} := t_k \cup \{\neg\varphi_k\}$.

Finally, we let $t^* = t_0 \cup t_1 \cup t_2 \cup \cdots$ Since every finite subset of t^* is contained in some t_k and hence consistent, t^* is consistent. t^* is complete, because every n-formula φ is equal to some φ_k, so we have either $\varphi \in t^*$ or $\neg\varphi_k \in t^*$.

3.3.6. Definition. We say that a type t is **consistent with** a theory Γ iff $t \cup \Gamma$ is consistent.

3.3.7. Fact. Let Γ be a theory, and t an n-type.

(0) Assume that $c_1, \ldots, c_n$ are new constants, i.e., constant symbols which do not occur in the formulas in Γ. Then t is consistent with Γ iff the theory $t(c_1, \ldots, c_n) \cup \Gamma$ is consistent.

(1) t is consistent with a theory Γ iff for all m, for all $\varphi_1, \ldots, \varphi_m$ in t, $\Gamma \cup \{\exists x_1 \cdots \exists x_n (\varphi_1 \wedge \cdots \wedge \varphi_m)\}$ is consistent.

(2) If Γ is a complete theory, then t is consistent with Γ iff for all m, for all $\varphi_1, \ldots, \varphi_m$ in t, $\Gamma \vdash \exists x_1 \cdots \exists x_n (\varphi_1 \wedge \cdots \wedge \varphi_m)$.

(3) If t is a complete n-type, then t is consistent with Γ iff $\Gamma \subseteq t$.

Proof: (0) Exercise. (Use the theorem on constants.)

(1) t is consistent with Γ iff every finite subset of t is consistent with Γ. So let $\{\varphi_1, \ldots, \varphi_m\}$ be a finite subset of t. We have

$$\begin{aligned}
\Gamma \cup \{\varphi_1, \ldots, \varphi_m\} \text{ is inconsistent } & \text{iff } \Gamma \cup \{\varphi_1 \wedge \cdots \wedge \varphi_m\} \text{ is inconsistent} \\
& \text{iff } \Gamma \vdash \neg(\varphi_1 \wedge \cdots \wedge \varphi_m) \\
& \text{iff } \Gamma \vdash \forall \mathrm{x}_1 \cdots \forall \mathrm{x}_n \neg(\varphi_1 \wedge \cdots \wedge \varphi_m) \\
& \text{iff } \Gamma \vdash \neg\exists \mathrm{x}_1 \cdots \exists \mathrm{x}_n (\varphi_1 \wedge \cdots \wedge \varphi_m) \\
& \text{iff } \Gamma \cup \{\exists \mathrm{x}_1 \cdots \exists \mathrm{x}_n (\varphi_1 \wedge \cdots \wedge \varphi_m)\} \\
& \quad \text{is inconsistent.}
\end{aligned}$$

(2) For any Γ, we have:

$$\Gamma \cup \{\exists \mathrm{x}_1 \cdots \exists \mathrm{x}_n (\varphi_1 \wedge \cdots \wedge \varphi_m)\} \text{ is consistent iff } \Gamma \nvdash \neg\exists \mathrm{x}_1 \cdots \exists \mathrm{x}_n (\varphi_1 \wedge \cdots \wedge \varphi_m)$$

For complete Γ, this is equivalent to

$$\Gamma \vdash \exists \mathrm{x}_1 \cdots \exists \mathrm{x}_n (\varphi_1 \wedge \cdots \wedge \varphi_m)$$

(3) If $\Gamma \subseteq t$, then clearly Γ is consistent with t. Conversely, since t is complete, we have for all $\varphi \in \Gamma$: Either $\varphi \in t$ or $\neg\varphi \in t$. Now $\neg\varphi \in t$ would imply that $t \cup \Gamma$ is inconsistent. Hence we have $\varphi \in t$. So $\Gamma \subseteq t$.

3.3.8. Definition. If t is an n-type, and $\tau_1, \ldots, \tau_n$ are terms (or $\mathcal{M}$-terms), we write $t(\tau_1, \ldots, \tau_n)$ for $\{\varphi(\tau_1, \ldots, \tau_n) : \varphi \in t\}$.
We write $\mathcal{M} \models t$ iff for all $\varphi \in t$, $\mathcal{M} \models \varphi$.
Let t be a 1-type, $\mathcal{M}$ a model. We say that $a \in \mathcal{M}$ **realizes** t if $t \subseteq \mathrm{type}_{\mathcal{M}}(a)$ or equivalently, if $\mathcal{M} \models t(a)$.
If there is $a \in \mathcal{M}$ realizing t, we say that $\mathcal{M}$ **realizes** t, otherwise we say that $\mathcal{M}$ **omits** t.
Similarly, if t is an n-type, we say that $\mathcal{M}$ realizes t iff there are $a_1, \ldots a_n \in M$ such that $t \subseteq \mathrm{type}_{\mathcal{M}}(a_1, \ldots, a_n)$, i.e., $\mathcal{M} \models t(a_1, \ldots, a_n)$.

3.3.9. Fact.

(1) Clearly, if $t \cup \mathrm{Th}(\mathcal{M})$ is inconsistent, then $\mathcal{M}$ omits t. (Otherwise, letting $\mathrm{c}_1, \ldots, \mathrm{c}_n$ be new constants, $\mathcal{M}$ could be expanded to a model of the inconsistent theory $t(\mathrm{c}_1, \ldots, \mathrm{c}_n) \cup \mathrm{Th}(\mathcal{M})$.)
(2) If $\mathcal{M}$ realizes t, and $\mathcal{M} \simeq \mathcal{N}$, then $\mathcal{N}$ realizes t.
(3) If $\mathcal{M}$ realizes t, and $\mathcal{M} \prec \mathcal{N}$ (or $\mathcal{M}$ can be elementarily embedded into $\mathcal{N}$), then $\mathcal{N}$ realizes t.

3.3.10. Example. In the language of number theory, the set

$$t := \{0 < x_1, S0 < x_1, \ldots\}$$

is a type. This type is consistent with Th(ℕ) (since every finite subset of Th(ℕ) $\cup\, t$ is consistent).
t (as well as $cl_{\vdash}(t)$) is incomplete — for example, neither $\exists y\, y + y = x_1$ nor $\neg\exists y\, y + y = x_1$ can be derived from t. t is not realized in ℕ (but it is realized in every nonstandard model of Th(ℕ), see Chapter 2.3).

3.3.11. Fact. Let $\mathcal{M}$ be a model, t a type, and assume $t \supseteq \mathrm{Th}(\mathcal{M})$. Then there is a model $\mathcal{N}$ such that
(1) $\mathcal{M} \prec \mathcal{N}$
(2) $\mathcal{N}$ realizes t.

Proof: Assume t is a type. We let c_1, ..., c_n be new constants and let $\Gamma := \mathit{Diag}(\mathcal{M}) \cup t(c_1, \ldots, c_n)$. We leave it as an exercise to check that Γ is consistent (again using the theorem on constants). Let $\mathcal{N}$ be a model of Γ, then $\mathcal{N}$ realizes t, and $\mathcal{M}$ can be elementarily embedded into $\mathcal{N}$ (by 3.1.16), so by 3.1.17 we may as well assume $\mathcal{M} \prec \mathcal{N}$.

Recall that a set X is called "enumerable" iff it is finite or there is a function $f : \mathbb{N} \to X$ **onto** X. X is called countable if X is enumerable and infinite, or equivalently, if there is a 1-1 onto function $f : \mathbb{N} \to X$. Instead of "enumerable" we also say "finite or countable" or "at most countable." A set is called uncountable if it is not enumerable, i.e., infinite but not countable.

One of the main problems in model theory is the following question:

How many nonisomorphic models can a complete theory Γ have?

We will give a few examples, dealing only with countable models.

(1) A complete theory with exactly one countable model (i.e., all of its countable models are isomorphic).
(2) A proof that there is no complete theory which has exactly two countable models.
(3) A complete theory which has exactly three nonisomorphic models.
(4) For every $n > 2$, a complete theory which has exactly n many nonisomorphic models.
(5) A complete theory which has countably many nonisomorphic models.

(6) A complete theory with uncountably many countable models.

3.3.12. Example. The theory of dense linear orderings without endpoints. We consider a language with one binary relation symbol $<$, and we let DLO be the theory consisting of the following closed formulas:

(1a) $\forall x \neg x < x$.

(1b) $\forall x \forall y \forall z (x < y < z \to x < z)$. (We write $x < y < z$ to abbreviate $(x < y \wedge y < z)$.)

(1c) $\forall x \forall y (x = y \vee x < y \vee y < x)$.

(2a) $\forall x \exists y \, x < y$.

(2b) $\forall x \exists y \, y < x$.

(3) $\forall x \forall y : (x < y \to \exists z : x < z < y)$.

Axiom (1a)–(1c) characterize linear orders. Axiom (2a)–(2b) insure that the linear order has no endpoints. Axiom (3) says that the ordering is dense, i.e., between any two distinct points a third point can be found.

3.3.13. Fact. Let $(\mathbb{Q}, <)$ be the set of rational numbers with the usual ordering. Then $(\mathbb{Q}, <) \models DLO$.

3.3.14. Theorem. Let $\mathcal{M} = (M, <)$ be a countable model of DLO. Then $\mathcal{M}$ is isomorphic to $\mathbb{Q}$.

To prove this theorem, we first need the following definition:

3.3.15. Definition. If $\mathcal{A}$, $\mathcal{B}$ are models of DLO, $A_0 \subseteq A$, $B_0 \subseteq B$, then a function $f : A_0 \to B_0$ is called a **partial isomorphism** if f is 1-1 and onto B_0, and for all $a, a' \in A_0$, $\mathcal{A} \models a < a'$ iff $\mathcal{B} \models f(a) < f(a')$.

3.3.16. Fact. Assume $\mathcal{A}$ and $\mathcal{B}$ are models of DLO, $A_0 \subseteq A$, $B_0 \subseteq B$ are finite sets, and $f : A_0 \to B_0$ is a partial isomorphism. Also assume that $a^* \in A$.

Then there is a partial isomorphism f^* extending f (i.e., $f^*(x) = f(x)$ for all x in the domain of f) such that a^* is in the domain of f^*.

Proof: If $a^* \in A_0$ then we can let $f^* = f$.

Otherwise, let $A_0 = A_0^+ \cup A_0^-$, where

$$A_0^- := \{a \in A_0 : a < a^*\} \quad A_0^+ := \{a \in A_0 : a > a^*\}$$

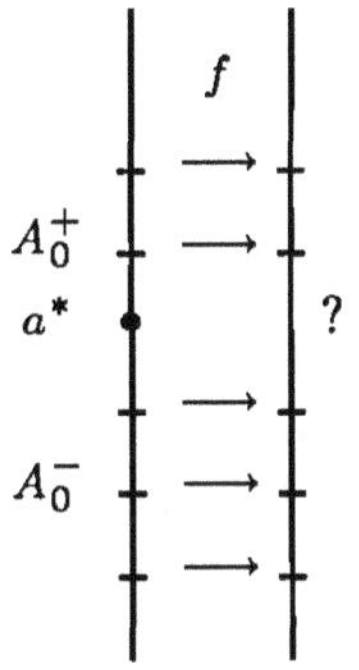

(It is possible that one of these sets is empty.)

Let $B_0^- := f_0(A_0^-)$, $B_0^+ := f(A_0^+)$. Then, since f is a partial isomorphism, all elements of B_0^- are less than all elements of B_0^+. We find an element b^* which is bigger than all elements of B_0^- and smaller than all elements of B_0^+, since $\mathcal{B} \models DLO$. Now define f^* by $f^*(a^*) = b^*$.

3.3.17. Fact. Assume $\mathcal{A}$ and $\mathcal{B}$ are models of DLO, $A_0 \subseteq A$, $B_0 \subseteq B$ are finite sets, and $f : A_0 \to B_0$ is a partial isomorphism. Also assume that $b^* \in B$.

Then there is a partial isomorphism f^* extending f (i.e., $f^*(x) = f(x)$ for all x in the domain of f) such that b^* is in the range of f^*.

Proof: Similar to the previous fact.

Proof of 3.3.14: We will actually prove that any two countable models $\mathcal{A}$ and $\mathcal{B}$ of DLO are isomorphic. The technique of this proof is called a "back-and-forth" argument.

Let $A = \{a_0, a_1, a_2, \dots\}$, $B = \{b_0, b_1, \dots\}$. We will construct finite sets $A_0 \subseteq A_1 \subseteq \cdots \subseteq A$ and $B_0 \subseteq B_1 \subseteq \cdots \subseteq B$ and partial isomorphisms $f_0 \subseteq f_1 \subseteq \cdots$.

f_n will be a 1-1 function from A_n onto B_n.

We start with $A_0 = \{a_0\}$, $B_0 = \{b_0\}$, and f_0 mapping a_0 to b_0. We will ensure $a_n \in A_{2n-1}$ and $b_n \in B_{2n}$ for all $n > 0$.

Let $A_1 = A_0 \cup \{a_1\}$. We can find a function f_1 extending f_0 (and an appropriate set B_1) by the first lemma. Now let $B_2 = B_1 \cup \{b_1\}$, then we can find a function f_2 with range B_2 by the second lemma, etc.

More formally, if we already have A_n, B_n and f_n, and n is even, say $n = 2k$, we use the first lemma to find sets $A_{n+1} = A_n \cup \{a_k\}$ and a set B_{n+1} and a function f_{n+1} (a partial isomorphism from A_{n+1} onto B_{n+1}) extending f_n. If n is odd, say $n = 2k + 1$ we use the second lemma to take care of b_k. Finally we obtain the desired isomorphism f by letting $f(a_n) = f_n(a_n)$.

3.3.18. Corollary.

(1) For any closed formula φ, $DLO \vdash \varphi$ iff $\mathbb{Q} \models \varphi$.

(2) DLO is a complete theory.

Proof: (1) If $DLO \vdash \varphi$, then $\mathbb{Q} \models \varphi$, since $\mathbb{Q}$ is a model of DLO. Conversely, if $DLO \not\vdash \varphi$, then $DLO \cup \{\neg\varphi\}$ is consistent, so it has a countable model. This model is isomorphic to $\mathbb{Q}$, so $\mathbb{Q} \models \neg\varphi$.

(2) $cl_\vdash(DLO) = \mathrm{Th}(\mathbb{Q})$.

3.3.19. Example. We consider a language $\mathcal{L}$ containing one binary relation symbol $<$ and infinitely many constant symbols $\mathtt{c}_1, \mathtt{c}_2, \ldots$ We let DLO^+ be the theory DLO, together with the following axioms:

(4) $\mathtt{c}_1 > \mathtt{c}_2$, $\mathtt{c}_2 > \mathtt{c}_3$, $\mathtt{c}_3 > \mathtt{c}_4$, ...

3.3.20. Fact. The following four models are models of DLO^+. (See figure 6.)

(1) $\mathcal{M}_1$ is the set of rational numbers with $c_n^{\mathcal{M}_1} = \frac{1}{n}$.

(1') $\mathcal{M}_1'$ is the set of those rational numbers which are either positive or ≤ -1, with $c_n^{\mathcal{M}_1'} = \frac{1}{n}$.

(2) $\mathcal{M}_2$ is the set of rational nonzero numbers with $c_n^{\mathcal{M}_1} = \frac{1}{n}$.

(3) $\mathcal{M}_3$ is the set of positive rational numbers with $c_n^{\mathcal{M}_1} = \frac{1}{n}$.

$\mathcal{M}_1$, $\mathcal{M}_2$, $\mathcal{M}_3$ are nonisomorphic to each other, and $\mathcal{M}_1$ is isomorphic to $\mathcal{M}_1'$.

Proof: It is clear that these structures satisfy the theory DLO^+. $\mathcal{M}_1$ is isomorphic to $\mathcal{M}_1'$ via the map

$$i(x) := \begin{cases} x & \text{if } x > 0 \\ x - 1 & \text{if } x \leq 0 \end{cases}$$

To see that $\mathcal{M}_3$ is not isomorphic to either $\mathcal{M}_1$ or $\mathcal{M}_2$, note that $\mathcal{M}_2$ and $\mathcal{M}_1$ realize the type $\{\mathtt{x} < \mathtt{c}_0, \mathtt{x} < \mathtt{c}_1, \mathtt{x} < \mathtt{c}_2, \ldots\}$, but this type is not realized in the model $\mathcal{M}_3$.

To see that $\mathcal{M}_1$ and $\mathcal{M}_2$ are not isomorphic, we argue indirectly: If f is an isomorphism from $\mathcal{M}_1$ to $\mathcal{M}_2$, let $a := f(0)$. Since $\mathcal{M}_1 \models 0 < \mathtt{c}_n$ for all n, we must have $\mathcal{M}_2 \models a < \mathtt{c}_n$ for all n, so $a < 0$. Let $b = a/2$, and $b = f(c)$. Then $\mathcal{M}_2 \models a < b < \mathtt{c}_n$ for all n, so we should have $\mathcal{M}_1 \models 0 < c < \frac{1}{n}$ for all n, which is impossible.

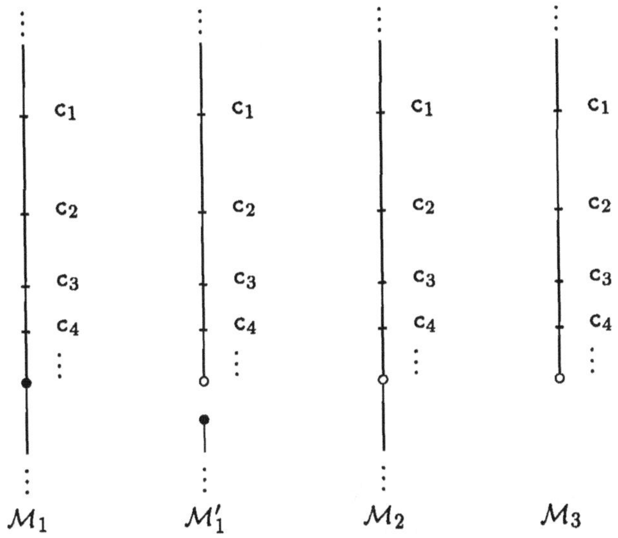

Figure 6.

3.3.21. Theorem. Let $\mathcal{M}_1$, $\mathcal{M}_2$, $\mathcal{M}_3$ be the models from 3.3.20. Then

(1) $\mathcal{M}_1$ and $\mathcal{M}_3$ can be elementarily embedded into $\mathcal{M}_2$.

(2) Every model of DLO^+ is isomorphic to $\mathcal{M}_1$, $\mathcal{M}_2$ or $\mathcal{M}_3$.

(3) DLO^+ is a a complete theory.

Proof: (1) $\mathcal{M}_3$ is a submodel of $\mathcal{M}_2$. To see that $\mathcal{M}_3 \prec \mathcal{M}_2$, we use 3.1.14. So let $A \subseteq \mathcal{M}_3$ be a finite set and $\mathcal{L}_0$ a language using only finitely many constant symbols, and let $b \in \mathcal{M}_2$. If $b \in \mathcal{M}_3$, then there is nothing to prove (we can choose the identity automorphism), so assume $b < 0$. Let $a_1 < a_2$ be positive rational numbers such that

(a) For all $q \in A$, $a_2 < q$.

(b) For all n: if $c_n \in \mathcal{L}_0$, then $a_2 < \frac{1}{n}$.

Let f_1 be an order-preserving map from the rational numbers $< b$ onto the rational nonzero numbers $< a_1$. (Both sets are dense linear orders, so by 3.3.14 there is such a map.) Similarly we can find an order-preserving map from the rational nonzero numbers between b and a_2 onto the rational numbers between a_1 and a_2.

Now we will define an automorphism f of $\mathcal{M}_2 \restriction \mathcal{L}_0$ (see figure 7):

$$f(x) = \begin{cases} x & \text{if } x \geq a_2 \\ f_2(x) & \text{if } b < x < a_2 \\ a_1 & \text{if } x = b \\ f_1(x) & \text{if } x < b \end{cases}$$

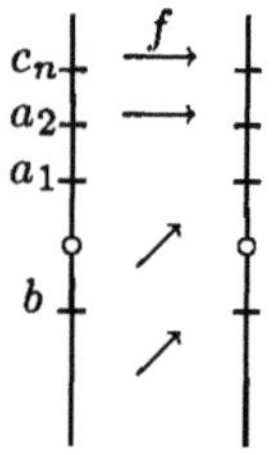

Figure 7.

We leave it to the reader to check that f is in fact a strictly increasing (continuous) function, leaving all $c_n^{\mathcal{M}_2}$ (for $c_n \in \mathcal{L}_0$) invariant, and $f(b) = a$.

Similarly we can show $\mathcal{M}_1' \prec \mathcal{M}_2$. We leave the details as an exercise.

(2) Let $\mathcal{M}$ be a model of DLO^+. Consider the set

$$C := \{c_1^{\mathcal{M}}, c_2^{\mathcal{M}}, \ldots\}.$$

Before we continue, we remark that for any element $a \in M$ and for all n, $a < c_n$ is an $\mathcal{L}(M)$-formula, and so the statement "for all n, $\mathcal{M} \models a < c_n$" is meaningful (and either true or false). However, $\mathcal{M} \models \forall n\ a < c_n$ is not well-defined, since $\forall n\ a < c_n$ is not in the language $\mathcal{L}(M)$.

There are three cases:

Case 1: The set C has a greatest lower bound, i.e., there is an element a_0 such that for all n, $\mathcal{M} \models a_0 < c_n$, but whenever $b > a_0$, there is an n such that $\mathcal{M} \models b > c_n$. In this case $\mathcal{M}$ is isomorphic to $\mathcal{M}_1$. The isomorphism has to map a_0 to 0.

Case 2: The set C has a lower bound, but no greatest lower bound. I.e., there is an element a such that for all n, $\mathcal{M} \models a < c_n$, but for any such a there is an $a' > a$ with the same property. In this case $\mathcal{M}$ is isomorphic to $\mathcal{M}_2$.

Case 3: The set C is unbounded in $\mathcal{M}$, i.e., there is no a satisfying $\mathcal{M} \models a < c_n$ for all n. In this case, $\mathcal{M}$ is isomorphic to $\mathcal{M}_3$.

The proofs for all these cases are back-and-forth arguments similar to 3.3.14. We leave the details as an exercise.

(3) Let $\Gamma := \text{Th}(\mathcal{M}_1)$. Then Γ is a complete theory, and we already know (from (1)) that $\Gamma = \text{Th}(\mathcal{M}_2) = \text{Th}(\mathcal{M}_3)$.

We claim that for any closed formula φ, $DLO^+ \vdash \varphi$ iff $\varphi \in \Gamma$.
Clearly if $DLO^+ \vdash \varphi$, then $\mathcal{M}_1 \models \varphi$, so $\varphi \in \Gamma$.
Conversely, if $DLO^+ \nvdash \varphi$, then $DLO^+ \cup \{\neg\varphi\}$ is consistent, so it has a countable model. This model must be isomorphic to one of $\mathcal{M}_1$, $\mathcal{M}_2$, $\mathcal{M}_3$, so $\neg\varphi \in \Gamma$.

3.3.22. Example. A theory with many models: We have already seen that Th($\mathbb{N}$) has uncountably many models. Using the language of types, we can rephrase the proof from section 2.3 as follows:

(1) Any set A of prime numbers determines a type t^A (consistent with Th($\mathbb{N}$)), namely the type "x is divisible by all members of A, but not divisible by prime numbers $\notin A$."
(2) If $A \neq B$, then $t^A \cup t^B$ is inconsistent.
(3) Thus, in any countable model of Th($\mathbb{N}$), only countably many of the t^A are realized.
(4) If we have only countably many countable models of Th($\mathbb{N}$), then only countably many of the types t^A are realized.
(5) Every t^A is realized in **some** countable model.
(6) Therefore, there must be uncountably many nonisomorphic models.

Saturated Models

In our discussion of atomic and saturated countable models, we fix a complete theory Γ. To simplify the notation, we will assume that for all closed formulas φ, $\Gamma \vdash \varphi$ iff $\varphi \in \Gamma$ or in other words, for all φ, either $\varphi \in \Gamma$ or $\neg\varphi \in \Gamma$.

3.3.23. Definition. Let $\mathcal{N}$ be a model for a language $\mathcal{L}$, $A \subseteq N$, and n a natural number. An **n-type over A** is a set of n-formulas in the language $\mathcal{L}(A)$. (So an n-type in the original language can be called an n-type over $\emptyset$.)
If t is an n-type over A, then we say that $\mathcal{N}$ realizes t iff $\mathcal{N}_A$ (see 3.1.6) realizes t.

3.3.24. Definition. $\mathcal{M}$ is called **saturated** (or, more precisely, ω-saturated) iff:

For every finite $A \subseteq M$, and for every 1-type t over A:
If t is consistent with $\mathrm{Th}(\mathcal{M}_A)$, then $\mathcal{M}_A$ realizes t.

3.3.25. Example. $\mathbb{N}$ is not saturated.

Proof: Consider the type (over $\emptyset$) $t := \{\mathtt{x} > \mathtt{0}, \mathtt{x} > \mathtt{S0}, \mathtt{x} > \mathtt{SS0}, \ldots\}$.

3.3.26. Example. $(\mathbb{Q}, <)$ is saturated. This will follow from corollary 3.3.33 below.

The next lemma shows that not only 1-types, but also all n-types are realized in saturated models.

3.3.27. Lemma. Assume $\mathcal{M}$ is saturated, $A \subseteq M$ is a finite set, and t is an n-type over A consistent with $\mathrm{Th}(\mathcal{M}_A)$. Then $\mathcal{M}_A$ realizes t.

Proof: We will proceed by induction on n, simultaneously for all A (i.e., we will prove $\forall n : \forall A \forall t \ldots$ by induction on n).

For $n = 1$ this is just the definition of "saturation."

Let $n = k + 1$, and let t be an n-type. The formulas in t have free variables among $\mathtt{x}_1, \ldots, \mathtt{x}_{k+1}$. Define

$$t' := \{\exists \mathtt{x}_{k+1} \varphi : t \vdash \varphi,\ \varphi \text{ is a } k+1\text{-formula}\}.$$

t' is a k-type over A, and $t' \subseteq t$, since $\vdash \varphi \rightarrow \exists \mathtt{x}_{k+1} \varphi$ for all φ.

So t' is consistent with $\mathrm{Th}(\mathcal{M}_A)$. By induction hypothesis we can find $m_1, \ldots, m_k$ realizing t', i.e., $\mathcal{M}_A \models t'(m_1, \ldots, m_k)$. Now let

$$t'' := \{\varphi(\underline{m}_1, \ldots, \underline{m}_k, \mathtt{x}_{k+1}) : t \vdash \varphi\}$$

t'' is a 1-type over $\mathcal{M}_B$, where $B := A \cup \{m_1, \ldots, m_k\}$. (Strictly speaking, t'' is not a 1-type, since its free variable is $\mathtt{x}_{k+1}$ rather than $\mathtt{x}_1$, so we should really look at $t''(\mathtt{x}_{k+1}/\mathtt{x}_1)$.)

We claim that for every formula φ:
if $t \vdash \varphi$, we have $\exists \mathtt{x}_{k+1} \varphi(\underline{m}_1, \ldots, \underline{m}_k, \mathtt{x}_{k+1}) \in \mathrm{Th}(\mathcal{M}_B)$. [**Proof:** Since $\exists \mathtt{x}_{k+1} \varphi(\mathtt{x}_1, \ldots, \mathtt{x}_k, \mathtt{x}_{k+1}) \in t'$, and $(m_1, \ldots, m_n)$ realizes t', we must have $\exists \mathtt{x}_{k+1} \varphi(\underline{m}_1, \ldots, \underline{m}_k, \mathtt{x}_{k+1}) \in \mathrm{Th}(\mathcal{M}_B)$.]

So t'' is consistent with $\mathrm{Th}(\mathcal{M}_B)$ (by 3.3.7). By assumption, there exists m_{k+1} realizing t''. Now check that $m_1, \ldots, m_{k+1}$ realizes t.

3.3.28. Fact. If $\mathcal{L}_1 \subseteq \mathcal{L}_2$, $\mathcal{L}_2$ has the same function and relation symbols as $\mathcal{L}_1$, and all but finitely many constant symbols of $\mathcal{L}_2$ are in $\mathcal{L}_1$, then: A model $\mathcal{M}$ for the language $\mathcal{L}_2$ is saturated (for the language $\mathcal{L}_2$) iff its restriction $\mathcal{M} \restriction \mathcal{L}_1$ is saturated (for the language $\mathcal{L}_1$).

Proof: If t is a type in $\mathcal{L}_1$, then t is also a type in $\mathcal{L}_2$. Moreover, if t is a type over $A \subseteq M$, and $\mathrm{Th}(\mathcal{M}\restriction\mathcal{L}_1) \cup t$ is consistent, then also $\mathrm{Th}(\mathcal{M}\restriction\mathcal{L}_2) \cup t$ is consistent. So if $\mathcal{M}$ is saturated, then $\mathcal{M}$ realizes t, hence $\mathcal{M}\restriction\mathcal{L}_1$ realizes t.

Conversely, let t be a type over A in the language $\mathcal{L}_2$. Let $\mathsf{c}_1, \ldots, \mathsf{c}_n$ be the set of constant symbols which are in $\mathcal{L}_2 - \mathcal{L}_1$, and let $B := \{\mathsf{c}_1^{\mathcal{M}}, \ldots, \mathsf{c}_n^{\mathcal{M}}\}$. By replacing each c_i by $\mathsf{c}_i^{\mathcal{M}}$ we obtain from t a new type t' in $\mathcal{L}_1(A \cup B)$. By 3.3.27, t' is realized in $\mathcal{M}\restriction\mathcal{L}_1$. So t is realized in $\mathcal{M}$.

3.3.29. Theorem (Saturated models are universal). If $\mathcal{N}$ is a saturated model of the complete theory Γ, and $\mathcal{M}$ is a countable model of Γ, then there is an elementary embedding of $\mathcal{M}$ into $\mathcal{N}$.

(A model $\mathcal{N}$ with the property that every countable model $\mathcal{M}$ of Γ embeds elementarily into it is called "universal" for the countable models of Γ)

Proof: Let $M = \{m_1, m_2, \ldots\}$, where all m_i are distinct. We will define a sequence of elements $\{n_1, n_2, \ldots\}$ of N satisfying

$$\text{For all } k,\ \mathrm{type}_{\mathcal{N}}(n_1, \ldots, n_k) = \mathrm{type}_{\mathcal{M}}(m_1, \ldots, m_k).$$

We start by letting n_1 be any element of $\mathcal{N}$ realizing $\mathrm{type}_{\mathcal{M}}(m_1)$. Given $n_1, \ldots, n_k$ with $\mathrm{type}_{\mathcal{N}}(n_1, \ldots, n_k) = \mathrm{type}_{\mathcal{M}}(m_1, \ldots, m_k) =: t$, we let $\mathsf{c}_1, \ldots, \mathsf{c}_k$ be new constants, and define expansions $\mathcal{M}'$ and $\mathcal{N}'$ of $\mathcal{M}$ and $\mathcal{N}$, respectively, by letting $\mathsf{c}_i^{\mathcal{M}'} = m_i$, $\mathsf{c}_i^{\mathcal{N}'} = n_i$. Then both $\mathcal{M}'$ and $\mathcal{N}'$ satisfy the theory $t(\mathsf{c}_1, \ldots, \mathsf{c}_k)$, and $\mathcal{N}'$ is a saturated model of this theory, by the previous fact. So we can find an element n_{k+1} such that $\mathrm{type}_{\mathcal{N}'}(n_{k+1}) = \mathrm{type}_{\mathcal{M}'}(m_{k+1})$, or equivalently,

$$\mathrm{type}_{\mathcal{N}}(n_1, \ldots, n_k, n_{k+1}) = \mathrm{type}_{\mathcal{M}}(m_1, \ldots, m_k, m_{k+1}).$$

This completes the induction step.

We now define a submodel $\mathcal{N}^* \subseteq \mathcal{N}$ with universe $\{n_1, n_2, \ldots\}$ which is isomorphic to $\mathcal{M}$ via the function $g : M \to N$ which maps m_i to n_i. Then for any k-formula φ, we have

$$\begin{aligned}\mathcal{N}^* \models \varphi(n_1, \ldots, n_k) &\leftrightarrow \mathcal{M} \models \varphi(m_1, \ldots, m_k) \text{ by definition of } \mathcal{N}^*\\ &\leftrightarrow \varphi \in \mathrm{type}_{\mathcal{M}}(m_1, \ldots, m_k)\\ &\leftrightarrow \varphi \in \mathrm{type}_{\mathcal{N}}(n_1, \ldots, n_k)\\ &\leftrightarrow \mathcal{N} \models \varphi(n_1, \ldots, n_k)\end{aligned}$$

so $\mathcal{N}^* \prec \mathcal{N}$. Similarly we can check that for for any i-formula ψ and any $j_1, \ldots, j_i$,

$$\mathcal{N}^* \models \psi(n_{j_1}, \ldots, n_{j_i}) \leftrightarrow \mathcal{M} \models \psi(m_{j_1}, \ldots, m_{j_i}) \leftrightarrow \mathcal{N} \models \psi(n_{j_1}, \ldots, n_{j_i}).$$

3.3.30. Theorem. Any two countable saturated models (of a complete theory Γ) are isomorphic.

Proof: This proof is again a back-and-forth argument, similar to 3.3.14. Each single step can be done as in 3.3.29.

3.3.31. Fact. If $\mathcal{M} \prec \mathcal{N}$, and $A \subseteq M$, then also $\mathcal{M}_A \prec \mathcal{N}_A$.

3.3.32. Theorem (Characterization of Theories with Saturated Models).
Assume that Γ is a complete theory. Then the following are equivalent:
(1) Γ has a countable saturated model.
(2) For all n there are at most countably many complete n-types $\supseteq \Gamma$.
(3) For all models $\mathcal{M}$ of Γ: Whenever $A \subseteq M$ is finite, there are at most countably many complete types over A extending $\mathrm{Th}(\mathcal{M}_A)$.

Proof of (1) → (2): If $\mathcal{M}$ is a countable saturated model, then every complete type is realized by some element (or n-tuple) from $\mathcal{M}$, and the same element (n-tuple) cannot realize two different complete types. Since $\mathcal{M}$ has only countably many elements, there can be only countably many complete types.
Proof of (2) → (3): If $A = \{a_1, \ldots, a_k\}$, then we can assign to each complete n-type over A a complete $n+k$-type over the empty set as follows: First, if θ is an n-formula in $\mathcal{L}(A)$, then there is an $n+k$-formula $\bar{\theta}$ in $\mathcal{L}$ such that

$$\mathcal{M} \models \theta(\mathrm{x}_1, \ldots, \mathrm{x}_n) \leftrightarrow \bar{\theta}(\mathrm{x}_1, \ldots, \mathrm{x}_n, a_1, \ldots, a_k)$$

Now if t is a complete n-type over A then the set $\bar{t}$ defined by

$$\bar{t} := \{\bar{\theta} : \theta \in t\}$$

is a complete $n+k$-type. [Proof: For any $n+k$-formula φ:

$$\bar{t} \vdash \varphi \quad \text{iff} \quad t \vdash \varphi(\mathrm{x}_1, \ldots, \mathrm{x}_n, \underline{a}_1, \ldots, \underline{a}_k)$$

by the lemma on constants. Since t is complete, $\bar{t}$ is also complete.] Clearly, for any two distinct complete types t_1 and t_2 we have $\bar{t}_1 \neq \bar{t}_2$. Since there are only countably many types over $\emptyset$, there are also only countably many types over A.

Proof of (3) → (1): First we claim the following:
Claim 1: If $\mathcal{M}$ is a countable model of Γ, $A \subseteq M$ finite, and t a complete 1-type over A (containing $\mathrm{Th}(\mathcal{M}_A)$), then there is a countable model $\mathcal{M}'$ realizing t such that $\mathcal{M} \prec \mathcal{M}'$.
Proof: We let c be a new constant symbol, and let

$$\Gamma' := Diag(\mathcal{M}_A) \cup t(\mathtt{x}_1/\mathtt{c}).$$

Check that Γ' is consistent. By 3.1.16, we can find a countable model $\mathcal{M}'$ satisfying Γ' such that $\mathcal{M}$ can be elementarily embedded into $\mathcal{M}' \upharpoonright \mathcal{L}$. So by 3.1.17 we can find such a countable model $\mathcal{M}'$ such that $\mathcal{M} \prec \mathcal{M}'$. Clearly $\mathcal{M}'$ realizes t.
Claim 2: If $\mathcal{M}$ is a countable model of Γ, A_1, A_2, ... are finite subsets of M, and $t_1, t_2, \ldots$ are complete 1-type over $A_1, A_2, \ldots$, respectively, (each containing the respective $\mathrm{Th}(\mathcal{M}_{A_n})$), then there is a countable model $\mathcal{M}^*$ such that $\mathcal{M} \prec \mathcal{M}^*$, and $\mathcal{M}^*$ realizes all t_n.
Proof: We construct a sequence $\mathcal{M} = \mathcal{M}_0 \prec \mathcal{M}_1 \prec \mathcal{M}_2 \prec \cdots$. At each step we apply claim 1.
First we get a countable model $\mathcal{M}_1$ realizing t_1. Since $\mathcal{M} \prec \mathcal{M}_1$, we also have $\mathcal{M}_{A_1} \prec (\mathcal{M}_1)_{A_1}$, so we have $t_2 \supseteq \mathrm{Th}(\mathcal{M}_{A_2}) = \mathrm{Th}((\mathcal{M}_1)_{A_2})$, and so we can apply the lemma again, etc.

If our list $A_1, A_2, \ldots$ is finite, say $A_1, \ldots, A_n$, then the model $\mathcal{M}_n$ will satisfy the requirement. If our list was infinite, then we let $\mathcal{M}^* = \lim_n \mathcal{M}_n$ and again get the required model.
Claim 3: If $\mathcal{M}$ is a countable model of Γ, then there is a countable saturated model $\mathcal{N}$ such that $\mathcal{M} \prec \mathcal{N}$.
Proof: For every finite $A \subseteq M$ there are only countably many complete types over A (which are consistent with $\mathrm{Th}(\mathcal{M}_A)$). Since there are only countably many finite subsets A, there are altogether only countably many pairs (t, A) such that A is a finite subset of M and t is a complete type over A (consistent with $\mathrm{Th}(\mathcal{M}_A)$).
Let (t_1, A_1), (t_2, A_2), ... be a list of all such pairs. By claim 2, we can find a countable model $\mathcal{M}_1$, $\mathcal{M} \prec \mathcal{M}_1$ such that all t_n are realized in $\mathcal{M}_1$.

Now we consider types over finite subsets of M_1: Let (t_1^1, A_1^1), (t_2^1, A_2^1), ... be a list of all pairs (t, A) such that A is a finite subset of M_1 and t is a complete type over A, consistent with $\mathrm{Th}(\mathcal{M}_A)$. We can find a countable model $\mathcal{M}_2$ realizing all these types.

We continue by induction, constructing a chain $\mathcal{M} \prec \mathcal{M}_1 \prec \mathcal{M}_2 \prec \cdots$ such that all types over finite subsets of $\mathcal{M}_n$ are realized in $\mathcal{M}_{n+1}$.

Finally, let $\mathcal{N} = \bigcup_k \mathcal{M}_k$. Every finite subset $A \subseteq N$ appears at some stage, i.e., as a subset of some M_k. If t is a type over A which is consistent

with $\mathrm{Th}(\mathcal{N}_A)$, then t is also consistent with $\mathrm{Th}((\mathcal{M}_k)_A)$. [$\mathcal{M}_k \prec \mathcal{N}$, so these two theories are in fact the same.]
So t is realized in $\mathcal{M}_{k+1}$ and hence in $\mathcal{N}$. So $\mathcal{N}$ is saturated.

3.3.33. Corollary. If there are only finitely many or countably many countable models of Γ, then there is a saturated countable model of Γ.

Proof: Every complete type consistent with Γ is realized in some countable model of Γ. Each countable model realizes only countably many complete types. So there can be only countably many complete types consistent with Γ.

Atomic Models

We have seen (in 3.3.29) that saturated models are, in a sense, big. At the other end of the scale we have atomic models, which are "small."

3.3.34. Definition. We will write $\varphi \vdash_\Gamma \psi$ for $\Gamma \cup \{\varphi\} \vdash \psi$.

3.3.35. Definition. A n-formula φ is called **complete** (or, more precisely, n-complete over Γ) if for every n-formula ψ we have either $\varphi \vdash_\Gamma \psi$ or $\varphi \vdash_\Gamma \neg\psi$, but not both.
In general, if $Z = \{\mathtt{z}_1, \ldots \mathtt{z}_k\}$ is a set of variables, then we say φ is complete for Z if the free variables of φ are contained in Z, and for every formula ψ with $\mathrm{Free}(\psi) \subseteq Z$ we have either $\varphi \vdash_\Gamma \psi$ or $\varphi \vdash_\Gamma \neg\psi$ (but not both). (So n-complete is really "complete for $\{\mathtt{x}_1, \ldots, \mathtt{x}_n\}$".)
An n-type is **atomic** (over Γ) if it contains an n-complete formula. (So every atomic type is a complete type.)

3.3.36. Example.

(1) If τ is a closed term, and Γ is a complete theory, then the formula $\mathtt{x}_1 = \tau$ is 1-complete.
(2) If $\Gamma = DLO$, then the formula $\mathtt{x}_1 < \mathtt{x}_2$ is complete.
(3) If $\Gamma = \mathrm{Th}(\mathbb{N})$, then the type $t := \{\mathtt{x} > \mathtt{0}, \mathtt{x} > \mathtt{S0}, \ldots\}$ contains no complete formula. Moreover, if t^* is a complete type extending $t \cup \Gamma$, then t^* is not atomic (see exercise 25).

3.3.37. Definition. If ψ is an n-formula, and if φ is a complete n-formula such that $\vdash_\Gamma \varphi \to \psi$, then φ is called a "completion" of ψ.
An n-formula ψ is called "incompletable" iff it has no completion. In other words, ψ is incompletable if there is no atomic n-type (consistent

with Γ) containing ψ. (Clearly, if φ is inconsistent with Γ, i.e., if $\Gamma \vdash \neg\varphi$, then φ is incompletable. So we will only consider formulas consistent with Γ.)

3.3.38. Definition. A model $\mathcal{M}$ is an **atomic model** of a theory Γ, if $\mathcal{M} \models \Gamma$, and for every $a_1, \ldots, a_k$, $\text{type}_{\mathcal{M}}(a_1, \ldots, a_k)$ is atomic.

3.3.39. Example.

(1) $\mathbb{N}$ is an atomic model. In general, if $\mathcal{M}$ is any model such that every element of M is of the form $\tau^{\mathcal{M}}$ for some term τ, then $\mathcal{M}$ is atomic.

(2) $(\mathbb{Q}, <)$ is an atomic model of DLO. (We will prove this below.)

3.3.40. Lemma (Atomic models are prime). Fix a theory Γ. If $\mathcal{M}$ is a countable atomic model of Γ, and $\mathcal{N}$ is any model of Γ, then there is an elementary embedding of $\mathcal{M}$ into $\mathcal{N}$.

(A model $\mathcal{M}$ with the property that it can be embedded into every model $\mathcal{N}$ of Γ is called "prime model of Γ.")

Proof: Let $M = \{m_1, m_2, \ldots\}$. For each k let φ_k be a complete k-formula satisfied by $(m_1, \ldots, m_k)$. The formula $\varphi_k \wedge \exists x_{k+1}\varphi_{k+1}$ is satisfied by $(m_1, \ldots, m_k)$, so $\varphi_k \wedge \exists x_{k+1}\varphi_{k+1}$ is consistent with Γ, and hence we cannot have $\varphi_k \vdash_\Gamma \neg\exists x_{k+1}\varphi_{k+1}$. φ_k is complete, so we must have

$$(*) \qquad \varphi_k \vdash_\Gamma \exists x_{k+1}\varphi_{k+1}.$$

Now let n_1 be any element of $\mathcal{N}$ satisfying φ_1, and define a sequence n_1, n_2, ... by requiring

$$\text{for all } k,\ \mathcal{N} \models \varphi_k(n_1, \ldots, n_k).$$

If $n_1, \ldots, n_k$ are given and satisfy φ_k, then by $(*)$ we also have

$$\mathcal{N} \models \exists x_{k+1}\varphi_{k+1}(n_1, \ldots, n_k, x_{k+1})$$

so we can find n_{k+1} as required.

Let $N^* = \{n_1, n_2, \ldots\}$. It is easy to see that $\mathcal{N}^*$ is a submodel of $\mathcal{N}$, $\mathcal{N}^* \prec \mathcal{N}$, and the map $g : M \to N$ defined by $g(m_i) = n_i$ is an elementary embedding.

Clearly if $\mathcal{M}$ is isomorphic to an atomic model, then $\mathcal{M}$ is itself atomic. Also the converse is true:

3.3.41. Lemma. Any two countable atomic models (of a complete theory Γ) are isomorphic.

Proof: The proof is similar to the proof that atomic models are prime (3.3.40), using a back-and-forth argument. We leave the details as an exercise.

We will now characterize those complete theories which have atomic models. First, we state a few easy facts:

3.3.42. Fact. Let Γ be a complete theory.

(1) If φ is n-complete, then $\Gamma \nvdash \neg\varphi$, and $\exists x_1 \cdots \exists x_n \varphi \in \Gamma$.
(2) If φ is $n+1$-complete, then $\exists x_{n+1} \varphi$ is n-complete.
(3) In general, if φ is $\{x_1, \ldots, x_n\}$-complete, and $\{x_1, \ldots, x_n\}$ is the disjoint union of the sets $\{y_1, \ldots, y_k\}$ and $\{z_1, \ldots, z_{n-k}\}$, then the formula $\exists y_1 \cdots \exists y_k \varphi$ is complete for $\{z_1, \ldots, z_{n-k}\}$.
(4) $\mathcal{M}$ is atomic if for all finite sets $A \subseteq M$ there exists a finite set $\{m_1, \ldots, m_n\} \supseteq A$ such that $(m_1, \ldots, m_n)$ satisfies some n-complete formula.
(5) Assume that φ_n and φ_{n+1} are n-complete and $n+1$-complete formulas, respectively, and that $\vdash_\Gamma \varphi_{n+1} \rightarrow \varphi_n$. Then $\vdash_\Gamma \varphi_n \rightarrow \exists x_{n+1} \varphi_{n+1}$.

Proof: (1) (Recall that we assume $\varphi \in \Gamma$ whenever $\Gamma \vdash \varphi$, and that Γ is complete.) We must have either $\exists x_1 \cdots \exists x_n \varphi \in \Gamma$ or $\forall x_1 \cdots \forall x_n \neg\varphi \in \Gamma$. The second alternative is impossible, since it would imply $\vdash_\Gamma \neg\varphi$, and hence $\vdash_\Gamma \varphi \rightarrow \psi$ for **all** ψ.

(2) Let ψ be an n-formula. Then ψ is also an $n+1$-formula, so either $\vdash_\Gamma \varphi \rightarrow \psi$ or $\vdash_\Gamma \varphi \rightarrow \neg\psi$. Since x_{n+1} is not free in ψ, we can use the principle of $\exists$-introduction (see 1.4.31), to get $\vdash_\Gamma \exists x_{n+1} \varphi \rightarrow \psi$ or $\vdash_\Gamma \exists x_{n+1} \varphi \rightarrow \neg\psi$.

(3) Similar to (2). We have to use $\exists$-introduction k many times.

(4) This follows from (3): Let $A = \{a_1, \ldots, a_i\} \subseteq \{m_1, \ldots, m_n\}$. So we can write $\{1, \ldots n\}$ as the union of two disjoint sets $\{k_1, \ldots, k_i\}$ and $\{l_1, \ldots, l_j\}$, where $A = \{m_{k_1}, \ldots, m_{k_i}\}$. We know that $(m_1, \ldots m_n)$ satisfies a $\{x_1, \ldots, x_n\}$-complete formula φ. So $(a_1, \ldots, a_i)$ satisfies the $\{x_{k_1}, \ldots, x_{k_i}\}$-complete formula $\exists x_{l_1} \cdots \exists x_{l_j} \varphi$.

(5) Assume not, then we must have $\vdash_\Gamma \varphi_n \rightarrow \forall x_{n+1} \neg\varphi_{n+1}$ and hence $\vdash_\Gamma \varphi_n \rightarrow \neg\varphi_{n+1}$. Hence $\vdash_\Gamma \varphi_{n+1} \rightarrow \neg\varphi_{n+1}$, so $\vdash_\Gamma \neg\varphi_{n+1}$, a contradiction.

3.3.43. Theorem. A complete theory Γ has an atomic model iff for every n, every n-formula which is consistent with Γ is completable (or equivalently, is contained in some atomic n-type).

Proof: First we prove the "easy" direction. Assume Γ has an atomic countable model, and let ψ be an n-formula consistent with Γ. Since we cannot have $\vdash_\Gamma \forall x_1 \cdots \forall x_n \neg\psi$, we must have $\vdash_\Gamma \exists x_1 \cdots \exists x_n \psi$. Let $\mathcal{M}$ be an atomic model of Γ, then there must be $a_1, \dots, a_n \in M$ such that $\mathcal{M} \models \psi(a_1, \dots, a_n)$. Now the type of $(a_1, \dots, a_n)$ contains a complete n-formula φ as well as the formula ψ. Since $\vdash_\Gamma \varphi \to \neg\psi$ is impossible, we have $\vdash_\Gamma \varphi \to \psi$, so φ is indeed a completion of ψ.

Now we prove the converse: Let $\psi_2, \psi_3, \dots$ be a list of formulas (see exercise 28) such that

(1) For all $n \geq 2$, ψ_n is an n-formula.
(2) Whenever ψ is a $k+1$-formula there is some $n \geq k$ such that $\psi(x_{k+1}/x_{n+1}) = \psi_{n+1}$.

Now let $\mathcal{N}$ be a model of Γ. We will construct an atomic model $\mathcal{M} \prec \mathcal{N}$.

We will define a sequence $\varphi_1, \varphi_2, \dots$ of formulas and a sequence $m_1, m_2, \dots$ of elements of N satisfying the following for each n

(a) φ_n is n-complete.
(b) $\mathcal{N} \models \varphi_n(m_1, \dots, m_n)$.
(c) $\mathcal{N} \models \exists x_{n+1} \psi_{n+1}(m_1, \dots, m_n, x_{n+1}) \to \psi_{n+1}(m_1, \dots, m_n, m_{n+1})$.

The inductive construction works as follows:

Step 1: Let φ_1 be any 1-complete formula. Since $\mathcal{N} \models \exists x_1 \varphi_1$ (by 3.3.42(1)), there is an element $m_1 \in N$ such that $\mathcal{N} \models \varphi_1(m_1)$.

Step $n+1$: We already have elements $m_1, \dots, m_n$ and a complete formula φ_n such that $\mathcal{N} \models \varphi_n(m_1, \dots, m_n)$.

Now consider the $n+1$-formula $[\varphi_n \wedge ((\exists x_{n+1}\psi_{n+1}) \to \psi_{n+1})]$. We claim that this formula is consistent with Γ.

[**Proof of the claim:** If not, then we have

1	$\Gamma \cup \{\varphi_n\} \vdash \neg [\varphi_n \wedge ((\exists x_{n+1}\psi_{n+1}) \to \psi_{n+1})]$	
2	$\Gamma \cup \{\varphi_n\} \vdash \varphi_n \to (\exists x_{n+1}\psi_{n+1}) \wedge \neg\psi_{n+1}$	(using tautology axioms)
3	$\Gamma \cup \{\varphi_n\} \vdash (\exists x_{n+1}\psi_{n+1}) \wedge \neg\psi_{n+1}$	(nonlogical axiom + MP)
4	$\Gamma \cup \{\varphi_n\} \vdash (\exists x_{n+1}\psi_{n+1})$	(using tautologies)
5	$\Gamma \cup \{\varphi_n\} \vdash \neg\psi_{n+1}$	(using tautologies, from 3)
6	$\Gamma \cup \{\varphi_n\} \vdash \forall x_{n+1} \neg\psi_{n+1}$	(Generalization theorem)

so (4) and (6) show that $\Gamma \cup \{\varphi_n\}$ is inconsistent, a contradiction. This proves the claim that our $n+1$-formula is consistent with Γ.]

So we can find a complete formula φ_{n+1} such that

$$\vdash_\Gamma \varphi_{n+1} \rightarrow [\varphi_n \wedge ((\exists x_{n+1}\psi_n) \rightarrow \psi_n)] \qquad (*)$$

Since $\vdash_\Gamma \varphi_{n+1} \rightarrow \varphi_n$, we have by 3.3.42(5):

$$\vdash_\Gamma \varphi_n \rightarrow \exists x_{n+1}\, \varphi_{n+1}$$

Since $\mathcal{N} \models \varphi_n(m_1, \ldots, m_n)$ and $\mathcal{N} \models \Gamma$ we get

$$\mathcal{N} \models \exists x_{n+1}\varphi_{n+1}(m_1, \ldots, m_n, x_{n+1})$$

and we can find m_{n+1} satisfying (b). m_{n+1} will also satisfy (c), because of $(*)$.

Now that we have completed the inductive construction, we let $M := \{m_1, m_2, \ldots\}$ and claim the following:

(1) There is a model $\mathcal{M} \subseteq \mathcal{N}$ with universe M.
(2) $\mathcal{M} \prec \mathcal{N}$.
(3) $\mathcal{M}$ is atomic.

We leave the proof of (1) to the reader. To prove (2), we use the Tarski-Vaught criterion 3.1.12.

So assume $\mathcal{N} \models \exists y\theta$, where θ is a formula in the language $\mathcal{L}(M)$. So θ is of the form $\varphi(m_1, \ldots, m_k, y)$ for some $k+1$-formula φ. We have $\mathcal{N} \models \exists x_{k+1}\varphi(m_1, \ldots, m_n, x_{k+1})$. By the definition of our list of formulas ψ_1, ψ_2, ... we can find some n such that $\psi_{n+1} = \varphi(x_{k+1}/x_{n+1})$. Since the variables x_{k+1}, ..., x_n are not free in ψ_n, we have

$$\mathcal{N} \models \exists x_{n+1}\, \psi_{n+1}(m_1, \ldots, m_n, x_{n+1}).$$

So we have $\mathcal{N} \models \psi_{n+1}(m_1, \ldots, m_{n+1})$, and we are done.

(3) follows from (2) and (b), since the φ_n are complete formulas.

We have seen that there are theories which have an atomic model but no saturated countable model, for example, the theory of the natural numbers. Contrast this with the following theorem:

3.3.44. Theorem. If Γ has a saturated countable model, then Γ has an atomic countable model.

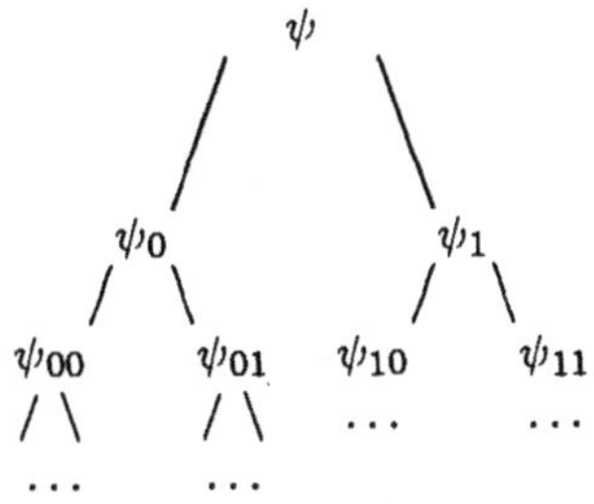

Figure 8.

Proof: (Sketch) Assume that ψ is an incompletable formula consistent with Γ. We will construct a tree of incompletable formulas such that every branch of this tree determines a different (complete) type. (See figure 8.)

Since ψ is incompletable, it is in particular incomplete, so there is an n-formula ψ_0 such that neither $\vdash_\Gamma \psi \to \psi_0$ nor $\vdash_\Gamma \psi \to \neg\psi_0$. Letting $\psi_1 := \neg\psi_0$, we have that both $\{\psi, \psi_0\}$ and $\{\psi, \psi_1\}$ are types consistent with Γ (and inconsistent with each other).

Also $\psi \wedge \psi_0$ is incomplete, so we can find a formula ψ_{00} and a formula $\psi_{01} = \neg\psi_{00}$ such that both $\{\psi, \psi_0, \psi_{00}\}$ and $\{\psi, \psi_0, \psi_{01}\}$ are consistent n-types, and similarly we can find formula ψ_{10} and ψ_{11} to get consistent n-types $\{\psi, \psi_1, \psi_{10}\}$ and $\{\psi, \psi_1, \psi_{11}\}$. We continue by induction.

Finally, every $A \subseteq \{1, 2, 3, \dots\}$ defines an infinite sequence of 0's and 1's and hence a branch through this tree.

(We do not want to formalize this. A hint for a formal proof without too much notation is given in exercise 29.)

For example, if $A = \{1, 3, 4, 6, 7, \dots\}$, then

$A = \{$ 1, 3, 4, 6, 7, $\dots\}$

corresponds to the sequence

1 0 1 1 0 1 1 ...

and thus to the type

$t_A = \{ \psi_1, \psi_{10}, \psi_{101}, \psi_{1011}, \psi_{10110}, \psi_{101101}, \psi_{1011011} \cdots \}$.

Since there are uncountably many sets A, we get uncountably many types.

See exercise 29 for a reformulation of this proof.

3.3.45. Fact. If $\mathcal{M}$ is atomic and saturated, then every countable model is isomorphic to $\mathcal{M}$, and conversely.

Proof: Exercise. (Combine the proofs of "Saturated models are universal" (3.3.29) and "Atomic models are prime" (3.3.40), constructing an isomorphism by a back-and-forth argument.)

ω-Categorical Theories

Theories which (like Th($\mathbb{Q}$)) have a unique countable model are called ω-categorical. In theorem 3.3.47 we will characterize such theories.

3.3.46. Definition. Let Γ be a theory, and φ and ψ n-formulas. We say that φ and ψ are Γ-equivalent iff $\vdash_\Gamma \varphi \leftrightarrow \psi$.

3.3.47. Theorem (Characterization of categorical theories). Let Γ be a complete consistent theory. Then the following statements are equivalent:

(1) All countable models of Γ are isomorphic.
(2) All countable models of Γ are atomic.
(3) All complete types $t \supseteq \Gamma$ are atomic.
(4) For every n there are only finitely many complete n-types $t \supseteq \Gamma$.
(5) For every n there is a list of finitely many n-formulas such that every n-formula is Γ-equivalent to one formula on the list, i.e., there is a natural number k_n and a finite set $\{\varphi_1^n, \dots, \varphi_{k_n}^n\}$ of n-formulas such that for every n-formula ψ there is $i \leq k_n$ such that $\vdash_\Gamma \psi \leftrightarrow \varphi_i^n$.

Proof: It is clear that (1), (2) and (3) are equivalent.

Now assume (3), and fix a natural number n. Every complete n-type $\supseteq \Gamma$ contains a (different) complete formula, so there are either finitely many or countably many such types. Assume, towards a contradiction, that there are infinitely many such types, say t_1, t_2, t_3, ... is a list of all complete types consistent with Γ. For each k, let φ_k be a complete n-formula in t_k. Then we have $\Gamma \vdash \varphi_k \rightarrow \neg\varphi_{k'}$ whenever $k \neq k'$, and hence $\neg\varphi_k \in t_{k'}$.

Now let

$$t^* := \{\neg\varphi_1, \neg\varphi_2, \dots\}.$$

For each k we have that the set $t_k^* := \Gamma \cup \{\neg\varphi_1, \dots, \neg\varphi_k\}$ is consistent, because $t_k^* \subseteq t_{k+1}$.

So t^* is consistent and can be extended to a complete type $\bar{t}^*$. But for all k we have $\varphi_k \in t_k$, $\neg\varphi_k \in \bar{t}^*$, so $\bar{t}^* \neq t_k$, a contradiction.

So (3) implies (4).

Now assume (4), and fix n. Assume there are only k many complete types consistent with Γ, say $\{t_1, \dots, t_k\}$. To each n-formula φ we assign the (finite) set A_φ of all complete types containing φ:

$$A_\varphi := \{t : t \supseteq \Gamma \text{ is a complete } n\text{-type}, \varphi \in t\} \subseteq \{t_1, \dots, t_k\}.$$

There are only at most 2^k many such sets A_φ, since each A_φ is a subset of a set of size k. We will show that if $A_\varphi = A_\psi$, then φ and ψ are Γ-equivalent.
Claim: If $A_\varphi \subseteq A_\psi$, then $\Gamma \vdash \varphi \to \psi$.
Proof of the claim: If $\Gamma \not\vdash \varphi \to \psi$, then $\Gamma \cup \{\varphi, \neg\psi\}$ is a consistent type. So there exists a complete n-type t_i containing this type, i.e., we have $\varphi \in t_i$, $\neg\psi \in t_i$. So $t_i \in A_\varphi$, $t_i \notin A_\psi$, hence $A_\varphi \not\subseteq A_\psi$.

From the claim we conclude that if $A_\varphi = A_\psi$, then φ and ψ are Γ-equivalent. So there is a list of (at most) 2^k many formulas such that every formula is equivalent to a formula on this list.

Finally, we show that (5) implies (3). Let $t \supseteq \Gamma$ be a complete n-type, and let $\varphi_1, \ldots, \varphi_k$ be a list such that every n-formula is equivalent to some φ_i.
For $i \leq k$ define

$$\psi_i := \begin{cases} \varphi_i & \text{if } \varphi_i \in t \\ \neg\varphi_i & \text{if } \varphi_i \notin t, \text{ i.e., } \neg\varphi_i \in t \end{cases}$$

and let $\psi^* := \psi_1 \wedge \cdots \wedge \psi_k$. Then $\psi^* \in t$. Conversely, for any $\varphi \in t$ the formula φ must be Γ-equivalent to some φ_i. Since $\Gamma \cup \{\varphi\} \subseteq t$ we also have $\varphi_i \in t$, so $\varphi_i = \psi_i$, so $\vdash_\Gamma \psi^* \to \varphi$. Hence t is atomic.

We conclude this section by showing the promised theorem (called "Vaught's Never Two" theorem):

3.3.48. Theorem. A complete theory cannot have exactly two nonisomorphic countable models.

Proof: Let Γ be a complete theory.

If Γ has only two (or finitely many) nonisomorphic countable models, then (by 3.3.33) Γ has a saturated countable model $\mathcal{N}$, and by 3.3.44 Γ has an atomic countable model $\mathcal{M}$. $\mathcal{M}$ and $\mathcal{N}$ cannot be isomorphic, because this would imply that Γ has a unique countable model (by 3.3.45).

We will find another countable model which is neither atomic nor saturated, hence not isomorphic to $\mathcal{N}$ and not isomorphic to $\mathcal{M}$.

First we notice that since $\mathcal{N}$ is not atomic, we can find $a_1, \ldots, a_n \in N$ such that $\text{type}_{\mathcal{N}}(a_1, \ldots, a_n)$ contains no n-complete formula.

Consider the language $\bar{\mathcal{L}} = \mathcal{L}(\underline{a}_1, \ldots, \underline{a}_n)$. $\mathcal{N}$ can be expanded to a countable model $\bar{\mathcal{N}}$ of $\mathcal{L}$, by interpreting the constant symbols $\underline{a}_i$ as elements a_i.

Let $\bar{\Gamma} := \text{Th}_{\bar{\mathcal{L}}}(\bar{\mathcal{N}})$. Then $\bar{\Gamma} \cap \mathcal{L} = \Gamma$, and $\bar{\mathcal{N}}$ is a saturated countable model of $\bar{\Gamma}$ (by 3.3.28).

Hence, there exists an atomic countable model $\bar{\mathcal{M}}$ of $\bar{\Gamma}$. We claim

(1) $\bar{\mathcal{M}}$ is not saturated.
(2) $\bar{\mathcal{M}} \restriction \mathcal{L}$ is not saturated (hence not isomorphic to $\mathcal{N}$).
(3) $\bar{\mathcal{M}} \restriction \mathcal{L}$ is not atomic (hence not isomorphic to $\mathcal{M}$).

Proof of (1): Since Γ has more than one countable model, there exists an n such that there are infinitely many n-formulas in $\mathcal{L}$ which are pairwise not Γ-equivalent. But if $\Gamma \not\vdash \psi \leftrightarrow \varphi$, then Γ (being a complete theory) proves the negation of the universal closure of $\psi \leftrightarrow \varphi$, so also $\bar{\Gamma} \not\vdash \psi \leftrightarrow \varphi$, so all these formulas are also pairwise not $\bar{\Gamma}$-equivalent. So $\bar{\Gamma}$ has at least two nonisomorphic countable models, and thus $\bar{\mathcal{M}}$, the atomic model, cannot be saturated, by 3.3.45.

Proof of (2): by 3.3.28.

Proof of (3): Note that although $\text{type}_{\bar{\mathcal{M}}}(a_1^{\bar{\mathcal{M}}}, \ldots, a_n^{\bar{\mathcal{M}}})$ is atomic (over the theory $\bar{\Gamma}$, because it contains the atomic formula $\mathrm{x}_1 = \underline{a}_1 \wedge \cdots \wedge \mathrm{x}_n = \underline{a}_n$), the type

$$\text{type}_{\bar{\mathcal{M}} \restriction \mathcal{L}}(a_1^{\bar{\mathcal{M}}}, \ldots, a_n^{\bar{\mathcal{M}}})$$

(which is $= \text{type}_{\mathcal{N}}(a_1, \ldots, a_n)$) is **not** atomic.

Exercises

Types

1. Show that $\text{type}_{\mathcal{M}}(a)$ is always a complete type.

2. Prove 3.3.7(0).

3. Assume that $\mathcal{M}$ is a model in which each element is of the form $\tau^{\mathcal{M}}$, for some closed term τ. [What is an example for such a model?] Find a type $t \supseteq \text{Th}(\mathcal{M})$ which is omitted by $\mathcal{M}$.

Examples

4. Let DLO^- be the theory containing only the following formulas:
(1a) $\forall \mathrm{x} \neg \mathrm{x} < \mathrm{x}$.
(1b) $\forall \mathrm{x} \forall \mathrm{y} \forall \mathrm{z} (\mathrm{x} < \mathrm{y} < \mathrm{z} \to \mathrm{x} < \mathrm{z})$.
(1c) $\forall \mathrm{x} \forall \mathrm{y} (\mathrm{x} = \mathrm{y} \vee \mathrm{x} < \mathrm{y} \vee \mathrm{y} < \mathrm{x})$.
(2) $\forall \mathrm{x} \exists \mathrm{y}\, \mathrm{x} < \mathrm{y}$.
(3) $\forall \mathrm{x} \forall \mathrm{y} : (\mathrm{x} < \mathrm{y} \to \exists \mathrm{z} : \mathrm{x} < \mathrm{z} < \mathrm{y})$. Show that DLO^- has exactly two nonisomorphic countable models. Does this contradict theorem 3.3.48?

5. Let Γ be a consistent theory all of whose countable models are isomorphic. Show that Γ is complete.

6. Which of the models $\mathcal{M}_1$, $\mathcal{M}_2$, $\mathcal{M}_3$ in 3.3.20 is atomic? Which is saturated? For the two models that are not saturated, find 1-types which are omitted. For the two models that are not atomic, find elements which do not satisfy complete formulas.

7. Find a complete theory with exactly four nonisomorphic countable models. (Hint: Use DLO^+, and add a unary predicate P to the language. Add axioms $P(c_n)$, and add axioms to ensure that the elements satisfying P as well as the elements not satisfying P are dense.)

8. Consider the language which has only countably many constant symbols $c_1, c_2, \ldots$ (no function or relation symbols). Let $\Gamma = \{c_1 \neq c_2, c_2 \neq c_3, c_1 \neq c_3, \ldots\}$.
(1) Show that Γ has countably many nonisomorphic countable models. (Hint: For any model $\mathcal{M}$, consider the set $M - \{c_1^{\mathcal{M}}, c_2^{\mathcal{M}}, \ldots\}$.)
(2) Show that there is a countable model of Γ into which all countable models can be elementarily embedded.
(3) Conclude that Γ is a complete theory.
(4) Describe the atomic and the saturated countable model of Γ.

9. Let $\Gamma = DLO^+$. Let $\mathcal{N}$ be a saturated countable model of Γ. Find an element a of N whose type is not atomic. Let $\bar{\mathcal{N}}$ and $\bar{\Gamma}$ be defined as in 3.3.48, and find an atomic countable model of $\bar{\Gamma}$.

10. Complete the proof of 3.3.21. That is, show that $\mathcal{M}_1' \prec \mathcal{M}_2$.

11. Let $\mathcal{L}_0$ be a language with only one binary relation symbol $<$. Let $\mathcal{L} \supset \mathcal{L}_0$ be the language that in addition has countably many constant symbols $c_1, c_2, \ldots$. Let Γ_0 be DLO, the theory of dense linear orders. Find a complete theory $\Gamma \supseteq \Gamma_0$ in the language $\mathcal{L}$ which has no countable saturated model.

Discrete Linear Orders

12. This exercise and the following refer to the theory DisLO. The theory DisLO (discrete linear orderings) has the following axioms:
(1a) $\forall x \neg x < x$.
(1b) $\forall x \forall y \forall z (x < y < z \rightarrow x < z)$. (We write $x < y < z$ to abbreviate $(x < y \wedge y < z)$.)

(1c) $\forall x \forall y(x = y \vee x < y \vee y < x)$.
(2a) $\forall x \exists y\,(x < y \wedge \neg\exists z(x < y < z)$.
(2b) $\forall x \exists y\, y < x \wedge \neg\exists z(y < z < x)$. Axioms (2a)–(2b) ensure that every element has an immediate successor and an immediate predecessor.
(a) Show that for any linear order $(A, <)$ the model $A \times \mathbb{Z}$ with the lexicographic order (i.e., $(a, z) < (a', z')$ iff $a < a'$ or $a = a' \wedge b < b'$) satisfies DisLO. ($\mathbb{Z}$ is the set of all integers.)
(b) Show that every model of DisLO is isomorphic to such a model $A \times \mathbb{Z}$, for some A.

13. We will write $x \leq y$ to abbreviate $x < y \vee x = y$. Define the formula $D(x, y)$ ("x is a direct successor of y") by

$$D(x,y) \;=\; (x < y \wedge (\neg\exists z : x < z < y)).$$

Let $D_0(x, y)$ be the formula $x = y$, and $D_{n+1}(x, y) = \exists z(D_n(x, z) \wedge D(z, y)$. Thus, $D_n(x, y)$ says that x is exactly n places before x.

Let $E_n(x, y)$ be the formula $\exists z\, D_n(x, z) \wedge z \leq y$, i.e., y is at least n places above x.

Show the following: If x, y, z are distinct variables, then $E_n(y, z)$ is (modulo DisLO) equivalent to:

$$\begin{aligned} & E_0(x,y) \wedge E_n(y,z) \\ \vee\ & D_1(y,x) \wedge E_{n-1}(x,z) \\ \vee\ & \quad \cdots \\ \vee\ & D_{n-1}(y,x) \wedge E_1(x,z) \\ \vee\ & E_n(y,x) \wedge E_0(x,z) \\ \vee\ & E_n(y,z) \wedge E_0(z,x). \end{aligned}$$

14. Let A be a set of at least 2 variables. We call a formula φ a "constellation for A" for A iff A can be enumerated (without repetitions) as $A = \{y_0, \ldots, y_n\}$, and φ is of the form

$$\varphi_1(y_0, y_1) \wedge \cdots \wedge \varphi_n(y_{n-1}, y_n)$$

and each φ_i is of the form E_{k_i} or D_{k_i} for some k_i.

If $A = \{x\}$, then the only constellation for A is the formula $x = x$.

Show the following:
(a) If φ is a constellation of A, and x is a variable not in A, then there is a finite disjunction of constellations ψ on $A \cup \{x\}$ such that DisLO$\vdash \varphi \leftrightarrow \psi$. (Use the previous exercise.)

(b) If φ_1 and φ_2 are two constellations on A, then
either DisLO$\vdash \neg(\varphi_1 \wedge \varphi_2)$,
or there is a constellation ψ such that DisLO$\vdash \psi \leftrightarrow \varphi_1 \wedge \varphi_2$.
(c) If φ is a constellation, then either $\neg\varphi$ is equivalent to a finite disjunction of constellations, or DisLO$\vdash \varphi$.
(d) Every formula which is consistent with DisLO is (modulo DisLO) equivalent to a finite disjunction of constellations, i.e., for every formula φ, either DisLO$\vdash \neg\varphi$, or there is a finite disjunction ψ of constellations such that DisLO$\vdash \varphi \leftrightarrow \psi$.
(d) Conclude that DisLO s a complete theory. (Use induction. The $\forall$x step can be reduced to the $\exists$ step, by (c).)

15. Show that $\mathbb{Z}$ can be elementarily embedded into every model of DisLO.

16.
(a) Let A be a linear order, $A' \subseteq A$. Show that $A' \times \mathbb{Z} \prec A \times \mathbb{Z}$. (See exercise 12.)
(b) Find two **nonisomorphic** models $\mathcal{M}_1$, $\mathcal{M}_2$ of DisLO such that $\mathcal{M}_1$ can be elementarily embedded into $\mathcal{M}_2$ and conversely.

17. Show that $\mathbb{Z}$ is the atomic model for DisLO. Does DisLO have a saturated countable model?

Saturated Models

18. Show: If $\mathcal{M}$ is an infinite model such that every element of M is of the form $\tau^{\mathcal{M}}$ for some closed term τ, then $\mathcal{M}$ cannot be saturated.

19. To show that the assumptions in 3.3.28 were necessary, find languages $\mathcal{L}_1 \subseteq \mathcal{L}_2$ and a nonsaturated model $\mathcal{N}$ for the language $\mathcal{L}_2$ such that $\mathcal{N} \upharpoonright \mathcal{L}_1$ is saturated, where
(a) $\mathcal{L}_1$ has the same function and relation symbols as $\mathcal{L}_2$. Repeat the exercise replacing condition (a) by
(b) $\mathcal{L}_2$ has the same relation and constant symbols as $\mathcal{L}_2$ and only one additional function symbol. Finally, repeat the exercise with the following condition:
(c) $\mathcal{L}_2$ has the same function and constant symbols as $\mathcal{L}_2$ and only one additional relation symbol.

20. Complete the proof of 3.3.29.

21. Show that any two countable saturated models are isomorphic. (Hint: If $f : \mathcal{M} \to \mathcal{N}$ is an isomorphism, then for all $a_1, a_2, \ldots \in \mathcal{M}$ we have

$$\begin{aligned}\mathrm{type}_{\mathcal{M}}(a_1) &= \mathrm{type}_{\mathcal{N}}(f(a_1))\\ \mathrm{type}_{\mathcal{M}_{\{a_1\}}}(a_2) &= \mathrm{type}_{\mathcal{N}_{\{f(a_1)\}}}(f(a_2))\\ &\vdots\end{aligned}$$

Use a back-and-forth argument.)

22. Find nonisomorphic countable models $\mathcal{M}$ and $\mathcal{N}$ such that $\mathcal{M}$ can be elementarily embedded into $\mathcal{N}$, and $\mathcal{N}$ can be elementarily embedded into $\mathcal{M}$.

23. Prove fact 3.3.31: If $\mathcal{M} \prec \mathcal{N}$, $A \subseteq M$, then also $\mathcal{M}_A \prec \mathcal{N}_A$.

Atomic Models

24. Show that every atomic type is a complete type.

25. Let $\Gamma = \mathrm{Th}(\mathbb{N})$, and let t be an atomic 1-type. Show that for some n, the formula $\mathtt{S}^n\mathtt{0} = \mathtt{x}_1$ is in t. (Hint: Recall that $\mathbb{N}$ is an atomic model.) Use this fact to show 3.3.36(3).

26. Let Γ be a theory with an infinite model. Let $\mathcal{M}$ be a model. Show that the following are equivalent:

(1) $\mathcal{M}$ is a prime model of Γ, i.e., $\mathcal{M}$ can be elementarily embedded into every model of Γ.

(2) $\mathcal{M}$ can be elementarily embedded into every countable model of Γ.

Show that the above conditions imply that Γ is complete.

27. Fix a theory Γ which has a countable atomic model. Show that any two countable atomic models are isomorphic. (Hint: If $f : \mathcal{M} \to \mathcal{N}$ is an isomorphism, then for all $a \in \mathcal{M}$ we have

$$\mathrm{type}_{\mathcal{M}}(a) = \mathrm{type}_{\mathcal{N}}(f(a)).$$

In particular, if for some complete formula φ we have $\mathcal{M} \models \varphi(a)$, then $\mathcal{N} \models \varphi(f(a))$. Use a back-and-forth argument to construct f.)

28. Fix any language $\mathcal{L}$. Find a list $\psi_2, \psi_3, \ldots$ of formulas such that
(1) For all $n \geq 2$, ψ_n is an n-formula.
(2) Whenever ψ is a $k+1$-formula then there is some n such that the formula $\psi(\mathbf{x}_{k+1}/\mathbf{x}_{n+1})$ is equal to φ_{n+1}.

29. (Alternative proof of 3.3.44) Fix a complete theory Γ, and assume that $t_1, t_2, \ldots,$ is a finite or countable set of complete n-types consistent with Γ. Assuming ψ is an incompletable n-formula, construct a (complete) type t^* such that $t^* \neq t_i$ for all i.
(Hint: Construct $t := \{\psi, \psi_1, \psi_2, \ldots\}$ by induction such that for all n:
(1) $\neg\psi_n \in t_n$.
(2) $\{\psi, \psi_1, \ldots, \psi_n\}$ is consistent with Γ. Then let t^* be a complete type extending t. Conclude that if a complete theory Γ has a saturated countable model, then Γ has an atomic countable model.)

30. Prove theorem 3.3.45: If $\mathcal{M}$ is atomic and saturated, then every countable model is isomorphic to $\mathcal{M}$, and conversely.

ω-Categorical Theories

31. Show 3.3.47(1) $\leftrightarrow$ 3.3.47(2) $\leftrightarrow$ 3.3.47(3).

32. Let Γ be DLO. How many complete 1-types are there? How many nonequivalent 1-formulas are there? How many complete 2-types? How many nonequivalent 2-formulas?

Chapter 4

The Incompleteness Theorem

4.0. Introduction

In this chapter we will present one of the most surprising and interesting theorems proved in this century. This theorem has deep influences in most branches of mathematics, as well as consequences for the epistemology and the methodology of science.

In a first approach, we can paraphrase this theorem by saying that the usual axioms for number theory, introduced by Peano and Dedekind (see section 4.2) are an incomplete axiomatic system. This means that there is a sentence φ in the language of number theory such that neither φ nor $\neg\varphi$ are provable from the Peano axioms.

This result in itself is very surprising if you know that all the usual results obtained in number theory are consequences of the Peano axioms. In other words, Peano Arithmetic is a very powerful set of axioms from the point of view of ordinary mathematics, and it was very natural to think that all possible questions about the natural numbers could be decided in this system.

But if we look more closely at the proof of the theorem, we notice that any "possible" axiomatic system for the natural numbers must be incomplete. We mean here by "possible" that the set of axioms must be, if not finite, at least a decidable set, or in other words, that there should be a finitary method that decides whether a given formula is an axiom or not.

As a consequence, we also see that the usual system of axioms for set theory is incomplete. In fact, any axiom system which is rich enough to simulate the number theoretic machinery given in the Peano system will be incomplete.

The central idea of the proof of the incompleteness theorem (proved by Kurt Gödel in 1931) is the following:

As we saw in the previous chapter, we can have a 1-1 correspondence between the symbols of an enumerable language $\mathcal{L}$ and the natural numbers.

Now we can use the structure of the Peano system to encode all the symbols, terms, formulas, and sequences of formulas of our language into natural numbers, and we can also encode the concept "$\varphi_1, \ldots, \varphi_n$ is a derivation from the Peano axioms." That is, we can find a formula $\varphi(\mathbf{x})$ such that $\varphi(n)$ expresses the fact that n encodes a sequence of formulas which is a derivation.

This process of encoding is the main idea in Gödel's proof of the incompleteness theorem. Given the encoding, the concepts described in the previous chapters, such as "free variable," "substitution," etc., are **definable** in the Peano arithmetical system.

Gödel's proof finishes by defining a formula φ that can be interpreted to mean "the formula φ is not provable in Peano Arithmetic." This implies that φ is not provable in Peano Arithmetic, and also that the negation of φ is not provable.

There is another possible interpretation of this incompleteness. We can view a computer as a logical system. An actual computer is always finite, so in principle we can know exactly how the computer behaves, i.e., we could make a finite list that describes its action on any program.

But we can also imagine an "ideal" computer, with potentially infinite storage space. Gödel's incompleteness theorem allows us to find a mathematical problem that this computer cannot solve.

In this chapter we will prove the weakest form of Gödel's incompleteness theorem. We will use much stronger assumptions than were used in the original proof. But the main idea of Gödel remains in this approach, which we think allows for a more readily grasped presentation of the incompleteness theorem.

The main difference is that we assume that the standard model for Peano arithmetic exists. This model is the set of natural numbers. We use this in order to give a semantical argument to show that PA proves φ. So we can avoid giving formal proofs in PA.

4.1. The Language of Peano Arithmetic

Peano Arithmetic (PA) is an axiomatic system that was developed to give formal proofs (derivations) of theorems about natural numbers. In this section we describe the language of PA.

Here is a picture of the natural numbers:

$$\begin{array}{ccccccccc} \cdot & \longrightarrow & \cdot & \longrightarrow & \cdot & \longrightarrow & \cdot & \longrightarrow & \cdots \\ 0 & & 1 & & 2 & & 3 & & \cdots \end{array}$$

There is a smallest number, 0, and after each number there is a next number, called its "successor." To reflect these notions within PA, we will have a constant symbol $\mathtt{0}$ and a one place function symbol $\mathtt{S}$.

4.1.1. Definition. $\mathtt{0}$ is a symbol for a constant. $\mathtt{S}$ is a symbol for a one place function.

There is no need for additional constants. For example, instead of having a constant for the number 1, we can use the term $\mathtt{S(0)}$, abbreviated $\mathtt{S0}$. For 2 we use $\mathtt{SS0}$, for 16 we use $\mathtt{SSSSSSSSSSSSSSSS0}$, etc.

It is important to notice that we will work on two different levels. One level is the common language, the language that we usually speak. In this language we have words for zero (0), one (1), etc. There is another level, the formal language of PA. This is the language defined in 4.1.1.

When you and I talk about PA, we use the first language. All the abbreviations are in the first language and they DENOTE elements of the second language.

4.1.2. Definition. Whenever n is a natural number, $\underline{n}$ is the term $\mathtt{S}\cdots\mathtt{S0}$, that contains exactly n $\mathtt{S}$'s. So $\underline{0} = \mathtt{0}$, $\underline{1} = \mathtt{S(0)}$, $\underline{2} = \mathtt{S(S(0))}$, etc.

Strictly speaking, we define a function t that maps (our) natural numbers to terms, by

$$t(0) = \mathtt{0}.$$
$$t(n+1) = \mathtt{S}(x), \text{ where } x = t(n),$$

and we will write $\underline{n}$ instead of $t(n)$.

There are three natural operations associated with the natural numbers. These are addition, multiplication and exponentiation. We will introduce a function symbol for each of these operations.

4.1.3. Definition. Add, Mult, Exp are two place function symbols in the language of PA.

The intended interpretation of these functions are as addition, multiplication, and exponentiation. Conforming to common usage, we will write $(\mathtt{x}+\mathtt{y})$, $(\mathtt{x}\cdot\mathtt{y})$ and $(\mathtt{x}^{\mathtt{y}})$ (or sometimes $\mathtt{x}\uparrow\mathtt{y}$) instead of Add(x, y), Mult(x, y), and Exp(x, y), respectively.

Thus, our language contains the nonlogical symbols 0, S, Add, Mult, and Exp.

Furthermore, we will have the equality symbol =, the logical connectives $\neg$ and $\wedge$, and the universal quantifier $\forall$ in our language, as well as a set $\mathtt{x}_0$, $\mathtt{x}_1$, ... of individual variables.

$A \vee B$ abbreviates $\neg(\neg A \wedge \neg B)$, $A \to B$ abbreviates $\neg(A \wedge (\neg B))$, $\exists\mathtt{x}(\cdots)$ stands for $\neg\forall\mathtt{x}\neg(\cdots)$, etc.

To make our formulas more readable, we will also use other letters for variables, such as $\mathtt{p}, \mathtt{q}, \mathtt{e}, \mathtt{x}, \mathtt{y}, \ldots$. It is understood that all these are really different variables, and that bound variables are always renamed to avoid clashes.

[E.g., if $\varphi = \varphi(\mathtt{x})$ is the formula $\exists\mathtt{y}(\mathtt{x} = \mathtt{Sy})$, then $\varphi(\mathtt{z})$ is the formula $\exists\mathtt{y}(\mathtt{z} = \mathtt{Sy})$, but $\varphi(\mathtt{y})$ is the formula $\exists\mathtt{t}(\mathtt{y} = \mathtt{St})$.]

The Standard Model

There are many models for the language given above. (For example, there is a model with only one element.) However, there is a canonical model that we had in mind when we described the language, and that we will also use to guide us when we formulate the axioms of PA in the next section: The natural numbers $\mathbb{N}$.

The universe of $\mathbb{N}$ is the set $\{0, 1, 2, \ldots\}$. The interpretation of 0 is the number 0, the interpretation of S is the function s that maps each number n to the number $n+1$, and the interpretations of Add, Mult, Exp are the addition (+), multiplication ($\cdot$) and exponentiation ($\uparrow$) of natural numbers.

We write $\mathbb{N}$ for the set $\{0, 1, 2, \ldots\}$ as well as the model $(\mathbb{N}, 0, s, +, \cdot, \uparrow)$.

Exercises

1. How many nonisomorphic models with exactly two elements are there for the language containing as the only nonlogical symbol the symbol S? For the language containing only Add? For the full language of PA?

4.2. The Axioms of Peano Arithmetic

In this section we give a formal axiomatic development of the arithmetic of natural numbers.

We begin by listing the axioms of PA.

PA1. $\forall x[Sx \neq 0]$.

PA2. $\forall x \forall y[Sx = Sy \rightarrow x = y]$.

Note that any model satisfying these two axioms must be infinite.

PA3. $\forall x[x + 0 = x]$.

PA4. $\forall x \forall y[x + Sy = S(x + y)]$.

These two axioms are sufficient to define addition, as the following examples show:

4.2.1. Example. $\vdash_{PA} \underline{2} + \underline{2} = \underline{4}$.

Proof:

1 $\vdash_{PA} SS0 + SS0 = S(SS0 + S0)$ (PA4, plus substitution axiom)
2 $\vdash_{PA} SS0 + S0 = S(SS0 + 0)$ (PA4, plus substitution axiom)
3 $\vdash_{PA} SS0 + 0 = SS0$ (PA3, plus substitution axiom)
4 $\vdash_{PA} S(SS0 + 0) = S(SS0)$ (from 3, using an equality axiom)
5 $\vdash_{PA} SS0 + S0 = S(SS0)$ (from 4 and 2, using equality axioms)
6 $\vdash_{PA} S(SS0 + S0) = S(S(SS0))$ (from 5, using equality axiom)
7 $\vdash_{PA} SS0 + SS0 = S(S(SS0))$ (from 6 and 1, using equality axiom).

4.2.2. Example. In general, for all natural numbers m, n we have

$$\vdash_{PA} \underline{m} + \underline{n} = \underline{m+n} \ldots$$

Proof: We fix m and use induction on n. Clearly

$$\vdash_{PA} \underline{m} + \underline{0} = \underline{m+0}$$

by axiom PA3.

Now note that $\underline{n+1}$ is $S(\underline{n})$, so assuming $\vdash_{PA} \underline{m} + \underline{n} = \underline{m+n}$,

1 $\vdash_{PA} \underline{m} + \underline{n} = \underline{m+n}$
2 $\vdash_{PA} \underline{m} + \underline{n+1} = S(\underline{m} + \underline{n})$ (by PA4)
3 $\vdash_{PA} S(\underline{m} + \underline{n}) = S(\underline{m+n})$ (by 1, using equality axioms)
4 $\vdash_{PA} \underline{m} + \underline{n+1} = S(\underline{m+n})$ (again using equality axioms)

so we are done.

Similarly, the following four axioms are sufficient to define multiplication and exponentiation:

PA5. $\forall x[x \cdot 0 = 0]$.

PA6. $\forall x \forall y[x \cdot Sy = (x \cdot y) + x]$.

PA7. $\forall x[x^0 = S(0)]$.

PA8. $\forall x \forall y[x^{S(y)} = x^y \cdot x]$.

We leave it to the reader to verify the following fact:

4.2.3. Fact. PA1–PA8 are valid in $\mathbb{N}$.

The axioms so far are sufficient for any arithmetical calculation. However, they are not sufficient to determine the structure of $\mathbb{N}$, as the following example shows:

4.2.4. Example. Let $\mathbb{N}$ be the set of natural numbers, and let A be any nonempty set. Fix an element $a_0 \in A$. Define a model $\mathcal{M}$ as follows:

The universe of $\mathcal{M}$ is the set $\mathbb{N} \times A$. So elements of our model are pairs $\langle n, a \rangle$, where $n \in \mathbb{N}$ and $a \in A$.

$0^{\mathcal{M}} = \langle 0, a_0 \rangle$.

The interpretation of the successor function is the function that maps $\langle n, a \rangle$ to $\langle n+1, a \rangle$.

$\langle n, a \rangle +^{\mathcal{M}} \langle n', a' \rangle = \langle n + n', a \rangle$.

Multiplication and exponentiation are defined similarly.

Then, $\mathcal{M}$ always satisfies the axioms PA1–PA8, but whenever A has more than one element, $\mathcal{M}$ is not isomorphic to $\mathbb{N}$ — in fact addition will not even be commutative.

Proof: Exercise.

For this reason, we introduce the **induction axioms**.

Any model $\mathcal{M}$ satisfying PA1 and PA2 must contain the infinite set $\{0^{\mathcal{M}}, (S0)^{\mathcal{M}}, \dots\}$, but we would also like to have an axiom that says that **every** element of the model is in this set. One approach would be to admit formulas of infinite length. Then we could postulate

$$\forall x(x = 0 \lor x = S0 \lor x = SS0 \lor \cdots).$$

Another approach would be to admit quantifiers ranging not only over elements, but also over subsets (so-called second order quantifiers), and enrich our language by the binary $\in$ relation symbol. Then we could write

$$\forall A\, [0 \in A \land \forall x\, (x \in A \rightarrow Sx \in A) \quad \rightarrow \quad \forall x\, (x \in A)]$$

We will not use either of these approaches, because there is no satisfactory proof system for either infinitary languages (as in the first approach) or second order languages (as in the second approach).

However, the second approach can be modified as follows: instead of all sets A we consider only those sets defined by formulas.

4.2.5. Definition. For any formula $\varphi(\mathrm{x})$ with free variable x, let IND_φ be the formula

$$\varphi(0) \wedge \forall \mathrm{x}(\varphi(\mathrm{x}) \rightarrow \varphi(\mathrm{Sx})) \quad \rightarrow \quad \forall \mathrm{x} \varphi(\mathrm{x}).$$

We can also formulate a more general version of IND_φ:

4.2.6. Definition. If $\varphi(\mathrm{x}_0, \ldots, \mathrm{x}_n)$ is a formula with free variables $\mathrm{x}_0, \ldots, \mathrm{x}_n$, we let IND_φ be the formula

$$\forall \mathrm{x}_1 \cdots \forall \mathrm{x}_n \Big[\varphi(0, \mathrm{x}_1, \ldots, \mathrm{x}_n) \wedge \forall \mathrm{x}_0 \Big(\varphi(\mathrm{x}_0, \mathrm{x}_1, \ldots, \mathrm{x}_n) \rightarrow \varphi(\mathrm{Sx}_0, \mathrm{x}_1, \ldots, \mathrm{x}_n) \Big) \\ \rightarrow \quad \forall \mathrm{x}_0 \varphi(\mathrm{x}_0, \mathrm{x}_1, \ldots, \mathrm{x}_n) \Big]$$

(or more formally, $\forall \mathrm{x}_1 \cdots \forall \mathrm{x}_n \big[\varphi(\mathrm{x}_0/0) \wedge \forall \mathrm{x}_0 \big(\varphi \rightarrow \varphi(\mathrm{x}_0/\mathrm{Sx}_0) \big) \rightarrow \forall \mathrm{x}_0 \varphi \big]$).

Now we are ready to formulate the last axiom of Peano Arithmetic. Actually, it is not a single axiom, but a list of infinitely many axioms.

4.2.7. Definition. PA9 is the list of all axioms IND_φ, for all formulas φ.

We will write PA^- for the (finite) set of axioms PA1–PA8, and we will write PA for the (infinite) set of all axioms in PA1–PA9.

4.2.8. Fact. $\mathbb{N}$ satisfies PA9 (and hence all of PA).

Exercises

1. Prove that any model for the language $\{0, S\}$ satisfying PA1 and PA2 must be infinite.

2. Give two models for the language $\{0, S\}$ satisfying PA1 and PA2 that are not isomorphic.

3. Show that there are two nonisomorphic models for the language $\{0, S\}$ with exactly two elements. Show that any two models with exactly two elements that satisfy PA1 must be isomorphic.

4. Show that $\vdash_{PA} \underline{k{\cdot}m} = \underline{n}$ iff $k{\cdot}m = n$ and $\vdash_{PA} \underline{k^m} = \underline{n}$ iff $k^m = n$

5. Prove that PA1–PA8 are valid in $\mathbb{N}$.

6. Show that if the set A contains only one element, then the model $\mathcal{M}$ from 4.2.4 is isomorphic to $\mathbb{N}$. Show that if A contains more than one element, addition in $\mathcal{M}$ is not commutative. (You can even find an element $m \in \mathcal{M}$ such that $\mathcal{M} \models 0 + m \neq m$.)

7. Show that the integer numbers $\mathbb{Z} = \{\ldots, -2, -1, 0, 1, 2, \ldots\}$, with the usual interpretation of 0, successor, addition and multiplication satisfy PA1–PA6. Find a formula φ such that $\mathbb{Z} \not\models IND_\varphi$.

8. Show that the nonnegative real numbers $\mathbb{R}_0^+$, with the usual interpretation of 0, successor ($Sx = x + 1$), addition, multiplication and exponentiation (where $0^0 = 0$) satisfy PA$^-$. Find a formula φ such that $\mathbb{R} \not\models IND_\varphi$.

9. Let $\mathcal{M} = (M, 0, s)$ be a model for the language $\{0, S\}$ satisfying PA1 and PA2, and assume that for any subset $A \subseteq M$:

> If A contains 0 and is closed under the successor function s (i.e., $a \in A$ implies $sa \in A$ for any a),
> then $A = M$.

Show that there is a unique isomorphism $f : \mathbb{N} \to \mathcal{M}$.

10. Show that for any formula φ with one free variable, $\mathbb{N} \models IND_\varphi$.

11. Show that for all φ, $\mathbb{N} \models IND_\varphi$.

12. (a) Let $\psi(x_0)$ be the formula $0 + x_0 = x_0$. Show that

$$\vdash_{PA^-} IND_\varphi \to \forall x\, 0 + x = x.$$

(b) Let $\varphi(x_0, x_1)$ be the formula $x_0 + x_1 = x_1 + x_0$. Show that

$$\vdash_{PA^-} IND_\varphi \to \forall x \forall y (x + y = y + x).$$

13. Show that $\mathbb{N}$ satisfies PA9.

4.3. Basic Theorems of Number Theory in PA

We will give examples of derivations in PA. First recall the following facts about derivations: (We write φ and ψ for formulas, Γ for any set of closed formulas.)

1. The deduction theorem (1.4.23):
If $\Gamma \cup \{\varphi\} \vdash \psi$, then $\Gamma \vdash (\varphi \to \psi)$.

2. The generalization theorem (1.4.27):
If $\Gamma \vdash \varphi$, and x is a variable not occurring free in any formula in Γ, then also $\Gamma \vdash \forall x\, \varphi$.

3. Introduction of existential quantifiers (1.4.31): (a): If $\Gamma \vdash \varphi \to \psi(x/\tau)$, where x is a variable and τ is a term (that may be substituted for x in φ), then $\Gamma \vdash \varphi \to \exists x \psi$.

(b) If $\Gamma \vdash \varphi \to \psi$, and x is a variable that is **not free in** ψ, then also $\Gamma \vdash \exists x\, \varphi \to \psi$.

4.3.1. Example. Every nonzero number has a predecessor:

$$\vdash_{PA} \forall x (x = 0 \lor \exists t\, x = St).$$

Proof: We will use "induction on x" for the formula $\varphi(x) = \forall x (x = 0 \lor \exists t\, x = St)$. That is, we will invoke the induction axiom once we have shown $\vdash_{PA} 0 = 0 \lor \exists t\, St = 0$ and $\vdash_{PA} x = 0 \lor \exists t\, St = x \to Sx = 0 \lor \exists t\, St = Sx$. Both are easy, since $\vdash_{PA} 0 = 0$ and $\vdash_{PA} Sx = Sx \to \exists t St = Sx$.

4.3.2. Example. $\vdash_{PA} x + r = 0 \to r = 0$

We will use a proof "by cases," according to whether $\mathtt{r}$ is 0 or a successor. Formally, we will prove $\vdash_{PA} \mathtt{r} = 0 \to (\mathtt{x} + \mathtt{r} = 0 \to \mathtt{r} = 0)$ (which is a tautology) and

$$(\star) \qquad \vdash_{PA} \exists \mathtt{t}\, \mathtt{r} = \mathtt{St} \to (\mathtt{x} + \mathtt{r} = 0 \to \mathtt{r} = 0)$$

and then use 4.3.1 and tautological reasoning to get the desired result.

1	$\mathtt{r} = \mathtt{St} \vdash_{PA} \mathtt{x} + \mathtt{r} = \mathtt{x} + \mathtt{St}$	(equality axioms)
2	$\mathtt{r} = \mathtt{St} \vdash_{PA} \mathtt{x} + \mathtt{r} = \mathtt{S}(\mathtt{x} + \mathtt{t})$	(PA4)
3	$\mathtt{r} = \mathtt{St} \vdash_{PA} \mathtt{S}(\mathtt{x} + \mathtt{t}) \neq 0$	(PA1)
4	$\mathtt{r} = \mathtt{St} \vdash_{PA} \mathtt{x} + \mathtt{r} \neq 0$	(2, 3 and equality axioms)
5	$\mathtt{r} = \mathtt{St} \vdash_{PA} \mathtt{x} + \mathtt{r} = 0 \to \mathtt{r} = 0$	(tautological reasoning)
6	$\vdash_{PA} \mathtt{r} = \mathtt{St} \to (\mathtt{x} + \mathtt{r} = 0 \to \mathtt{r} = 0)$	(deduction)
7	$\vdash_{PA} (\exists \mathtt{t}\, \mathtt{r} = \mathtt{St}) \to (\mathtt{x} + \mathtt{r} = 0 \to \mathtt{r} = 0)$	(∃-introduction)

4.3.3. Definition. We write $\mathtt{x} \leq \mathtt{y}$ for $\exists \mathtt{r}\, \mathtt{x} + \mathtt{r} = \mathtt{y}$, and $\mathtt{x} < \mathtt{y}$ for $\mathtt{x} \leq \mathtt{y} \land \mathtt{x} \neq \mathtt{y}$.

4.3.4. Example. The following two theorems characterize the relation $\leq$:
(1) $\vdash_{PA} \mathtt{x} \leq 0 \leftrightarrow \mathtt{x} = 0$.
(2) $\vdash_{PA} \mathtt{x} \leq \mathtt{Sy} \leftrightarrow \mathtt{x} \leq \mathtt{y} \lor \mathtt{x} = \mathtt{Sy}$.

Proof: (1)

1	$\mathtt{x} + \mathtt{r} = 0 \vdash_{PA} \mathtt{r} = 0$	(by 4.3.2)
2	$\mathtt{x} + \mathtt{r} = 0 \vdash_{PA} \mathtt{x} = 0$	(From (1), using PA3)
3	$\vdash_{PA} \mathtt{x} + \mathtt{r} = 0 \to \mathtt{x} = 0$	(Deduction theorem)
4	$\vdash_{PA} (\exists \mathtt{r}\mathtt{x} + \mathtt{r} = 0) \to \mathtt{x} = 0$	(∃-introduction)
5	$\vdash_{PA} \mathtt{x} \leq 0 \to \mathtt{x} = 0$	(definition of ≤)
6	$\vdash_{PA} \mathtt{x} = 0 \to \mathtt{x} \leq 0$	(easy exercise)
7	$\vdash_{PA} \mathtt{x} \leq 0 \leftrightarrow \mathtt{x} = 0$	(tautologies from 5 and 6)

(2): By induction on $\mathtt{x}$. The induction basis $\vdash_{PA} 0 \leq \mathtt{Sy} \leftrightarrow 0 \leq \mathtt{y} \lor 0 = \mathtt{Sy}$ is easy and left to the reader. For the induction step, we again use a case distinction: $\mathtt{r}$ can be zero ...

1	$\mathtt{Sx+r=Sy, r=0} \vdash_{PA} \mathtt{Sx=Sy}$	(using PA3)
2	$\mathtt{Sx+r=Sy} \vdash_{PA} \mathtt{r=0 \rightarrow Sx=Sy}$	(ded.th.)

... or a successor:

3	$\mathtt{Sx+r=Sy, r=St} \vdash_{PA} \mathtt{Sx+St=Sy}$	(equality axioms)
4	$\mathtt{Sx+r=Sy, r=St} \vdash_{PA} \mathtt{S(Sx+t)=S(y)}$	(using PA4)
5	$\mathtt{Sx+r=Sy, r=St} \vdash_{PA} \mathtt{Sx+t=y}$	(using PA2)
6	$\mathtt{Sx+r=Sy, r=St} \vdash_{PA} \mathtt{Sx \leq y}$	($\exists$-intro.)
7	$\mathtt{Sx+r=Sy} \vdash_{PA} \mathtt{r=St \rightarrow Sx \leq y}$	(ded.th.)
8	$\mathtt{Sx+r=Sy} \vdash_{PA} \mathtt{(\exists t\, r=St) \rightarrow Sx \leq y}$	($\exists$-intro.)

Putting lines 2 and 8 together we get

9	$\mathtt{Sx+r=Sy} \vdash_{PA} \mathtt{r=0 \vee (\exists t\, r=St) \rightarrow Sx=Sy \vee Sx \leq y}$	(taut., from (2) and (8))
10	$\mathtt{Sx+r=Sy} \vdash_{PA} \mathtt{Sx=Sy \vee Sx \leq y}$	(using 4.3.2)
11	$\vdash_{PA} \mathtt{Sx+r=Sy \rightarrow Sx=Sy \vee Sx \leq y}$	(ded.th.)
12	$\vdash_{PA} \mathtt{Sx \leq Sy \rightarrow Sx=Sy \vee Sx \leq y}$	($\exists$-intro.)

This ends the proof of the "induction step."

4.3.5. Example.

$$\mathtt{x \leq 0 \leftrightarrow x=0}$$
$$\mathtt{x \leq \underline{1} \leftrightarrow x=0 \vee x=\underline{1}}$$
$$\mathtt{x \leq \underline{2} \leftrightarrow x=0 \vee x=\underline{1} \vee x=\underline{2}}$$
$$\mathtt{x \leq \underline{3} \leftrightarrow x=0 \vee x=\underline{1} \vee x=\underline{2} \vee x=\underline{3}}$$

and in general for all **n**:

$$\mathtt{x \leq \underline{n} \leftrightarrow x=0 \vee x=\underline{1} \vee \cdots \vee x=\underline{n}}.$$

Proof: Note that the quantifier "for all n" quantifies over "our" natural numbers, and cannot be written as a quantifier $\forall \mathtt{n}$ inside PA. So to prove this theorem, we will use "outer induction," that is, induction on the usual natural numbers.

The case $n=0$ was already proved in 4.3.4(1). So let $n=k+1$. Since $\underline{n}=\underline{k+1}=\mathtt{S}(\underline{k})$, we can apply 4.3.4(2) to get

$$\vdash_{PA} \mathtt{x} \leq \underline{k+1} \leftrightarrow \mathtt{x} \leq \underline{k} \vee \mathtt{x} = \underline{k+1}.$$

So by induction hypothesis we get

$$\vdash_{PA} x \leq \underline{k+1} \leftrightarrow (x = 0 \vee \cdots \vee x = \underline{k}) \vee x = \underline{k+1}.$$

4.3.6. Example. The associative law of addition:

$$\vdash_{PA} \forall x \forall y \forall z\, (x + y) + z = x + (y + z).$$

We will prove this in Peano Arithmetic "by induction on **z**," i.e., we will prove

(1) $\vdash_{PA} (x + y) + 0 = x + (y + 0)$.

(2) $\vdash_{PA} \left[(x + y) + z = x + (y + z) \to (x + y) + Sz = x + (y + Sz)\right]$.

(1) follows easily from axiom PA1 and the equality axioms:

1	$\vdash_{PA}(x + y) + 0 = x + y$	(PA1)
2	$\vdash_{PA} x + (y + 0) = x + y$	(PA1 and equality axioms)
3	$\vdash_{PA}(x + y) + 0 = x + (y + 0)$	(using 1, 2, and equality axioms).

For (2), we need axiom PA2 (plus equality axioms, but from now on we will not mention explicitly where we use equality axioms):

1	$(x + y) + z = x + (y + z) \vdash_{PA}(x + y) + Sz = S((x + y) + z)$	(PA2)
2	$(x + y) + z = x + (y + z) \vdash_{PA} x + (y + Sz) = x + S(y + z)$	(PA2)
3	$(x + y) + z = x + (y + z) \vdash_{PA} x + S(y + z) = S(x + (y + z))$	(PA2)
4	$(x + y) + z = x + (y + z) \vdash_{PA} S(x + (y + z)) = S((x + y) + z)$	(ind.hyp.)
5	$(x + y) + z = x + (y + z) \vdash_{PA}(x + y) + Sz = x + (y + Sz)$	(1,4,3,2)

Now, using the deduction theorem and the generalization theorem, we get the desired result (2).

4.3.7. Example.

$$\vdash_{PA} x \leq y \wedge y \leq z \to x \leq z$$

Proof: Exercise 1.

4.3.8. Fact. $\vdash_{PA} \neg x < 0$, and

$$\vdash_{PA} x < Sy \leftrightarrow x < y \vee x = y.$$

Proof: Exercise 2.

We will write $\forall x{<}y\,\varphi$ as an abbreviation for $\forall x(x{<}y \to \varphi)$, and $\exists x{<}y\,\varphi$ to abbreviate $\exists x(x{<}y \wedge \varphi)$.

4.3.9. Fact.

(1) $\vdash_{PA} \forall x{<}0\,\varphi$.

(2) $\vdash_{PA} \forall y{<}Sx\,\varphi(y) \leftrightarrow [\forall y{<}x\,\varphi(y)] \wedge \varphi(x)$.

Proof: (1) By 4.3.8, using tautological reasoning, we get $\vdash_{PA} x{<}0 \to \varphi$.

(2) Again using 4.3.8 and tautological reasoning, we get

$$\vdash_{PA} (y < Sx \to \varphi(y)) \quad \leftrightarrow \quad (y < x \to \varphi(y)) \wedge (y = x \to \varphi(y))$$

which is equivalent to

$$\vdash_{PA} (y < Sx \to \varphi(y)) \quad \leftrightarrow \quad (y < x \to \varphi(y)) \wedge \varphi(x).$$

The following is sometimes called the "Course-of-values" induction principle:

4.3.10. Lemma. Assume that φ is a formula with only one free variable x. Then:

$$\vdash_{PA} \forall x\,\big[(\forall y{<}x\,\varphi(y)) \to \varphi(x)\big] \to \forall x\,\varphi(x).$$

[Writing ψ for $\neg\varphi$, we can use tautological reasoning to transform this formula into the following form, often called the "minimum principle:"

$$\vdash_{PA} \exists x\,\psi(x) \to \exists x\big[\psi(x) \wedge \forall y{<}x\,\neg\psi(y)\big].$$

In other words: If there exists a number with a certain property ψ, then there exists a minimal such number.]

Instead of the formula given above, we will prove

$$\vdash_{PA} \forall x\big[(\forall y{<}x\,\varphi(y)) \to \varphi(x)\big] \to \forall x \forall y{<}x\,\varphi(y)$$

and

$$\vdash_{PA} \forall x \forall y{<}x\,\varphi(y) \to \forall x\,\varphi(x).$$

The first formula will be proved by induction on x. So we need

(1) $\forall x\big[(\forall y{<}x\,\varphi(y) \to \varphi(x))\big] \vdash_{PA} \forall y{<}0\,\varphi(y)$.

(2) $\forall x\big[(\forall y{<}x\,\varphi(y) \to \varphi(x))\big] \vdash_{PA} \forall y{<}x\,\varphi(y) \to \forall y{<}Sx\,\varphi(y)$.

Both follow easily from 4.3.9.

The second formula follows from

$$\forall x \forall y{<}x\, \varphi(y) \vdash_{PA} \forall y{<}Sx \varphi(y)$$

and $\vdash_{PA} x < Sx$.

4.3.11. Example. We will sketch how to derive "there are infinitely many prime numbers."

Formally, the statement we want to prove is

$$\forall x\, \exists p(x{<}p \wedge \mathit{prime}(p))$$

where *prime*(p) abbreviates

$$\neg(\exists x \exists y(x > \underline{1} \wedge y > \underline{1} \wedge x{\cdot}y = p)$$

Informally, the proof goes as follows: Fix a number x. Let y be the product of all numbers $\leq x$ (other than 0). So y is divisible by all numbers $\leq x$ (except 0). But then

$$y + 1 \text{ is not divisible by any number } z \leq x.$$

Now find a prime number p dividing $y + 1$. Clearly, p cannot be $\leq x$. So p must be $> x$.

How can we find p? We need the following fact:

Whenever $x > 1$, then x has a prime factor p

To prove this fact, consider the set $A := \{n : n > 1, n \text{ divides } x\}$. A is nonempty, as $x \in A$. So A has a minimal element p. We claim that p is prime. For if $p = r{\cdot}s$, with both $r, s > 1$, then r and s are divisors of x that are less than p but > 1, so $r, s \in A$, which contradicts the definition of p.

Now we will give a formalized version of this proof. We will write $x|y$ for $\exists z\, x{\cdot}z = y$, i.e., "x divides y."

As a first step towards a derivation, we see that we need the following:

(A) $\vdash_{PA} \forall x \exists y[y > \underline{1} \wedge \forall z(z \leq x \wedge z > \underline{1} \rightarrow z|y)]$

(B) $\vdash_{PA} \forall x \exists y[y > \underline{1} \wedge \forall z(z \leq x \wedge z > \underline{1} \rightarrow \neg z|y)]$

(C) $\vdash_{PA} \forall y(y > \underline{1} \rightarrow \exists p\, \mathit{prime}(p) \wedge p|y)$

(D) $\vdash_{PA} \forall x \exists p(\mathit{prime}(p) \wedge p \geq x)$.

Proof of (A): We will use induction, i.e., we have to prove

(A1) $\vdash_{PA} \exists y[y > \underline{1} \wedge \forall z(z \leq 0 \wedge z > \underline{1} \rightarrow z|y)]$
(A2) $\exists y[y > \underline{1} \wedge \forall z(z \leq x \wedge z > \underline{1} \rightarrow \neg z|y)] \vdash_{PA}$
$$\exists y[y > \underline{1} \wedge \forall z(z \leq Sx \wedge z > \underline{1} \rightarrow \neg z|y)].$$

We leave (A1) to the reader.
As for (A2), it will be enough to show

$$\vdash_{PA} \big(\forall z \leq x : z > \underline{1} \rightarrow z|y\big) \rightarrow \big(\forall z \leq Sx : z > \underline{1} \rightarrow z|y{\cdot}Sx\big)$$

Again, we leave this as an exercise.

From the formula above, using first the second $\exists$-introduction rule, and then the first, we get

$$\vdash_{PA} \exists y[y > \underline{1} \wedge \forall z(z \leq x \wedge z > \underline{1} \rightarrow z|y)] \rightarrow$$
$$\rightarrow \exists y[y > \underline{1} \wedge \forall z(z \leq Sx \wedge z > \underline{1} \rightarrow z|y)].$$

(B) can be obtained from (A) using again $\exists$-introduction, plus the fact that

$$\vdash_{PA} z|y \wedge z > \underline{1} \rightarrow \neg z|Sy$$

For (C), we have to use induction again, this time in the form of the minimum principle: Write $\varphi(x, p)$ for the formula $p|x \wedge p > \underline{1} \wedge \forall z{<}p : z > \underline{1} \rightarrow \neg z|x$, i.e., "p is the smallest nontrivial divisor of x." We want to show $\varphi(x, p) \rightarrow prime(p)$.

(C1) $\varphi(x, p), r{\cdot}s = p, r > \underline{1} \vdash_{PA} r|x$ (because $\vdash_{PA} p|x \wedge r{\cdot}s = p \rightarrow r|x$)
(C2) $\varphi(x, p), r{\cdot}s = p, r > \underline{1} \vdash_{PA} r \geq p$ (using the definition of φ)
(C3) $\varphi(x, p), r{\cdot}s = p, r > \underline{1} \vdash_{PA} s = \underline{1}$
(because $\vdash_{PA} r{\cdot}s = p \wedge r \geq p > \underline{1} \rightarrow s = \underline{1}$).

Using the derivation theorem and tautological reasoning, we get
(C4) $\varphi(x, p) \vdash_{PA} \neg(r{\cdot}s = p \wedge r > \underline{1} \wedge s > \underline{1})$
(C5) $\varphi(x, p) \vdash_{PA} \neg(r{\cdot}s = p \wedge r > \underline{1} \wedge s > \underline{1}) \wedge p > \underline{1} \wedge p|x$ (nonlogical axioms)
(C6) $\varphi(x, p) \vdash_{PA} prime(p)$ (by definition of *prime*).

By the deduction theorem, we get
(C7) $\vdash_{PA} \varphi(x, p) \rightarrow prime(p) \wedge p|x$
(C8) $\vdash_{PA} \varphi(x, p) \rightarrow \exists p\,[prime(p) \wedge p|x]$
(C9) $\vdash_{PA} \exists p \varphi(x, p) \rightarrow \exists p\, prime(p) \wedge p|x$

The minimum principle gives us
(C10) $\vdash_{PA} \exists p(p > \underline{1} \wedge p|x) \rightarrow \exists p \varphi(x, p)$
(C11) $x > \underline{1} \vdash_{PA} (x > \underline{1} \wedge x|x)$ (easy)
(C12) $x > \underline{1} \vdash_{PA} \exists p \varphi(x, p)$ (from 11)
(C13) $x > \underline{1} \vdash_{PA} \exists p\, prime(p) \wedge p|x$.

(D) now follows easily.

It turns out that the axioms of Peano Arithmetic are sufficiently strong to prove almost any theorem that number theorists have proved. (Of course, nobody actually writes down full derivations. Proofs are often given more informally. Thus they become shorter and more readable.)

Thus, one could conjecture the following:

For every closed formula φ,

$$¿\quad \mathbb{N} \models \varphi \leftrightarrow \vdash_{\mathrm{PA}} \varphi \quad ?$$

The fact that the implication $\leftarrow$ is true is often called the "soundness" of PA: since all axioms of PA (as well as all logical axioms, of course) are valid in $\mathbb{N}$ (by 4.2.3 and 4.2.8), and modus ponens, applied to formulas valid in a model, again yields a formula valid in this model (1.3.44), all theorems of PA are valid in $\mathbb{N}$.

The implication $\rightarrow$ is true for practically all statements φ that are proved in number theory. However, Gödel's incompleteness theorem states in one version that

there is a closed formula φ which is valid in $\mathbb{N}$, but not provable in PA

or equivalently,

there is a closed formula φ such that neither φ nor $\neg\varphi$ are provable in PA.

(See exercise 9.)

The formula φ in question is of course in the language of PA, and thus talks about natural numbers. So it can be called a number theoretical statement. However, its true (i.e., intended) meaning emerges only by translating it into a "metamathematical" statement — i.e., a statement not about numbers, but about formulas.

In fact, the "meaning" of φ will be "φ is not provable in PA." From this translation it is easy to see that φ must be true (i.e., valid in our model $\mathbb{N}$), so it is not provable. For if φ were false, its negation "φ is provable in PA" would be true. By the soundness property of PA, φ would have to be true, which would be a contradiction.

In the sections ahead we will explain how to interpret statements about numbers as statements about formulas. To this end, we will assign to each formula φ a number $\ulcorner\varphi\urcorner$, called the "code" of φ.

Then for many properties P of formulas φ (for examples "φ is a closed formula," or "φ is an axiom of PA") we can find a formula ψ representing P, i.e. such that $\mathbb{N} \models \psi(n)$ iff $n = \ulcorner\phi\urcorner$ for some formula φ having the property P.

First, each symbol of our language will be coded by a natural number. Since a formula is a sequence of symbols, a formula will then correspond to a sequence of natural numbers. In the following section we will show how to code any sequence of natural numbers into a single natural number.

Exercises

1. Use ∃-introduction to show that

$$\vdash_{PA} x \leq y \wedge y \leq z \rightarrow x \leq z.$$

2. Show that $\vdash_{PA} \neg x < 0$, and

$$\vdash_{PA} x < Sy \leftrightarrow x < y \vee x = y$$

3. Show that $\vdash \exists x{<}y\, \varphi \leftrightarrow \neg\forall x{<}y\, \neg\varphi$.

4. Show that for any formula $\varphi(x)$ with free variable x:

$\forall x\, \left[(\forall y{<}x\, \varphi(y)) \rightarrow \varphi(x)\right] \rightarrow \forall x\, \varphi(x) \dashv\vdash \exists x\, \neg\varphi(x) \rightarrow \exists x\left[\neg\varphi(x) \wedge \forall y{<}x\, \varphi(y)\right]$

5. Show that

$$\vdash_{PA} \exists y[y > \underline{1} \wedge \forall z(z \leq 0 \wedge z > \underline{1} \rightarrow z|y)]$$

(Hint: Show that

$$\vdash_{PA} [\underline{2} > \underline{0} \wedge \forall z(z \leq \underline{0} \wedge z > \underline{1} \rightarrow z|\underline{2})],$$

and then use ∃-introduction(1).)

6. Show that

$$\vdash_{PA} (\forall z \leq x : z > \underline{1} \rightarrow z|y) \rightarrow (\forall z \leq Sx : z > \underline{1} \rightarrow z|y{\cdot}Sx)$$

7. Show that $\vdash_{PA} p|x \wedge r{\cdot}s = p \rightarrow r|x$.

8. Show that $\vdash_{PA} r{\cdot}s = p \wedge r \geq p > \underline{1} \rightarrow s = \underline{1}$

9. Show that the two versions of Gödel's incompleteness theorem given on the previous page are in fact equivalent.

4.4. Encoding Finite Sequences of Numbers

We will have to code finite sequences of integers by single integers. We will use the fact that every number can be uniquely written as a product of prime powers.

4.4.1. Definition.

(1) Let $p_1 := 2$, $p_2 := 3$, $p_3 := 5$, $p_4 := 7$, $p_5 := 11$, ..., $p_k :=$ the k-th prime number.

(2) We will code a sequence $\langle a_1, \ldots, a_n \rangle$ of positive natural numbers by the single natural number $\#(a_1, \ldots, a_n)$, defined by

$$\#(a_1, \ldots, a_n) := 2^{a_1} \cdot 3^{a_2} \cdots p_n^{a_n}.$$

(3) We call $\#(a_1, \ldots, a_n)$ the "code" of the sequence $\langle a_1, \ldots, a_n \rangle$.

(4) For any number c and for any $k > 0$, we let $(c)_k := \max\{l : p_k^l | c\}$, where $x|y \leftrightarrow \exists z\, x{\cdot}z = y \leftrightarrow x$ divides y.

4.4.2. Example. $\#(3,4,3,1) = 2^3{\cdot}3^4{\cdot}5^3{\cdot}7^1 = 567000$. $(56700)_1 = 3$, $(56700)_2 = 4$, $(56700)_3 = 3$, $(56700)_4 = 1$, $(56700)_n = 0$ for $n > 4$.

To represent these relations within the language of Peano Arithmetic, we define the following formulas:
(Recall that $\mathsf{x}|\mathsf{y}$ abbreviates $\exists \mathsf{z}\, \mathsf{x}{\cdot}\mathsf{z} = \mathsf{y}$.)

4.4.3. Definition.

$prime(\mathsf{x}) :=(\mathsf{x} > \underline{1}) \wedge \neg\big(\exists \mathsf{y}\, \exists \mathsf{z}\ \mathsf{x} = \mathsf{y}{\cdot}\mathsf{z} \wedge \mathsf{y} > \underline{1} \wedge \mathsf{z} > \underline{1}\big).$

[Of course, formally this means that we define the formula *prime* with one free variable x_0 to be

$$(\mathsf{x}_0 > \mathsf{S0}) \wedge \neg(\exists \mathsf{x}_1\, \exists \mathsf{x}_2\ \mathsf{x}_0 = \mathsf{x}_1{\cdot}\mathsf{x}_2 \wedge \mathsf{x}_1 > \mathsf{S0} \wedge \mathsf{x}_2 > \mathsf{S0}),]$$

4.4.4. Fact. The corresponding set $prime := \{p : \mathbb{N} \models prime(p)\}$ is the set of all prime numbers.

4.4.5. Definition.

$$nextprime(\mathsf{x},\mathsf{y}) := prime(\mathsf{x}) \wedge prime(\mathsf{y}) \wedge \mathsf{x} < \mathsf{y} \wedge \forall \mathsf{z}\big(prime(\mathsf{z}) \rightarrow \mathsf{z} \leq \mathsf{x} \vee \mathsf{z} \geq \mathsf{y}\big).$$

4.4.6. Fact. $\mathbb{N} \models \mathit{nextprime}(p,q)$ iff p and q are adjacent primes, i.e., for some i, $p = p_i$ and $q = p_{i+1}$.

Not all numbers are of the form $\#(a_1,\ldots,a_k)$ for some k (since we required the a_i to be > 0):

$$\begin{aligned} \#(1) &= 2 \\ \#(2) &= 4 \\ \#(1,1) &= 6 \\ \#(3) &= 8 \\ \#(2,1) &= 12 \\ &\ldots \end{aligned}$$

There is a formula in the language of PA that selects exactly the sequence numbers, i.e., the numbers that are of this type.

We define the set of all codes of sequences by

$$seq := \{p_1^{e_1}\cdots p_n^{e_n} : n \in \mathbb{N}, e_1,\ldots,e_n > 0\}.$$

(For example, $12 \in seq$, as $12 = 2^2{\cdot}3^1 = p_1^2{\cdot}p_2^1$, but $15 \notin seq$, because $15 = 3{\cdot}5 = 2^0{\cdot}3^1{\cdot}5^1 = p_1^0{\cdot}p_2^1{\cdot}p_3^1$.)

4.4.7. Definition.

$$\mathit{seq}(\mathsf{c}) := \forall \mathsf{p},\mathsf{q} : \big(\mathit{prime}(\mathsf{p}) \land \mathit{prime}(\mathsf{q}) \land \mathsf{p} < \mathsf{q} \land \mathsf{q}|\mathsf{c} \rightarrow \mathsf{p}|\mathsf{c}\big) \ \land \mathsf{c} > \underline{1}.$$

4.4.8. Fact. $\mathbb{N} \models \mathit{seq}(n)$ iff $n \in seq$.

To obtain a formula $\varphi(x,y)$ that decides whether x is the y-th prime number, we first consider the following set of numbers:

Let

$$\begin{aligned} products &:= \{\#(1), \#(1,2), \#(1,2,3), \ldots\} \\ &= \{2^1, 2^1{\cdot}3^2, 2^1{\cdot}3^2{\cdot}5^3, 2^1{\cdot}3^2{\cdot}5^3{\cdot}7^4, \ldots\} = \\ &= \{2, 18, 2250, 5402250, \ldots\}. \end{aligned}$$

The corresponding formula is given by

4.4.9. Definition.

$$\begin{aligned} \mathit{products}(\mathsf{x}) := {} & 2|\mathsf{x} \land \neg 4|\mathsf{x} \land \mathit{seq}(\mathsf{x}) \land \forall \mathsf{p}\ \forall \mathsf{q}\ \forall \mathsf{e} \\ & \big(\mathit{nextprime}(\mathsf{q},\mathsf{p}) \land \mathsf{p}|\mathsf{x} \ \rightarrow \ (\mathsf{q}^{\mathsf{e}}|\mathsf{x} \leftrightarrow \mathsf{p}^{\mathsf{Se}}|\mathsf{x})\big). \end{aligned}$$

4.4.10. Fact.

$$products = \{n \in \mathbb{N} : \mathbb{N} \models products(n)\}.$$

Proof: It is clear that if $k \in products$, then $\mathbb{N} \models products(k)$.

Conversely, let $\mathbb{N} \models products(k)$. We can represent k uniquely as

$$k = p_1^{e_1} \cdot p_2^{e_2} \cdots p_n^{e_n} \qquad (e_n > 0).$$

We claim that for all $i < n$, $e_{i+1} = e_i + 1$.

Proof of this claim: Since $\mathbb{N} \models nextprime(p_i, p_{i+1})$, we have for all e:

$$\mathbb{N} \models p_i^e | k \leftrightarrow p_{i+1}^{e+1} | k.$$

Thus for all e, we have

$$\mathbb{N} \models e \leq e_i \leftrightarrow e + 1 \leq e_{i+1}.$$

Letting $e := e_i$, we get $e_i + 1 \leq e_{i+1}$, and letting $e := e_{i+1} - 1$ we get $e_{i+1} - 1 \leq e_i$. Thus $e_{i+1} = e_i + 1$.

Since $e_1 = 1$, we have for all $i \leq n$, $e_i = i$.

Now we define:

4.4.11. Definition.

$nthprime(\mathtt{p}, \mathtt{k}) := prime(\mathtt{p}) \wedge \exists \mathtt{w}\, products(\mathtt{w}) \wedge \mathtt{p}^{\mathtt{k}} | \mathtt{w} \wedge \neg \mathtt{p}^{\mathtt{k}+\underline{1}} | \mathtt{w}.$

4.4.12. Fact. $\mathbb{N} \models nthprime(p, k)$ iff $p = p_k$, the k-th prime number.

Proof: Exercise.

(Recall that $e = (c)_k$ iff $p_k^e | \mathtt{c}$ and but not $p_k^{e+1} | \mathtt{c}$.) We define $entry := \{(c, k, e) : c \in seq, e = (c)_k\}$. Informally, $(c, k, e) \in entry$ iff c codes a sequence whose k-th entry is e.

4.4.13. Definition.

$entry(\mathtt{c}, \mathtt{k}, \mathtt{e}) := seq(\mathtt{c}) \wedge (\mathtt{k} > 0) \wedge (\mathtt{e} > 0) \wedge \forall \mathtt{p}$
$\quad (prime(\mathtt{p}) \wedge nthprime(\mathtt{p}, \mathtt{k}) \rightarrow \mathtt{p}^{\mathtt{e}} | \mathtt{c} \wedge \neg \mathtt{p}^{\mathtt{e}+1} | \mathtt{c}).$

4.4.14. Fact.

$\mathbb{N} \models entry(c, k, e)$ iff c codes a sequence, $(c)_k = e$ and $e > 0$.

Proof: Exercise.

4.4.15. Fact. Whenever $\langle e_1, \ldots, e_n \rangle$ is a finite sequence of positive integers, there is a code $c \in seq$ such that for all $k \leq n$: $\mathbb{N} \models entry(c, k, e_k)$.

Proof: Let $c := p_1^{e_1} \cdots p_n^{e_n}$.

The next step is to code terms and formulas of Peano Arithmetic as natural numbers. To each term t (to each formula φ) we will associate a number $\ulcorner t \urcorner$ ($\ulcorner \varphi \urcorner$), its so-called "Gödel number" in a way that represents the structure of t (φ).

First we define $\ulcorner x \urcorner$ for all symbols x in our language according to the following table:

x	$\ulcorner x \urcorner$
$\mathtt{0}$	2
$\mathtt{S}$	4
$=$	6
$+$	8
$\cdot$	10
$\uparrow$	12
$\forall$	14
$\wedge$	16
$\neg$	18
$\mathtt{x}_n$	$2n+1$

(We do not need codes for the parentheses or the comma, since our way of coding will reflect the structure of the formulas as given by its syntax tree.)

Now we define $\ulcorner t \urcorner$ by induction on terms as follows:

If t is a variable or the constant $\mathtt{0}$, then $\ulcorner t \urcorner$ is defined in the table above.

If $t = \mathtt{S}t_1$, we let $\ulcorner t \urcorner := 2^{\ulcorner \mathtt{S} \urcorner} \cdot 3^{\ulcorner t_1 \urcorner}$.

If $t = t_1 + t_2$, we let $\ulcorner t \urcorner := 2^{\ulcorner + \urcorner} \cdot 3^{\ulcorner t_1 \urcorner} \cdot 5^{\ulcorner t_2 \urcorner}$.

If $t = t_1 \cdot t_2$, we let $\ulcorner t \urcorner := 2^{\ulcorner \cdot \urcorner} \cdot 3^{\ulcorner t_1 \urcorner} \cdot 5^{\ulcorner t_2 \urcorner}$.

If $t = t_1 \uparrow t_2$, we let $\ulcorner t \urcorner := 2^{\ulcorner \uparrow \urcorner} \cdot 3^{\ulcorner t_1 \urcorner} \cdot 5^{\ulcorner t_2 \urcorner}$.

For example, $\ulcorner \mathtt{0} \urcorner = 2$, so $\ulcorner \mathtt{S0} \urcorner = 2^4 \cdot 3^2 = 144$, and $\ulcorner \mathtt{0} + \mathtt{S0} \urcorner = 2^8 \cdot 3^2 \cdot 5^{144}$.

4.4.16. Fact. The map $t \to \ulcorner t \urcorner$ is well-defined and 1-1 on the set of terms. (That means that whenever t_1 and t_2 are different terms, then $\ulcorner t_1 \urcorner$ and $\ulcorner t_2 \urcorner$ will be different numbers.)

Proof: By the unique readability theorem, 1.3.12.

4.4.17. Remark. To each term t we have now associated a natural number $\ulcorner t \urcorner$. This assignment depended on some arbitrary choices, e.g. the table given above.

Recall that to each number n we have associated a term $\underline{n}$, e.g., $\underline{3} = \mathtt{SSS0}$. These two mappings — Terms $\rightarrow \mathbb{N}$ and $\mathbb{N} \rightarrow$ Terms — have nothing to do with each other.

We can also compose these mappings: for example, $\ulcorner \mathtt{0} \urcorner = 2$, $\underline{2} = \mathtt{SS0}$, $\ulcorner \mathtt{SS0} \urcorner = 2^4 . 3^{\ulcorner \mathtt{S0} \urcorner} = 2^4 . 3^{2^4 . 3^2} = 2^4 . 3^{144}$, etc.

Now we define $\ulcorner \varphi \urcorner$ for formulas φ as follows:

If φ is the formula $t_1 = t_2$, we let $\ulcorner \varphi \urcorner := 2^{\ulcorner = \urcorner} . 3^{\ulcorner t_1 \urcorner} . 5^{\ulcorner t_2 \urcorner}$.

If $\varphi = \neg \psi$, we let $\ulcorner \varphi \urcorner := 2^{\ulcorner \neg \urcorner} . 3^{\ulcorner \psi \urcorner}$.

If $\varphi = \psi_1 \wedge \psi_2$, we let $\ulcorner \varphi \urcorner := 2^{\ulcorner \wedge \urcorner} . 3^{\ulcorner \psi_1 \urcorner} . 5^{\ulcorner \psi_2 \urcorner}$.

If $\varphi = \forall \mathtt{x}\, \psi$, we let $\ulcorner \varphi \urcorner := 2^{\ulcorner \forall \urcorner} . 3^{\ulcorner x \urcorner} . 5^{\ulcorner \psi \urcorner}$.

For example, since $\ulcorner \mathtt{x}_1 \urcorner = 3$, $\ulcorner \mathtt{0} \urcorner = 2$, $\ulcorner = \urcorner = 6$, we have $\ulcorner \mathtt{x}_1 = \mathtt{0} \urcorner = 2^6 . 3^3 . 5^2 = 43200$, and $\ulcorner \forall \mathtt{x}_1\, \mathtt{x}_1 = \mathtt{0} \urcorner = 2^{14} . 3^3 . 5^{43200}$.

The next step is to find a formula $\varphi_1(x)$ such that for every number n, $\mathbb{N} \models \varphi_1(n)$ iff $n = \ulcorner t \urcorner$ for some term t.

Similarly, we will find a formula $\varphi_2(x)$ such that that for every number n, $\mathbb{N} \models \varphi_2(n)$ iff $n = \ulcorner \psi \urcorner$ for some formula ψ.

Exercises

1. Prove 4.4.12: $\mathbb{N} \models nthprime(p, k)$ iff p is $= p_k$, the k-th prime.

2. Prove 4.4.14:

$\mathbb{N} \models entry(c, k, e)$ iff c codes a sequence, $(c)_k = e$ and $e > 0$.

4.5. Gödel Numbers

Below we will define a formula $term(\mathtt{t})$ such that $\mathbb{N} \models term(n)$ iff $n = \ulcorner t \urcorner$ for some term t.

We start by looking at atomic terms:

4.5.1. Definition.

$$var(\mathtt{x}) := (\exists \mathtt{y}\ \mathtt{x} = \underline{2} \cdot \mathtt{y} + \underline{1}). \qquad zero(\mathtt{x}) := (\mathtt{x} = \underline{\ulcorner 0 \urcorner}).$$

Clearly $\mathbb{N} \models zero(n)$ iff $n = \ulcorner 0 \urcorner$ $(= 2)$, and $\mathbb{N} \models var(n)$ iff for some k, $n = \ulcorner \mathtt{x}_k \urcorner$ is the code of the k-th variable.

Note the difference between the closed formula $var(\underline{\ulcorner \mathtt{x}_n \urcorner})$ and the formula $var(\mathtt{x}_n)$. Also note the difference between

- $\mathtt{0}$: a constant symbol in the language of PA, and hence also a term
- $\ulcorner 0 \urcorner$: the natural number 2 (the Gödel number of $\mathtt{0}$)
- $\underline{\ulcorner 0 \urcorner}$: the term $\mathtt{SS0}$, more accurately $\mathtt{S(S(0))}$, the term corresponding to 2.
- $\mathtt{o}$: one of the variables $\mathtt{x}_n$ in out language.
- o : a (meta-)variable, denoting a natural number.

We can express the fact that a term t is obtained by joining terms t_1 and t_2 with and operation o by the following formula $yieldtm(\mathtt{t}_1, \mathtt{t}_2, \mathtt{o}, \mathtt{t})$:

4.5.2. Definition.

$$yieldtm(\mathtt{t}_1, \mathtt{t}_2, \mathtt{o}, \mathtt{t}) := \left[(\mathtt{o} = \underline{\ulcorner + \urcorner} \vee \mathtt{o} = \underline{\ulcorner \cdot \urcorner} \vee \mathtt{o} = \underline{\ulcorner \uparrow \urcorner}) \wedge \left(\mathtt{t} = \underline{2}^{\mathtt{o}} \cdot \underline{3}^{\mathtt{t}_1} \cdot \underline{5}^{\mathtt{t}_2}\right)\right] \vee$$
$$\vee \left(\mathtt{o} = \underline{\ulcorner S \urcorner} \wedge \mathtt{t} = \underline{2}^{\mathtt{o}} \cdot \underline{3}^{\mathtt{t}_1}\right).$$

Informally, $yieldtm(m, n, o, p)$ says: "p codes the term obtained by joining the two terms coded by m and n, using the operation symbol coded by o (assuming m and n really code terms)."

If c is the code of a sequence, $obtained(c, k, l, o, n)$ will express the fact that the n-th element of the sequence was obtained from the k-th and the l-th element, which were joined by the symbol coded by o:

4.5.3. Definition.

$$obtained(\mathtt{c}, \mathtt{k}, \mathtt{l}, \mathtt{o}, \mathtt{n}) := \exists \mathtt{e}, \mathtt{f}, \mathtt{g} : entry(\mathtt{c}, \mathtt{k}, \mathtt{e}) \wedge entry(\mathtt{c}, \mathtt{l}, \mathtt{f}) \wedge$$
$$\wedge entry(\mathtt{c}, \mathtt{n}, \mathtt{g}) \wedge yieldtm(\mathtt{e}, \mathtt{f}, \mathtt{o}, \mathtt{g}).$$

The statement "the n-th number in the sequence c codes an atomic term" is expressed by the following formula:

4.5.4. Definition.

$$atomicterm(\mathtt{c},\mathtt{n}) :=\exists \mathtt{e}\, entry(\mathtt{c},\mathtt{n},\mathtt{e}) \wedge \big(zero(\mathtt{e}) \vee var(\mathtt{e})\big).$$

We say that a sequence $\langle t_1, t_2, \ldots, t_n\rangle$ "builds" the term t iff t appears in the sequence, and for all $k \leq n$ t_k is

either the constant term $\mathtt{0}$, or a variable $\mathtt{x}_n$,
or is of the form $t_i + t_j$ or $t_i{\cdot}t_j$ or $t_i^{t_j}$ for some $i, j < k$,
or $t_k = \mathtt{S}t_j$ for some $j < k$.

Before we define a formula $term(\mathtt{n})$ (expressing that n is a number coding a term), we define a formula $buildTerm(\mathtt{c},\mathtt{t})$ which expresses the fact that c is the code of a sequence $\langle \ulcorner t_1\urcorner, \ldots, \ulcorner t_n\urcorner\rangle$, where $\langle t_1, \ldots, t_n\rangle$ builds t.

4.5.5. Definition.

$$buildTerm(\mathtt{c},\mathtt{t}) :=seq(\mathtt{c}) \wedge \exists \mathtt{n}\, entry(\mathtt{c},\mathtt{n},\mathtt{t}) \wedge \forall \mathtt{n} \forall \mathtt{e} : \;\; entry(\mathtt{c},\mathtt{n},\mathtt{e}) \rightarrow$$
$$atomicterm(\mathtt{c},\mathtt{n}) \vee \exists \mathtt{k},\mathtt{l}{<}\mathtt{n}\, \exists \mathtt{o}\, obtained(\mathtt{c},\mathtt{k},\mathtt{l},\mathtt{o},\mathtt{n}).$$

($\exists \mathtt{k},\mathtt{l}{<}\mathtt{m} \ldots$ is the customary abbreviation for $\exists \mathtt{k} \exists \mathtt{l} (\mathtt{k}{<}\mathtt{n} \wedge \mathtt{l} < \mathtt{n} \wedge \ldots)$.).

Finally we define:

4.5.6. Definition.

$$term(\mathtt{t}) :=\exists \mathtt{c}\; buildTerm(\mathtt{c},\mathtt{t}).$$

4.5.7. Fact. $\mathbb{N} \models term(n)$ iff there is a term t such that $n = \ulcorner t\urcorner$.

For example, $\mathbb{N} \models term(144)$, because $144 = 2^4{\cdot}3^2 = 2^{\underline{\ulcorner \mathtt{S}\urcorner}}{\cdot}3^{\underline{\ulcorner \mathtt{0}\urcorner}}$, so $144 = \underline{\ulcorner \mathtt{S0}\urcorner}$.

Proof: We have to show that if $\mathbb{N} \models buildTerm(c, t)$, then for all $k > 0$: Either $(c)_k = 0$, or $(c)_k$ codes a term. The proof is a straightforward induction on k. (Remember that if for codes of sequences c, $(c)_k > 0$, then $(c)_l > 0$ for all $l < k$.)

The definition of the formula $formula(\mathtt{x})$ is very similar to the definition of $term(\mathtt{x})$. First we define *yieldfm* similar to *yieldtm*:

4.5.8. Definition.

$$yieldfm(\mathtt{f}_1,\mathtt{f}_2,\mathtt{o},\mathtt{v},\mathtt{f}) :=\big(\mathtt{o} = \underline{\ulcorner \wedge \urcorner} \wedge \mathtt{f} = \underline{2}^{\mathtt{o}}{\cdot}\underline{3}^{\mathtt{f}_1}{\cdot}\underline{5}^{\mathtt{f}_2}\big) \vee$$
$$\vee \big(\mathtt{o} = \underline{\ulcorner \neg \urcorner} \wedge \mathtt{f} = \underline{2}^{\mathtt{o}}{\cdot}\underline{3}^{\mathtt{f}_1}\big) \vee$$
$$\vee \big(\mathtt{o} = \underline{\ulcorner \forall \urcorner} \wedge var(\mathtt{v}) \wedge \mathtt{t} = \underline{2}^{\mathtt{o}}{\cdot}\underline{3}^{\mathtt{v}}{\cdot}\underline{5}^{\mathtt{f}_1}\big).$$

Informally, *yieldfm*(m,n,o,p,q) says: "If o codes $\wedge$, then q codes the conjunction of the formulas coded by m and n. If o codes $\exists$, then p codes a variable x, and q codes the formula $\forall x\psi$, where m codes ψ. If o codes $\neg$, then q codes the formula $\neg\psi$, where m codes ψ."

Now we look at atomic formulas.

4.5.9. Definition.

$$\mathit{equation}(\mathtt{f},\mathtt{t}_1,\mathtt{t}_2) := \mathit{term}(\mathtt{t}_1) \wedge \mathit{term}(\mathtt{t}_2) \wedge \mathtt{f} = \underline{2}^{\ulcorner = \urcorner}\cdot\underline{3}^{\mathtt{t}_1}\cdot\underline{5}^{\mathtt{t}_2}.$$

4.5.10. Definition.

$$\mathit{atomicfm}(\mathtt{c},\mathtt{n}) := \mathit{seq}(\mathtt{c}) \wedge \exists\mathtt{e}\,\exists\mathtt{t}_1\,\exists\mathtt{t}_2\,\mathit{entry}(\mathtt{c},\mathtt{n},\mathtt{e}) \wedge \mathit{equation}(\mathtt{e},\mathtt{t}_1,\mathtt{t}_2).$$

atomicfm(c,n) says that c is a sequence whose n-th entry codes an atomic formula.

If c is the code of a sequence, *fobtained*(c,k,l,o,v,n) will express the fact that the n-th element of the sequence codes either the conjunction of the formulas coded by the k-th and the l-th element (if $o = \ulcorner\wedge\urcorner$), or is the negation or an existential quantification of the formula coded by the k-th element (if $o = \ulcorner\neg\urcorner$ or $o = \ulcorner\forall\urcorner$, respectively).

4.5.11. Definition.

$$\mathit{fobtained}(\mathtt{c},\mathtt{k},\mathtt{l},\mathtt{o},\mathtt{v},\mathtt{n}) := \exists\mathtt{e},\mathtt{f},\mathtt{g} : \mathit{entry}(\mathtt{c},\mathtt{k},\mathtt{e}) \wedge \mathit{entry}(\mathtt{c},\mathtt{l},\mathtt{f}) \wedge \\ \wedge \mathit{entry}(\mathtt{c},\mathtt{n},\mathtt{g}) \wedge \mathit{yieldfm}(\mathtt{e},\mathtt{f},\mathtt{o},\mathtt{v},\mathtt{g}).$$

We say that a sequence $\langle\varphi_1,\varphi_2,\ldots,\varphi_n\rangle$ "builds" the formula φ iff φ appears in the sequence, and for all $k \leq n$ φ_k is either an atomic formula ($t_1 = t_2$ for terms t_1, t_2), or $\varphi_k = \varphi_i \wedge \varphi_j$ for some $i,j < k$, or $\varphi_k = \neg\varphi_j$ or $\varphi_k = \exists x_n\varphi_j$ for some $j < k$ and some variable x_n.

Before we define a formula *formula*(n) (expressing that n is a number coding a formula), we define a formula *buildFormula*(c, f) which expresses the fact that c is the code of a sequence $\langle\ulcorner\varphi_1\urcorner,\ldots,\ulcorner\varphi_n\urcorner\rangle$, where $\langle\varphi_1,\ldots,\varphi_n\rangle$ builds φ.

4.5.12. Definition.

$$\mathit{buildFormula}(\mathtt{c},\mathtt{f}) := \mathit{seq}(\mathtt{c}) \wedge \exists\mathtt{n}\,\mathit{entry}(\mathtt{c},\mathtt{n},\mathtt{f}) \wedge \forall\mathtt{n}\,\forall\mathtt{e} : \big[\mathit{entry}(\mathtt{c},\mathtt{n},\mathtt{e}) \rightarrow \\ \rightarrow \mathit{atomicfm}(\mathtt{c},\mathtt{n}) \vee \exists\mathtt{k},\mathtt{l}{<}\mathtt{n}\,\exists\mathtt{o}\,\exists\mathtt{v}\,\mathit{fobtained}(\mathtt{c},\mathtt{k},\mathtt{l},\mathtt{o},\mathtt{v},\mathtt{n})\big].$$

Finally we define:

4.5.13. Definition.

$formula(\mathtt{f}) :=\exists \mathtt{c}\ buildFormula(\mathtt{c}, \mathtt{f})$.

4.5.14. Fact. $\mathbb{N} \models formula(n)$ iff there is a formula φ such that $n = \ulcorner\varphi\urcorner$.

Proof: We have to show that if $\mathbb{N} \models buildFormula(c, f)$, then for all $k > 0$: Either $(c)_k = 0$, or $(c)_k$ codes a formula. The proof is a straightforward induction on k.

Finally, we formally (i.e., in PA) define what a "proof" (derivation) is, and what a theorem is.

A derivation is a sequence $\langle\varphi_1, \ldots, \varphi_m\rangle$ of formulas such that each φ_n either is an axiom, or can be obtained from two previous formulas by modus ponens, i.e. there are $k, l < n$ such that the formula φ_l is equal to $\varphi_k \to \varphi_n$.

Remember that we have not included the implication sign in our language. Thus, $\varphi \to \psi$ is really an abbreviation for $\neg(\varphi \wedge \neg\psi)$.

This can be straightforwardly translated into PA:

4.5.15. Definition.

$$implies(\mathtt{f}_1, \mathtt{f}_2, \mathtt{f}) :=formula(\mathtt{f}_1) \wedge formula(\mathtt{f}_2) \wedge \mathtt{f} = \underline{2}^{\underline{\ulcorner\neg\urcorner}} \cdot \underline{3}^{\underline{2}^{\underline{\ulcorner\wedge\urcorner}} \cdot \underline{3}^{\mathtt{f}_1} \cdot \underline{5}^{\underline{2}^{\underline{\ulcorner\neg\urcorner}} \cdot \underline{3}^{\mathtt{f}_2}}}.$$

Again, $\mathbb{N} \models implies(f_1, f_2, f)$ iff there are formulas φ_1 and φ_2 such that $f_1 = \ulcorner\varphi_1\urcorner$, $f_2 = \ulcorner\varphi_2\urcorner$, and $f = \ulcorner(\varphi_1 \to \varphi_2)\urcorner$.

First we have to define a formula in PA that recognizes if a number is the code of a (logical or nonlogical) axiom. Recall that there was an infinite list of logical axioms, which we classified into 5 groups. For example whenever φ was a formula in which the variable $\mathtt{x}$ was not free, $\varphi \to \forall \mathtt{x}\varphi$ was a group II axiom. We will call these axioms that were listed in I–V "pure" axioms.

Also every generalization $\forall \mathtt{y}(\varphi \to \forall \mathtt{x}\varphi)$, $\forall z\forall y(\varphi \to \forall x\varphi)$ (where $\mathtt{x}$ may or may not be different from $\mathtt{y}$ and $\mathtt{z}$) was an axiom.

Furthermore, we have to consider our nonlogical axioms, the axioms PA1–PA9. We leave these to the reader.

4.5.16. Definition.

$axiom(\mathtt{e}) :=logicalaxiom(\mathtt{e}) \vee PAaxiom(\mathtt{e})$.

A logical axiom can be obtained from a pure logical axiom by repeatedly applying the universal quantifier:

4.5.17. Definition.

$$\begin{aligned} \mathit{logicalaxiom}(\mathtt{f}) :=& \exists \mathtt{c} \exists \mathtt{p}\, \mathit{seq}(\mathtt{c}) \wedge \mathit{pureAxiom}(\mathtt{p}) \wedge \mathit{entry}(\mathtt{c}, \underline{1}, \mathtt{p}) \wedge \\ & \exists \mathtt{n} \mathit{entry}(\mathtt{c}, \mathtt{n}, \mathtt{f}) \wedge \\ & \forall \mathtt{k} \forall \mathtt{e} \forall \mathtt{e}' \mathit{entry}(\mathtt{c}, \mathtt{k}, \mathtt{e}) \wedge \mathit{entry}(\mathtt{c}, \mathtt{k} + \underline{1}, \mathtt{e}') \rightarrow \\ & \quad \exists \mathtt{v} \mathit{var}(\mathtt{v}) \wedge \mathtt{e}' = \underline{2}^{\underline{\forall}} \cdot \underline{3}^{\mathtt{v}} \cdot 5^{\mathtt{e}}. \end{aligned}$$

4.5.18. Definition.

$$\begin{aligned} \mathit{pureAxiom}(\mathtt{p}) :=& \mathit{GroupI}(\mathtt{p}) \vee \mathit{GroupII}(\mathtt{p}) \vee \mathit{GroupIII}(\mathtt{p}) \vee \mathit{GroupIV}(\mathtt{p}) \vee \\ & \vee \mathit{GroupV}(\mathtt{p}). \end{aligned}$$

We leave the formulation of the formulas *GroupI*, ..., *GroupV* to the reader. We will give only one example:

4.5.19. Definition.

$$\begin{aligned} \mathit{GroupII}(\mathtt{p}) :=& \big[\exists \mathtt{f}_1 \exists \mathtt{f}_2 \exists \mathtt{x}\, \mathit{implies}(\mathtt{f}_1, \mathtt{f}_2, \mathtt{f}) \wedge \mathit{var}(\mathtt{x}) \wedge \\ & \mathit{formula}(\mathtt{f}_1) \wedge \mathit{formula}(\mathtt{f}_2) \wedge \mathit{notfree}(\mathtt{x}, \mathtt{f}_1)\big]. \end{aligned}$$

(We also leave the definition of *notfree* to the reader.)

Finally, we can define:

4.5.20. Definition.

$$\begin{aligned} \mathit{derivation}(\mathtt{c}) :=& \mathit{seq}(\mathtt{c}) \wedge \forall \mathtt{n} \forall \mathtt{e} \mathit{entry}(\mathtt{c}, \mathtt{n}, \mathtt{e}) \rightarrow \\ & \mathit{axiom}(\mathtt{e}) \vee \exists \mathtt{k}, \mathtt{l} < \mathtt{n} \exists \mathtt{f} \exists \mathtt{g}\, \mathit{entry}(\mathtt{c}, \mathtt{k}, \mathtt{f}) \wedge \mathit{entry}(\mathtt{c}, \mathtt{l}, \mathtt{g}) \wedge \\ & \qquad \mathit{implies}(\mathtt{f}, \mathtt{e}, \mathtt{g}). \end{aligned}$$

Then $\mathbb{N} \models \mathit{derivation}(c)$ iff there exists a derivation

$$\begin{aligned} & \vdash_{\mathrm{PA}} \varphi_1 \\ & \dots \\ & \vdash_{\mathrm{PA}} \varphi_n \end{aligned}$$

such that $c = 2^{\ulcorner \varphi_1 \urcorner} \cdot 3^{\ulcorner \varphi_2 \urcorner} \cdots p_n^{\ulcorner \varphi_n \urcorner}$.

4.5.21. Definition.

$$\mathit{theorem}(\mathtt{f}) := \mathit{formula}(\mathtt{f}) \wedge \exists \mathtt{c} \exists \mathtt{n} \mathit{derivation}(\mathtt{c}) \wedge \mathit{entry}(\mathtt{c}, \mathtt{n}, \mathtt{f}).$$

So we have the following

4.5.22. Theorem. *theorem*$(\mathtt{f})$ has the property that for every formula φ,

$$\mathbb{N} \models \mathit{theorem}(\ulcorner \varphi \urcorner) \quad \text{iff} \quad \vdash_{\mathrm{PA}} \varphi$$

4.6. Substitution

We want to find a formula corresponding to the relation "$t' = t(x/s)$," i.e., "t' is the term obtained from the term t by substituting the term s for all occurrences of the variable x."

For example, if t is the term $\mathtt{S}(\mathtt{x}_1 + \mathtt{x}_2)$, and s is the term $\mathtt{S}(\mathtt{x}_0)$, then $t' = t(\mathtt{x}_1/s)$ will be the term $\mathtt{S}(\mathtt{S}(\mathtt{x}_0) + \mathtt{x}_2)$.

To describe the term t' within PA, we have to look at the way that t was built. First we consider a sequence c building t:

$$c = \langle\ \mathtt{x}_1\ ,\ \mathtt{x}_2\ ,\ \mathtt{x}_1 + \mathtt{x}_2\ ,\ \mathtt{S}(\mathtt{x}_1 + \mathtt{x}_2)\ \rangle$$

Now construct a sequence c' obeying the following rules:

(1) If c_k, the k-th entry in c, is the constant symbol $\mathtt{0}$, then also $c'_k = \mathtt{0}$.
(2) Similarly, if c_k is a variable other than $\mathtt{x}_1$ (i.e., not the variable that is to be substituted), then again $c'_k = c_k$.
(3) If $c_k = \mathtt{x}_1$, then $c'_k = s$.
(4) If $c_k = \mathtt{S}c_l$ for some $l < k$, then also $c'_k = \mathtt{S}c'_l$.
(5) Similarly, if $c_k = (c_l + c_m)$ for some $l, m < k$, then $c'_k = (c'_l + c'_m)$, etc.

So we obtain the sequence

$$c' = \langle\ \mathtt{S}(\mathtt{x}_0)\ ,\ \mathtt{x}_2\ ,\ \mathtt{S}(\mathtt{x}_0) + \mathtt{x}_2\ ,\ \mathtt{S}(\mathtt{S}(\mathtt{x}_0) + \mathtt{x}_2)\ \rangle.$$

Note that this is not a term-building sequence any more. However, all its entries are terms. In fact, each entry c'_k is exactly $= c_k(\mathtt{x}_1/s)$. (In our example, this can be checked by a simple inspection of all terms. In general, the proof is by induction on k.)

We can formalize this construction with the following formula:

4.6.1. Definition.

$$\begin{aligned}
\mathit{termsub}(\mathtt{t},\mathtt{v},\mathtt{s},\mathtt{t}') :=\ & \mathit{term}(\mathtt{t}) \wedge \mathit{var}(\mathtt{v}) \wedge \mathit{term}(\mathtt{s}) \wedge \mathit{term}(\mathtt{t}') \wedge \\
& \exists \mathtt{c} \exists \mathtt{c}' \Big[\mathit{buildTerm}(\mathtt{c},\mathtt{t}) \wedge \mathit{seq}(\mathtt{c}') \wedge \\
& \quad [\forall \mathtt{k} \forall \mathtt{l} \forall \mathtt{m} \forall \mathtt{o} : \mathit{obtained}(\mathtt{c},\mathtt{k},\mathtt{l},\mathtt{o},\mathtt{m}) \rightarrow \\
& \qquad\qquad \mathit{obtained}(\mathtt{c}',\mathtt{k},\mathtt{l},\mathtt{o},\mathtt{m})] \wedge \\
& \quad [\forall \mathtt{k} \forall \mathtt{e} : \mathit{entry}(\mathtt{c},\mathtt{k},\mathtt{e}) \rightarrow \\
& \qquad ([\mathtt{e} = \ulcorner \mathtt{0} \urcorner \rightarrow \mathit{entry}(\mathtt{c}',\mathtt{k},\mathtt{e})] \wedge \\
& \qquad [\mathtt{e} = \mathtt{v} \rightarrow \mathit{entry}(\mathtt{c}',\mathtt{k},\mathtt{s})] \wedge \\
& \qquad [\mathit{var}(\mathtt{e}) \wedge \mathtt{e} \neq \mathtt{v} \rightarrow \mathit{entry}(\mathtt{c}',\mathtt{k},\mathtt{e})])]\Big].
\end{aligned}$$

We can deal with formulas similarly: Given a formula φ, a variable x and a term s, we want to construct the formula $\varphi' = \varphi(x/s)$ as follows.

First find a sequence c that builds the formula φ. Then replace each entry c_k by $c_k(x/s)$, obeying the following rules:

(1) If c_k is an equation $t_1 = t_2$, where t_1 and t_2 are terms, then c'_k will be the equation $t_1(x/s) = t_2(x/s)$.

(2) If $c_k = \neg c_l$ for some $l < k$, then also $c'_k = \neg c'_l$. Similarly for $c_k = c_l \wedge c_m$.

(3) If $c_k = \forall y\, c_l$ where y is a variable other than x, $c'_k = \forall y c'_l$.

(4) If $c_k = \forall x\, c_l$, then x is not free in c_k, so c'_k should be $= c_k$.

We formalize this construction as follows. First we deal with atomic formulas:

4.6.2. Definition.

$$\begin{aligned}
\mathit{atomicsub}(\mathtt{f},\mathtt{v},\mathtt{s},\mathtt{f}') :=\exists \mathtt{t}_1 \exists \mathtt{t}_2 \exists \mathtt{t}'_1 \exists \mathtt{t}'_2 :\ & \mathit{equation}(\mathtt{f},\mathtt{t}_1,\mathtt{t}_2)\wedge \\
& \wedge \mathit{equation}(\mathtt{f}',\mathtt{t}'_1,\mathtt{t}'_2)\wedge \\
& \wedge \mathit{termsub}(\mathtt{t}_1,\mathtt{v},\mathtt{s},\mathtt{t}'_1)\wedge \\
& \wedge \mathit{termsub}(\mathtt{t}_2,\mathtt{v},\mathtt{s},\mathtt{t}'_2).
\end{aligned}$$

4.6.3. Definition.

$$\begin{aligned}
&\mathit{fmsub}(\mathtt{f},\mathtt{v},\mathtt{s},\mathtt{f}') := \mathit{formula}(\mathtt{f}) \wedge \mathit{formula}(\mathtt{f}') \wedge \mathit{var}(\mathtt{v}) \wedge \mathit{term}(\mathtt{s})\wedge \\
&\quad \exists \mathtt{c} \exists \mathtt{c}' \Bigg(\mathit{buildFormula}(\mathtt{c},\mathtt{f}) \wedge \mathit{seq}(\mathtt{c}')\wedge \\
&\qquad \forall \mathtt{k} \forall \mathtt{l} \forall \mathtt{m} \forall \mathtt{o} \forall \mathtt{x} :(\mathtt{o} = \underline{\ulcorner\neg\urcorner} \vee \mathtt{o} = \underline{\ulcorner\wedge\urcorner}) \rightarrow \\
&\qquad\qquad (\mathit{fobtained}(\mathtt{c},\mathtt{k},\mathtt{l},\mathtt{o},\mathtt{x},\mathtt{m}) \rightarrow \\
&\qquad\qquad \rightarrow \mathit{fobtained}(\mathtt{c}',\mathtt{k},\mathtt{l},\mathtt{o},\mathtt{x},\mathtt{m}))\ \wedge \\
&\qquad \big[\forall \mathtt{k} \forall \mathtt{e}\colon \mathit{entry}(\mathtt{c},\mathtt{k},\mathtt{e}) \wedge \mathit{atomicfm}(\mathtt{c},\mathtt{k},\mathtt{e}) \rightarrow \\
&\qquad\qquad \exists \mathtt{e}' :\mathit{entry}(\mathtt{c}',\mathtt{k},\mathtt{e}') \wedge \mathit{atomicsub}(\mathtt{e},\mathtt{v},\mathtt{s},\mathtt{e}')\big]\wedge \\
&\qquad \forall \mathtt{k} \forall \mathtt{l} \forall \mathtt{m} \forall \mathtt{x} : (\mathit{fobtained}(\mathtt{c},\mathtt{k},\mathtt{l},\underline{\ulcorner\forall\urcorner},\mathtt{x},\mathtt{m}) \wedge (\mathtt{x} \neq \mathtt{v}) \rightarrow \\
&\qquad\qquad \rightarrow \mathit{fobtained}(\mathtt{c}',\mathtt{o},\mathtt{k},\mathtt{l},\mathtt{x},\mathtt{m}))\ \wedge \\
&\qquad \forall \mathtt{k} \forall \mathtt{l} \forall \mathtt{m} \forall \mathtt{x} : (\mathit{fobtained}(\mathtt{c},\mathtt{k},\mathtt{l},\underline{\ulcorner\forall\urcorner},\mathtt{x},\mathtt{m}) \wedge (\mathtt{x} = \mathtt{v}) \rightarrow \\
&\qquad\qquad \rightarrow \exists \mathtt{e}(\mathit{entry}(\mathtt{c},\mathtt{m},\mathtt{e}) \wedge \mathit{entry}(\mathtt{c}',\mathtt{m},\mathtt{e}))) \Bigg).
\end{aligned}$$

It is tedious but not hard to check the following:

4.6.4. Theorem. The formula *fmsub* has the property that for any formula φ, any variable $\mathbf{x}_k$ and any term τ, for any natural number n:

$$\mathbb{N} \models \mathit{fmsub}(\ulcorner\varphi\urcorner, \ulcorner\mathbf{x}_k\urcorner, \ulcorner\tau\urcorner, n) \leftrightarrow n = \ulcorner\varphi(\mathbf{x}_k/\tau)\urcorner.$$

Note that this formula does not deal with "forbidden" substitution, i.e., it does not check whether some free variable of τ becomes bound after substitution.

Exercises

1. Check that for any formula φ, any variable $\mathbf{x}_k$ and any term τ, for any natural number n:

$$\mathbb{N} \models \mathit{fmsub}(\ulcorner\varphi\urcorner, \ulcorner\mathbf{x}_k\urcorner, \ulcorner\tau\urcorner, n) \leftrightarrow n = \ulcorner\varphi(\mathbf{x}_k/\tau)\urcorner.$$

2. Modify the definition of *fmsub* such that

$$\mathbb{N} \models \mathit{fmsub}(\ulcorner\varphi\urcorner, \ulcorner\mathbf{x}_k\urcorner, \ulcorner\tau\urcorner, n) \leftrightarrow n = \ulcorner\varphi(\mathbf{x}_k/\tau)\urcorner \text{ and } \mathit{allow}(\varphi, \tau, \mathbf{x}_k).$$

3. Using the modified definition of *fmsub* from the previous exercise, define *GroupIII*(p). (Recall that the pure axioms in group III were of the form $\forall \mathbf{x}\, \varphi \to \varphi(\mathbf{x}/\tau)$.)

4.7. The Incompleteness Theorem

4.7.1. Definition. If φ is a formula, we define $\mathcal{D}(\varphi)$ (the **diagonalisation** of φ) by

$$\mathcal{D}(\varphi) := \varphi(\mathbf{x}_0/\underline{\ulcorner\varphi\urcorner})$$

i.e., we substitute the term for $\ulcorner\varphi\urcorner$ for the free variable $\mathbf{x}_0$.

In this section we will show that there is a formula $\mathit{diag}(m, n)$ expressing the relationship "n is the code of diagonalisation of the formula coded by m."

First we have to translate the computation of the function $\ulcorner\urcorner$ into the language of Peano Arithmetic.

More precisely, we want to find a formula $\psi(x, y)$ such that $\mathbb{N} \models \psi(n, m)$ iff $m = \underline{\ulcorner n \urcorner}$.

We say that a sequence is a "successor sequence," if it has the form

$$\langle \ulcorner \mathtt{SO} \urcorner, \ulcorner \mathtt{SSO} \urcorner, \ldots, \ulcorner \mathtt{S}^n \mathtt{0} \urcorner \rangle, \quad \text{i.e.,} \quad = \langle \ulcorner \underline{1} \urcorner, \ulcorner \underline{2} \urcorner, \ldots, \ulcorner \underline{n} \urcorner \rangle.$$

4.7.2. Definition.

$$\mathit{succSeq}(\mathtt{c}) := \mathit{seq}(\mathtt{c}) \land \mathit{entry}(\mathtt{c}, \underline{1}, \underline{\ulcorner \mathtt{SO} \urcorner}) \land \forall \mathtt{n} \forall \mathtt{e} \forall \mathtt{e}' \; \mathit{entry}(\mathtt{c}, \mathtt{n}, \mathtt{e}) \land \mathit{entry}(\mathtt{c}, \mathtt{Sn}, \mathtt{e}') \rightarrow \mathtt{e}' = \underline{2}^{\underline{\ulcorner \mathtt{S} \urcorner}} \cdot \underline{3}^{\mathtt{e}}.$$

4.7.3. Definition.

$$\mathit{number}(\mathtt{n}, \mathtt{m}) := [\exists \mathtt{c} \; \mathit{succSeq}(\mathtt{c}) \land \mathit{entry}(\mathtt{c}, \mathtt{n}, \mathtt{m})] \lor [\mathtt{n} = 0 \land \mathtt{m} = \underline{\ulcorner 0 \urcorner}].$$

4.7.4. Fact.

$$\{\langle n, m \rangle : \mathbb{N} \models \mathit{number}(n, m)\} = \{\langle 0, \ulcorner 0 \urcorner \rangle, \langle 1, \ulcorner \mathtt{SO} \urcorner \rangle, \langle 2, \ulcorner \mathtt{SSO} \urcorner \rangle, \ldots\} = \{\langle 0, 2 \rangle, \langle 1, 144 \rangle, \ldots\}$$

Now we define the diagonalisation formula:

4.7.5. Definition.

$$\mathit{diag}(\mathtt{f}, \mathtt{f}') := \exists \mathtt{t} (\mathit{number}(\mathtt{f}, \mathtt{t}) \land \mathit{fmsub}(\mathtt{f}, \underline{\ulcorner \mathtt{x_0} \urcorner}, \mathtt{t}, \mathtt{f}')).$$

4.7.6. Fact. $\mathbb{N} \models \mathit{diag}(\ulcorner \varphi \urcorner, n) \leftrightarrow n = \ulcorner \mathcal{D}(\varphi) \urcorner$.

Proof: Let $\psi := \mathcal{D}(\varphi) = \varphi(\mathtt{x_0} / \underline{\ulcorner \varphi \urcorner})$. Then $\mathbb{N} \models \mathit{fmsub}(\ulcorner \varphi \urcorner, \ulcorner \mathtt{x_0} \urcorner, \ulcorner \underline{\ulcorner \varphi \urcorner} \urcorner, \ulcorner \psi \urcorner)$. So

$$\begin{aligned}
\mathit{diag}(\ulcorner \varphi \urcorner, n) &\leftrightarrow \exists t \; \mathit{number}(\ulcorner \varphi \urcorner, t) \land \mathit{fmsub}(\ulcorner \varphi \urcorner, \ulcorner \mathtt{x_0} \urcorner, t, n) \\
&\leftrightarrow \exists t \; (t = \ulcorner \underline{\ulcorner \varphi \urcorner} \urcorner) \land \mathit{fmsub}(\ulcorner \varphi \urcorner, \ulcorner \mathtt{x_0} \urcorner, t, n) \\
&\leftrightarrow \mathit{fmsub}(\ulcorner \varphi \urcorner, \ulcorner \mathtt{x_0} \urcorner, \ulcorner \underline{\ulcorner \varphi \urcorner} \urcorner, n) \\
&\leftrightarrow n = \ulcorner \psi \urcorner.
\end{aligned}$$

Now we will find a formula $\varphi(\mathrm{x}_0)$ that "says"

x_0 codes a formula ψ with the property: $\mathcal{D}(\psi)$ is not a theorem.

We let

$$\varphi(\mathrm{x}_0) := \mathit{formula}(\mathrm{x}_0) \wedge \exists \mathrm{y}[\mathit{diag}(\mathrm{x}_0, \mathrm{y}) \wedge \neg \mathit{theorem}(\mathrm{y})].$$

Now let

$$\sigma := \mathcal{D}(\varphi) = \varphi(\mathrm{x}_0 / \underline{\ulcorner \varphi \urcorner}).$$

φ has only one free variable (x_0), so σ is a closed formula.

σ says:

$$\ulcorner \varphi \urcorner \text{ codes a formula } \psi, \text{ and } \nvdash_{\mathrm{PA}} \mathcal{D}(\psi)$$

Since $\ulcorner \varphi \urcorner$ codes the formula φ, σ really says: $\nvdash_{\mathrm{PA}} \mathcal{D}(\varphi)$. Since $\mathcal{D}(\varphi)$ is exactly σ, σ says: "I am not a theorem of PA."

Now if σ were a theorem of PA, σ would be true (= valid in $\mathbb{N}$), so by its own testimony it would **not** be a theorem of PA.

So σ is not a theorem of PA. Since this is exactly what σ claims, σ is valid in $\mathbb{N}$.

In other words, there is a closed formula of PA which is true but not provable in PA.

Formally, the above argument can be phrased as follows:

(Note that if $\psi(x)$ is a formula with one free variable x, then $\mathbb{N} \models \psi(n)$ iff $\mathbb{N} \models \psi(\underline{n})$.)

Let $\sigma = \mathcal{D}(\varphi)$. So

$$\sigma \quad = \quad \mathit{formula}(\underline{\ulcorner \varphi \urcorner}) \wedge \exists \mathrm{y}(\mathit{diag}(\underline{\ulcorner \varphi \urcorner}, \mathrm{y}) \wedge \neg \mathit{theorem}(\mathrm{y})).$$

Since φ is a formula, $\mathbb{N} \models \mathit{formula}(\ulcorner \varphi \urcorner)$. So

$$\mathbb{N} \models \sigma \quad \leftrightarrow \quad \exists \mathrm{y}\, \mathit{diag}(\ulcorner \varphi \urcorner, \mathrm{y}) \wedge \neg \mathit{theorem}(\mathrm{y}).$$

We already remarked that $\mathbb{N} \models \mathit{diag}(\ulcorner \varphi \urcorner, \mathrm{y}) \leftrightarrow \mathrm{y} = \ulcorner \mathcal{D}(\varphi) \urcorner$ i.e., iff $\mathrm{y} = \ulcorner \sigma \urcorner$. So we get

$$\mathbb{N} \models \sigma \leftrightarrow \exists \mathrm{y}[\mathrm{y} = \ulcorner \sigma \urcorner \wedge \neg \mathit{theorem}(\mathrm{y})].$$

and therefore

$$(*) \qquad \mathbb{N} \models \sigma \leftrightarrow \neg \mathit{theorem}(\ulcorner \sigma \urcorner)$$

or equivalently

$$(\star) \qquad \mathbb{N} \models \neg \sigma \leftrightarrow \mathit{theorem}(\ulcorner \sigma \urcorner).$$

Now assume $\vdash_{\mathrm{PA}} \sigma$. Then $\mathbb{N} \models \mathit{theorem}(\sigma)$. So by ($\star$), $\mathbb{N} \models \neg \sigma$ which is impossible because $\vdash_{\mathrm{PA}} \sigma$ implies $\mathbb{N} \models \sigma$.

So $\nvdash_{\mathrm{PA}} \sigma$. This implies $\mathbb{N} \models \neg \mathit{theorem}(\sigma)$, hence by ($*$), $\mathbb{N} \models \sigma$.

So we finished the proof of the incompleteness theorem.

4.8. Other Axiom Systems

We have seen that the axioms of Peano Arithmetic do not capture all truths about the natural numbers, i.e., although all formulas provable in PA are valid in the model $\mathbb{N}$, there are formulas which are valid but not provable.

Is this an indication that we have chosen a "wrong" set of axioms? Should we have included more axioms? Maybe we could add the true but unprovable formula that we found in 3.7 to our axiom system and then get a complete theory? We will show in this section that this is not possible: Whatever "reasonable" way we choose to extend our system of PA, the same incompleteness phenomenon will occur.

What does "reasonable" mean here? There are two possible demands: the most natural requirement to put on an axiom system is that is should be "decidable." The point is that we only want to admit axiom systems which have the property that there is a finite decision procedure which, given any formula φ as input, after a finite time will either produce the output "φ is an axiom" or "φ is not an axiom." PA has this property, and PA$\cup\{\varphi\}$ has this property for any formula φ, but, for example, the set of all valid formulas does not have this property.

"Decidability" is an informal concept. The usual formalization of this concept is the notion of "recursiveness," which we will give in 4.9.11.

This motivates another, much weaker, demand on our axiom system: at least the set of axioms should be "definable." Again "definable" is an informal concept, which we formalize by the notion of "expressible" (by a formula), sometimes also called "first order definable." That is, we want a single formula $\chi(\mathrm{x})$ which "expresses" the fact that x is the Gödel number of an axiom, i.e., such that a formula φ is a nonlogical axiom of our system iff $\mathbb{N} \models \chi(\ulcorner\varphi\urcorner)$. We will see below that no expressible and hence also no recursive set of axioms can capture all truths about the natural numbers, i.e., for any such axiom set there will be a true but unprovable formula. Then we will show that Gödel's incompleteness theorem applies for any recursive set of axioms which includes the axioms of PA.

The method from the previous section actually allows us to prove the following result:

4.8.1. Theorem (Gödel's self-referential lemma). If $\varphi(\mathrm{x}_0)$ is a formula with free variable x_0, then there is a closed formula σ such that

$$\mathbb{N} \models \sigma \leftrightarrow \varphi(\ulcorner\sigma\urcorner).$$

Proof: We define a formula $\psi(x_0)$ by

$$\psi(x_0) := \mathit{formula}(x_0) \wedge \exists y\big(\mathit{Diag}(x_0, y) \wedge \varphi(y)\big).$$

We have shown in the previous section that $\mathbb{N} \models \mathit{Diag}(\ulcorner\chi\urcorner, y) \leftrightarrow y = \ulcorner\mathcal{D}(\chi)\urcorner$, so for any formula χ we get $\mathbb{N} \models \psi(\ulcorner\chi\urcorner) \leftrightarrow \varphi(\ulcorner\mathcal{D}(\chi)\urcorner)$. In particular, we get

$$\mathbb{N} \models \psi(\ulcorner\psi\urcorner) \leftrightarrow \varphi(\ulcorner\mathcal{D}(\psi)\urcorner).$$

Writing σ for the closed formula $\psi(\ulcorner\psi\urcorner)$ (i.e., for $\mathcal{D}(\psi)$), we get

$$\mathbb{N} \models \sigma \leftrightarrow \varphi(\ulcorner\sigma\urcorner).$$

4.8.2. Corollary (the Undefinability of Truth). The set of codes of closed formulas which are valid in $\mathbb{N}$ cannot be expressed by a formula. That is: there is no formula $\mathit{True}(x)$ such that for all closed formulas φ, $\mathbb{N} \models \mathit{True}(\ulcorner\varphi\urcorner) \leftrightarrow \varphi$. (See also exercise 6 in the next section.)

Proof: Apply the self-referential lemma to the formula $\neg\mathit{True}(x_0)$. This will yield a closed formula σ for which $\mathbb{N} \models \sigma \leftrightarrow \neg\mathit{True}(\ulcorner\sigma\urcorner)$. Hence we get $\mathbb{N} \models \sigma \leftrightarrow \neg\sigma$, which is impossible.

The next theorem will show us that also no set of axioms that is definable by a formula will allow us to derive exactly the valid formulas.

4.8.3. Theorem. Let A be a set of formulas which are valid in $\mathbb{N}$, and assume that $\mathit{Axiom}_A(x_0)$ is a formula such that $\varphi \in A$ iff $\mathbb{N} \models \mathit{Axiom}_A(\ulcorner\varphi\urcorner)$. Then

(1) Every formula which can be derived from A is valid in $\mathbb{N}$.
(2) There is a closed formula which is valid in $\mathbb{N}$ but cannot be derived from A.

Proof: (1) is true for any set A of formulas all of which are valid in $\mathbb{N}$, by the soundness of our derivation system (see 1.4.20).

To prove (2), we define the following formula (compare 4.5.20).

4.8.4. Definition.

$$\begin{aligned}\mathit{Derivation}_A(c) :=& \mathit{seq}(c) \wedge \forall n \forall e\ \mathit{entry}(c, n, e) \rightarrow \\ &\big[\mathit{logicalaxiom}(e) \vee \mathit{Axiom}_A(e) \vee \\ &\exists k, l{<}n \exists f \exists g\ \mathit{entry}(c, k, f) \wedge \mathit{entry}(c, l, g) \wedge \\ &\mathit{implies}(f, e, g)\big].\end{aligned}$$

(*logicalaxiom* was defined in 4.5.17.)

Clearly, $\mathbb{N} \models Derivation_A(n)$ iff $n = \#(\ulcorner\varphi_1\urcorner, \ulcorner\varphi_2\urcorner, \ldots, \ulcorner\varphi_k\urcorner)$ (see 4.4.1) for some sequence $(\varphi_1, \ldots, \varphi_k)$ of formulas which is a derivation from A.

4.8.5. Definition.
$Theorem_A(\mathtt{f}) :=formula(\mathtt{f}) \wedge \exists \mathtt{c} \exists \mathtt{n} Derivation_A(\mathtt{c}) \wedge entry(\mathtt{c}, \mathtt{n}, \mathtt{f})$.

So we easily get:

4.8.6. Fact. $A \vdash \varphi$ iff $\mathbb{N} \models Theorem_A(\varphi)$.

So we can apply the self-referential lemma again and get a formula σ such that

$$\mathbb{N} \models \sigma \leftrightarrow \neg Theorem_A(\ulcorner\sigma\urcorner).$$

The same proof as in the previous section now shows that we must have $\mathbb{N} \models \sigma$, but σ cannot be derived from the axioms A. This finishes the proof of 4.8.3.

4.8.7. Example. Let σ_0 be a formula which is valid in $\mathbb{N}$ but not provable in PA. Let A be $\mathrm{PA} \cup \{\sigma_0\}$. Then there is a closed formula σ_1 which is valid in $\mathbb{N}$ but not provable from A.

Proof: Let $n := \ulcorner\sigma_0\urcorner$. Then A is definable by the formula $axiom(\mathtt{x}) \vee \mathtt{x} = \underline{n}$, so we can apply 4.8.3 and get the desired conclusion.

4.9. Bounded Formulas

See section 3.1 for a definition of the language of Peano Arithmetic. Remember that we treat all logical connectives other than $\wedge$ and $\neg$ as abbreviations only (so as to shorten inductive proofs), and that $\exists \mathtt{x} \cdots$ abbreviates $\neg \forall \mathtt{x} \neg(\cdots)$.

Often we will replace a formula φ by an "equivalent" formula φ'. "Equivalent" can mean either equivalence in $\mathbb{N}$, i.e., $\mathbb{N} \models \varphi \leftrightarrow \varphi'$, or provable equivalence, i.e, $\vdash_{\mathrm{PA}} \varphi \leftrightarrow \varphi'$. Unless otherwise indicated, we are only interested in the first (weaker) concept of equivalence — although in most of the cases we will consider, a little more work would show that in fact the second (stronger) equivalence holds also.

For the next definition, recall that $\forall \mathtt{x} {\leq} \tau\, \varphi$ abbreviates for $\forall \mathtt{x}(\mathtt{x} {\leq} \tau \rightarrow \varphi)$, and $\exists \mathtt{x} {\leq} \tau\, \varphi$ abbreviates $\exists \mathtt{x}(\mathtt{x} {\leq} \tau \wedge \varphi)$.

4.9.1. Definition. The set of **bounded formulas** is the inductive structure defined as follows:

(1) Every atomic formula is a bounded formula.

(2) If ψ is a bounded formula, x a variable, and τ a term not containing the variable x, then the formula

$$\forall x{\le}\tau\ \psi$$

is a bounded formula.

(3) If ψ is a bounded formula, then $\neg\psi$ is a bounded formula.

(4) If ψ_1 and ψ_2 are bounded formulas, then $\psi_1 \wedge \psi_2$ is a bounded formula.

In other words, the blocks are the atomic formulas, the functions $F_\neg$ and $F_\wedge$ from 1.1.12 are operators, and for every variable x and every term τ not containing x we have a unary operator $A_{x,\tau}$ defined by $A_{x,\tau}(\psi) = (\forall x{\le}\tau\ \psi)$.

4.9.2. Example. The formula

$$[\underline{2} \le p \wedge \forall x{\le}p\, \forall y{\le}p\, [x{\cdot}y = p \rightarrow (x = p \vee y = p)]]$$

is a bounded formula.

4.9.3. Definition. A Σ^0_1-formula is a formula of the form $\exists x\psi$, where x is a variable and ψ is a bounded formula. A Π^0_1-formula is a formula of the form $\forall x\psi$, where x is a variable and ψ is a bounded formula.

Note that although, strictly speaking, bounded formulas are not Σ^0_1 formulas, every bounded formula is equivalent to a Σ^0_1 formula: $\mathbb{N} \models \psi \leftrightarrow \exists y\, \psi$, where y is any variable which does not appear in ψ.

4.9.4. Lemma (Closure properties of Σ^0_1 formulas). If φ_1 and φ_2 are Σ^0_1 formulas, and if x is a variable not free in the term τ, then the following are equivalent to Σ^0_1-formulas:

(a) $\exists x\, \varphi_1$

(b) $\varphi_1 \wedge \varphi_2$

(c) $\varphi_1 \vee \varphi_2$

(d) $\exists x{\le}\tau\, \varphi_1$

(e) $\forall x{\le}\tau\, \varphi_1$

We sometimes express this more sloppily (treating equivalent formulas as identical) as "Σ^0_1 formulas are closed under $\wedge$, $\vee$, existential quantification and bounded quantification."

Proof:

(a) Let φ_1 be $\exists y\psi$, where ψ is a bounded formula. We have to show that

$$\exists x(\exists y\, \psi)$$

is equivalent to a Σ_1^0 formula. Let $\mathbf{z}$ be any variable not appearing in φ_1. Then $\exists \mathbf{x} \exists \mathbf{y} \psi$ is equivalent to $\exists \mathbf{z}\, \exists \mathbf{x} \leq \mathbf{z}\, \exists \mathbf{y} \leq \mathbf{z}\, \psi$.
(b) Let $\varphi_1 = \exists \mathbf{x}_1\, \psi_1$, and $\varphi_2 = \exists \mathbf{x}_2\, \psi_2$. We have to show that

$$(\exists \mathbf{x}_1\, \psi_1) \wedge (\exists \mathbf{x}_2\, \psi_2)$$

is equivalent to a Σ_1^0 formula.

We may assume that the variable $\mathbf{x}_1$ does not appear in φ_2 — otherwise we can find a variable $\mathbf{z}_1$ not appearing in φ_1 and φ_2 at all, and use the fact that $\exists \mathbf{x}_1 \psi_1$ is equivalent to $\exists \mathbf{z}_1 (\psi_1(\mathbf{x}_1/\mathbf{z}_1))$. Similarly, we may assume that $\mathbf{x}_2$ does not appear in φ_1.

Now $\varphi_1 \wedge \varphi_2$ is equivalent to the formula $\exists \mathbf{x}_1\, \exists \mathbf{x}_2\, (\psi_1 \wedge \psi_2)$. By (a), this formula is equivalent to a Σ_1^0 formula.
(c) We can deal with

$$(\exists \mathbf{x}_1\, \psi_1) \vee (\exists \mathbf{x}_2\, \psi_2)$$

in a similar way.
(d) The formula

$$\exists \mathbf{x} \leq \tau (\exists \mathbf{y} \psi)$$

is really $\exists \mathbf{x}(\mathbf{x} \leq \tau \wedge \exists \mathbf{y} \psi)$, so by (a) and (b) it is equivalent to a Σ_1^0 formula.
(e) Let φ_1 be of the form $\exists \mathbf{y} \psi$ for some bounded formula ψ. So we have to show that

$$\forall \mathbf{x} \leq \tau (\exists \mathbf{y}\, \psi)$$

is equivalent to a Σ_1^0 formula. Let $\mathbf{z}$ be a variable not occurring in $\forall \mathbf{x} \leq \tau\, \varphi_1$. Since $\mathbb{N} \models \forall \mathbf{x} \leq \tau\, \exists \mathbf{y} \leq \mathbf{z}\, \psi \rightarrow \forall \mathbf{x} \leq \tau \exists \mathbf{y}\, \psi$, we have

$$\mathbb{N} \models \exists \mathbf{z}\, \forall \mathbf{x} \leq \tau\, \exists \mathbf{y} \leq \mathbf{z}\, \psi \rightarrow \forall \mathbf{x} \leq \tau \exists \mathbf{y}\, \psi.$$

Now for the converse direction. Assume that $\mathbf{x}_1, \ldots, \mathbf{x}_k$ are the free variables of $\varphi = \varphi(\mathbf{x}_1, \ldots, \mathbf{x}_k)$, and let $\psi = \psi(\mathbf{x}, \mathbf{y}, \mathbf{x}_1, \ldots, \mathbf{x}_k)$. Let $n_1, \ldots, n_k$ be natural numbers. We have to show that if

$$(\star) \qquad \mathbb{N} \models \forall \mathbf{x} \leq \tau(n_1, \ldots, n_k)\, \exists \mathbf{y}\, \psi(\mathbf{x}, \mathbf{y}, n_1, \ldots, n_k)$$

holds, then also

$$(\star\star) \qquad \mathbb{N} \models \exists \mathbf{z}\, \forall \mathbf{x} \leq \tau(n_1, \ldots, n_k)\, \exists \mathbf{y} \leq \mathbf{z}\, \psi(\mathbf{x}, \mathbf{y}, n_1, \ldots, n_k)$$

will hold. Now the $\mathbb{N}$-term $\tau(\mathbf{x}_1/n_1, \ldots, \mathbf{x}_k/n_k)$ is closed, so it has a value, which we will call n. Then $(\star)$ says that there are natural numbers m_0, m_1, $\ldots$, m_n such that for $i = 0, 1, \ldots, n$ we have

$$\mathbb{N} \models \psi(i, m_i, n_1, \ldots, n_k).$$

Now let m be the maximum of $m_0, \ldots, m_i$. Then

$$\mathbb{N} \models \forall \mathsf{x}{\leq}n\, \exists \mathsf{y}{\leq}m\, \psi(\mathsf{x}, \mathsf{y}, n_1, \ldots, n_k)$$

which implies ($\star\star$). (See also exercise 3(b) for an alternative proof.)

4.9.5. Remark. We shall see later that neither the **negation** of a Σ_1^0 formula nor the universal (unbounded) quantification of a Σ_1^0 formula is in general equivalent to a Σ_1^0 formula.

4.9.6. Lemma. For every bounded closed formula φ there is a quantifier-free closed equivalent formula $\hat{\varphi}$. Moreover, $\vdash_{\mathsf{PA}} \varphi \leftrightarrow \hat{\varphi}$.

Proof: We will use induction on φ (using the inductive system given in 4.9.1). The only nontrivial step is the following:

Let $\varphi = \forall \mathsf{x}{\leq}\tau\, \psi$ be a closed formula. So $\psi = \psi(\mathsf{x})$ contains no free variables except possibly x. τ is a closed term, so it has a value $n := \tau^{\mathsf{N}}$.

Now we have $\vdash_{\mathsf{PA}} \tau = \underline{n}$, so $\vdash_{\mathsf{PA}} \mathsf{x}{\leq}\tau \leftrightarrow \mathsf{x}{\leq}\underline{n}$. By 4.3.5 we also have

$$\vdash_{\mathsf{PA}} \mathsf{x}{\leq}\underline{n} \quad \leftrightarrow \quad \mathsf{x} = \mathsf{0} \vee \mathsf{x} = \mathsf{S0} \vee \cdots \vee \mathsf{x} = \underline{n}.$$

Now PA can prove that the following are equivalent:

$$\forall \mathsf{x}{\leq}\tau\, \psi(\mathsf{x})$$
$$\forall \mathsf{x}\big(\mathsf{x}{\leq}\underline{n} \rightarrow \psi(\mathsf{x})\big)$$
$$\forall \mathsf{x}\big[(\mathsf{x} = \mathsf{0} \rightarrow \psi(\mathsf{x})) \wedge (\mathsf{x} = \mathsf{S0} \rightarrow \psi(\mathsf{x})) \wedge \cdots \wedge (\mathsf{x} = \underline{n} \rightarrow \psi(\mathsf{x}))\big]$$
$$\forall \mathsf{x}(\mathsf{x} = \mathsf{0} \rightarrow \psi(\mathsf{x})) \wedge \forall \mathsf{x}(\mathsf{x} = \mathsf{S0} \rightarrow \psi(\mathsf{x})) \wedge \cdots \wedge \forall \mathsf{x}(\mathsf{x} = \underline{n} \rightarrow \psi(\mathsf{x}))$$
$$\psi(\mathsf{0}) \wedge \psi(\mathsf{S0}) \wedge \cdots \wedge \psi(\underline{n}).$$

Since each of the formulas $\psi(\underline{k})$ is a closed bounded formula, each of these formulas is equivalent to a closed formula $\widehat{\psi(\underline{k})}$ without quantifiers, and φ is equivalent to the conjunction of these formulas.

4.9.7. Explanation. If φ is a *closed* bounded formula, then it is easy to find out whether φ holds in the natural model or not, since there are only finitely many values to check.

If a set A of natural numbers is defined by a bounded formula, say

$$A = \{n : \mathbb{N} \models \psi(n)\}$$

where $\psi(\mathsf{x})$ is bounded and has only one free variable x, then we can easily (informally) see that A is decidable, because for each number n there are only finitely many values to check until we know if n is in A or not. For example, if $A = \{n : \mathbb{N} \models \exists x{\le}n\ x{\cdot}x + x = x^n + n\}$, then to find out if $2 \in A$, we only have to check 3 possible values of x:

(1) Is $0{\cdot}0 + 0 = 0^2 + 2$?
(2) Is $1{\cdot}1 + 1 = 1^2 + 2$?
(3) Is $2{\cdot}2 + 2 = 2^2 + 2$?

Similarly, if $B = \{n : \mathbb{N} \models \forall y{\le}n\ \exists y{\le}x\ \psi(x, y, n)\}$ (ψ quantifier-free) and we want to know if 2 is in B, we have to evaluate ψ 6 times:

(1) Does $\mathbb{N} \models \exists y{\le}0\ \psi(0, y, n)$? (1 evaluation of φ, with $y = 0$)
(2) Does $\mathbb{N} \models \exists y{\le}1\ \psi(1, y, n)$? (2 evaluations of φ, with $y = 0, 1$)
(3) Does $\mathbb{N} \models \exists y{\le}2\ \psi(2, y, n)$? (3 evaluations of φ, with $y = 0, 1, 2$)

4.9.8. Lemma.

(1) If τ is a closed term, $\tau^{\mathbb{N}} = n$, then $\vdash_{PA} \tau = \underline{n}$.
(2) If φ is a closed quantifier-free formula, then $\mathbb{N} \models \varphi$ iff $\vdash_{PA} \varphi$.
(3) If φ is a closed bounded formula, then $\mathbb{N} \models \varphi$ iff $\vdash_{PA} \varphi$.
(4) If φ is a closed Σ_1^0 formula, then $\mathbb{N} \models \varphi$ iff $\vdash_{PA} \varphi$.

Proof:

(1) By induction on τ.

(2) By the soundness theorem, we know that "$\vdash_{PA} \varphi$" implies "$\mathbb{N} \models \varphi$." For the other direction, use induction on φ.

(3) We can find a quantifier-free closed formula $\hat{\varphi}$ such that $\vdash_{PA} \varphi \leftrightarrow \hat{\varphi}$. So

$$\vdash_{PA} \varphi \text{ iff } \vdash_{PA} \varphi' \text{ iff } \mathbb{N} \models \varphi' \text{ iff } \mathbb{N} \models \varphi.$$

(4) Again, one direction follows from the soundness theorem. For the other direction, let $\varphi = \exists x\psi$, where $\psi = \psi(x)$ is a bounded formula with no free variables except x. Since $\mathbb{N} \models \exists x\, \psi(x)$, there is a natural number such that $\mathbb{N} \models \psi(n)$, i.e., $\mathbb{N} \models \psi(\underline{n})$. So by (3), $\vdash_{PA} \psi(\underline{n})$, which easily implies (see 1.4.31) $\vdash_{PA} \exists x\psi(x)$.

4.9.9. Definition. We say that a subset $A \subseteq \mathbb{N}$ is **expressed** by the formula $\varphi(x)$ if $A = \{n \in \mathbb{N} : \mathbb{N} \models \varphi(n)\}$. If the set A is defined by a property P then we say that this property P is expressed by φ.

Similarly, a set $A \subseteq \mathbb{N}^k$ is expressed by a formula $\varphi(\mathrm{x}_1, \ldots, \mathrm{x}_k)$ (with free variables among $\mathrm{x}_1, \ldots, \mathrm{x}_k$) if

$$A = \{(n_1, \ldots, n_k) \in \mathbb{N}^k : \mathbb{N} \models \varphi(n_1, \ldots, n_k)\}.$$

4.9.10. Examples. (1) The property "even" is expressed by the formula $\exists \mathrm{y}\, \mathrm{y} + \mathrm{y} = \mathrm{x}$, since the set of even numbers is

$$\{0, 2, 4, \ldots, \} \;=\; \{n \in \mathbb{N} : \mathbb{N} \models \exists \mathrm{y}\, \mathrm{y} + \mathrm{y} = n\}.$$

(2) The property "divides" is expressed by the formula $\varphi(\mathrm{x}, \mathrm{y}) = (\exists \mathrm{z}\, \mathrm{x}{\cdot}\mathrm{z} = \mathrm{y})$.

Note: we have already seen a set which is not expressed by any formula, namely, "n is the Gödel number of a true sentence." However, we shall see below that most of the sets occurring in "traditional" number theory are expressible by formulas, often even by bounded or Σ_1^0 formulas.

4.9.11. Definition. A set $A \subseteq \mathbb{N}$ is called **recursively enumerable** (or "r.e." for short) if it can be expressed by a Σ_1^0-formula, i.e., if there is a Σ_1^0 formula $\varphi(\mathrm{x})$ with one free variable x such that

$$A = \{n \in \mathbb{N} : \mathbb{N} \models \varphi(n)\}.$$

Similarly, a set $A \subseteq \mathbb{N}^k$, i.e., a k-ary relation, is called recursively enumerable if there is a Σ_1^0 formula φ with k free variables $\mathrm{x}_1, \ldots, \mathrm{x}_k$ such that

$$A = \{(n_1, \ldots, n_k) \in \mathbb{N} : \mathbb{N} \models \varphi(\mathrm{x}_1/n_1, \ldots, \mathrm{x}_k/n_k)\}.$$

A set $A \subseteq \mathbb{N}^k$ is called **recursive** if both A and $\mathbb{N}^k - A$ (i.e., the complement of A are r.e.

Membership for an r.e. set is in general not decidable. Let $A = \{n : \exists \mathrm{x}\, \psi(\mathrm{x}, n)\}$, where ψ is bounded. Then we could try out all possible values for x to find out if $n \in A$. Of course, if n IS in A, this method will work, because we will eventually find a witness, i.e., a number k such that $\mathbb{N} \models \psi(k, n)$. But if n is not in A, we may never find out.

Recursive sets, on the other hand, ARE decidable. Indeed, let $A = \{n : \exists \mathrm{x}\, \psi_1(\mathrm{x}, n)\}$, where ψ_1 is a bounded formula. Since A is recursive, there is also another bounded formula ψ_2 with the property that $\mathbb{N} - A = \{n : \exists \mathrm{x}_2\, \psi_2(\mathrm{x}_2, n)\}$. Now the following procedure will decide if a given number n is in A:

If $\mathbb{N} \models \psi_1(0, n)$, then $n \in A$. If $\mathbb{N} \models \psi_2(0, n)$, then $n \notin A$.
If neither of the above hold, we continue:
If $\mathbb{N} \models \psi_1(1, n)$, then $n \in A$. If $\mathbb{N} \models \psi_2(1, n)$, then $n \notin A$.
If we are still not done, we try again with 2, then with 3, etc.

Since one of $\exists \mathrm{x}\, \psi_1(\mathrm{x}, n)$ and $\exists \mathrm{x}\, \psi_2(\mathrm{x}, n)$ must hold, we will be done after a finite number of steps.

4.9.12. Definition. A function $f : \mathbb{N} \to \mathbb{N}$ is called **recursive** iff the graph of f (i.e., the set $\{(m, f(m)) : m \in \mathbb{N}\}$) is recursive. Similarly, a function $g : \mathbb{N}^k \to \mathbb{N}$ is recursive if the set $\{(m_1, \ldots, m_k, f(m_1, \ldots, m_k)) : m_1, \ldots, m_k \in \mathbb{N}\}$ is recursive.

4.9.13. Example. Addition is a recursive function from $\mathbb{N}^2$ to $\mathbb{N}$. Moreover, any function given by a term in the language of PA is recursive.

4.9.14. Example. (1) Every set defined by a bounded formula is recursive.
(2) If $f : \mathbb{N} \to \mathbb{N}$ is a recursive function and $\psi(\mathrm{x}, \mathrm{y})$ is a bounded formula, then the set

$$A = \{n \in \mathbb{N} : \mathbb{N} \models \exists \mathrm{y}(\mathrm{y} \leq f(n) \wedge \psi(n, \mathrm{y}))\}$$

is recursive.

Proof: (1) is clear.

For (2), first we see that

$$n \in A \quad \text{iff} \quad \mathbb{N} \models \exists \mathrm{z}\, (\mathrm{z} = f(n) \wedge \exists \mathrm{y} \leq \mathrm{z}\ \psi(n, \mathrm{y})$$

(where we consider the subformula $\mathrm{z} = f(n)$ as an abbreviation for the Σ^0_1 formula $\varphi(n, \mathrm{z})$ which describes the graph of f). So A is a Σ^0_1 set.
But we also have

$$n \notin A \quad \text{iff} \quad \mathbb{N} \models \exists \mathrm{z}\, (\mathrm{z} = f(n) \wedge \forall \mathrm{y} \leq \mathrm{z}\ \neg \psi(n, \mathrm{y}).$$

so also the complement of A is Σ^0_1. So A is recursive.

It turns out that for (total) functions, the concepts of "r.e." and "recursive" coincide. That is, if the graph of a function from $\mathbb{N}$ to $\mathbb{N}$ is r.e., then also its complement in $\mathbb{N} \times \mathbb{N}$ will be r.e.:

4.9.15. Lemma. Let $f : \mathbb{N} \to \mathbb{N}$ be a (total) function, and assume that the set $A := \{(m, f(m)) : m \in \mathbb{N}\}$ is recursively enumerable. Then f is recursive. (We sometimes write this as "f is recursive iff the property $n = f(m)$ can be expressed by a Σ_1^0-formula.")
Similarly, if $f : \mathbb{N}^k \to \mathbb{N}$ is a function, and the set

$$A := \{(m_1, \ldots, m_k, f(m_1, \ldots, m_k)) : m_1, \ldots, m_k \in \mathbb{N}\}$$

is recursively enumerable, then f is recursive.

Proof: We have to show that A is recursive. We already know that A is r.e., say $A = \{(m, n) : \mathbb{N} \models \exists \mathrm{y}\varphi(m, n, \mathrm{y})\}$, where φ is a bounded formula. We have to show that the complement of A is also r.e. Note that $(m, n) \notin A$ iff n is not equal to $f(m)$, i.e., if there is $n' \neq n$ such that $n' = f(m)$. Hence

$$(m, n) \notin A \quad \leftrightarrow \quad \mathbb{N} \models \exists \mathrm{z}(\mathrm{z} \neq n \wedge \exists \mathrm{y}\, \varphi(m, \mathrm{z}, \mathrm{y})).$$

(We will write this more informally as $n \neq f(m) \leftrightarrow \exists \mathrm{z}(\mathrm{z} \neq n \wedge \mathrm{z} = f(m))$). By 4.9.4, this formula is equivalent to a Σ_1^0 formula.

Exercises

1. Show that if ψ_1 and ψ_2 are bounded formulas, then also $\psi_1 @ \psi_2$ is a bounded formula, for all @ in $\{\wedge, \vee, \rightarrow, \leftrightarrow, |\}$. Also, if τ is a term not containing the free variable x, and ψ is bounded, then $\exists \mathrm{x} {\leq} \tau\, \psi$ is also bounded.

2. Define the set of "bounded formulas in prefix form" to be the following inductive structure: blocks are all quantifier-free formulas, and for every term τ and every variable x not in $\mathrm{Fr}(\tau)$ there are two operators $A_{\mathrm{x},\tau}$ and $E_{\mathrm{x},\tau}$, defined by $A_{\mathrm{x},\tau}(\varphi) = \forall \mathrm{x} {\leq} \tau\, \varphi$ and $E_{\mathrm{x},\tau}(\varphi) = \exists \mathrm{x} {\leq} \tau\, \varphi$. Show that every bounded formula is equivalent to a bounded formula in prefix form.

3. Let ψ be a formula (not necessarily bounded), x, y, z, t distinct variables, and τ a term in which x does not occur. Also assume that **z** occurs neither in τ nor in ψ.
(a) Show that

$$\vdash_{\mathrm{PA}} \forall \mathrm{x} {\leq} \mathrm{t}\, \exists \mathrm{y}\, \psi \quad \leftrightarrow \quad \exists \mathrm{z}\, \forall \mathrm{x} {\leq} \mathrm{t}\, \exists \mathrm{y} {\leq} \mathrm{z}\, \psi$$

(b) Show that

$$\vdash_{\mathrm{PA}} \forall \mathrm{x} {\leq} \tau\, \exists \mathrm{y}\, \psi \quad \leftrightarrow \quad \exists \mathrm{z}\, \forall \mathrm{x} {\leq} \tau\, \exists \mathrm{y} {\leq} \mathrm{z}\, \psi$$

(Hint: The $\leftarrow$ direction in (a) should be easy. For the $\rightarrow$ direction, use induction on $\mathtt{t}$, i.e., writing $\varphi(\mathtt{t})$ for the formula $\forall \mathtt{x}{\le}\mathtt{t}\, \exists \mathtt{y}\, \psi \;\rightarrow\; \exists \mathtt{z}\, \forall \mathtt{x}{\le}\mathtt{t}\, \exists \mathtt{y}{\le}\mathtt{z}\, \psi$, show that $\vdash_{\mathrm{PA}} \varphi(0)$ and $\vdash_{\mathrm{PA}} \varphi(\mathtt{t}) \rightarrow \varphi(\mathtt{St})$. Hint: Use the fact that

$$\vdash_{\mathrm{PA}} \forall \mathtt{x}{\le}\mathtt{St}\varphi \leftrightarrow \forall \mathtt{x}{\le}\mathtt{t}\, \varphi \wedge \varphi(\mathtt{x}/\mathtt{St})$$

(b) follows from (a).)

4. Prove 4.9.8(2).

5. Prove lemma 4.9.8(1).

Truth Definitions

6. Let A be a set of closed formulas. We say that a formula T_A is a truth definition for A if for all $\varphi \in A$:

$$\mathbb{N} \models \varphi \leftrightarrow T_A(\ulcorner\varphi\urcorner).$$

(1) Show that *theorem* (see 4.5.21) is a truth definition for the set of closed Σ^0_1 formulas.
(2) Find a truth definition for Π^0_1 formulas.

7. For $n \ge 0$, the set of Σ^0_{n+1}-formulas is the set of formulas of the form $\exists \mathtt{x}\, \psi$, where ψ is a Π^0_n-formula(and $\mathtt{x}$ a variable), and the set of Π^0_{n+1} formulas is the set of those ψ which are of the form $\forall \mathtt{x}\psi$, where ψ is Σ^0_n. Thus, Σ^0_2 formulas are of the form $\exists \mathtt{x}\forall \mathtt{y}\, \psi$, with ψ bounded, Σ^0_3 formulas are of the form $\exists \mathtt{x}\forall \mathtt{y}\exists \mathtt{z}\, \psi$ with ψ bounded, etc.

For each n, give a truth definition for the set of closed Σ^0_n formulas, and give a truth definition for the set of closed Π^0_n formulas.

(Hint: First construct a formula $\mathit{instanceA}(\mathtt{f}, \mathtt{v}, \mathtt{x}, \mathtt{f}')$ such that

$$\mathbb{N} \models \mathit{instanceA}(\ulcorner\varphi\urcorner, \ulcorner\mathtt{x}_k\urcorner, n, \ulcorner\varphi'\urcorner)$$

holds iff φ is of the form $\forall \mathtt{x}_k\, \psi$ (for some ψ) and φ' is the formula $\psi(\mathtt{x}_k/\underline{n})$. (I.e., φ' is an "instance" of the universally quantified formula φ.) Then show that if TrueE_n is a truth definition for Σ^0_n formulas, then TrueA_{n+1}, defined by

$$\mathit{TrueA}_{n+1}(\mathtt{f}) := \;\exists \mathtt{v}\, \forall \mathtt{x}\, \exists \mathtt{f}' \big(\mathit{instanceA}(\mathtt{f}, \mathtt{v}, \mathtt{x}, \mathtt{f}') \wedge \mathit{TrueE}_n(\mathtt{f}')\big)$$

is a truth definition for Π^0_{n+1}-formulas.)

4.10. A Finer Analysis of 4.4 and 4.5

We will now review 4.4 and check that all formulas appearing there are in fact equivalent to bounded formulas. For example, consider the formula $\mathsf{x}|\mathsf{y}$ (x divides y). Originally it was defined as an abbreviation for $\exists \mathsf{z}\, \mathsf{x}\cdot\mathsf{z} = \mathsf{y}$. However, it is easy to see that this formula is equivalent to $\exists \mathsf{z}\leq\mathsf{y}\, \mathsf{x}\cdot\mathsf{z} = \mathsf{y}$. Indeed, if $n|m$, so for some k, $n\cdot k = m$, then we must have $k\leq m$, unless $m = 0$. If $m = 0$ then we have $n|m$ for all n, witnessed by $k = 0$. So in either case we get: If n divides m, then there is a k less than or equal to m such that $n\cdot k = m$. Similarly, we can prove the following facts:

4.10.1. Fact.

(1) If $c \geq 1$, then $p\cdot k = c$ iff $p\leq c \wedge k\leq c \wedge p\cdot k = c$.

(2) If $p > 1$ (in particular if p is prime), then: $p^k|c$ iff $k\leq c \wedge p\leq c \wedge p^k|c$.

(3) If $\mathbb{N} \models \mathit{entry}(c, k, e)$, then $k\leq c \wedge e\leq c$. (See 4.4.13.)

(4) For all $k \geq 1$ we have $p_k\leq 2^{k^k}$. (p_k is the k-th prime number: $p_1 = 2$, $p_2 = 3$, etc.)

(5) If $m_1, \ldots, m_k$ are numbers $\leq m$, then $\#(m_1, \ldots, m_n)\leq 2^{k^k\cdot m\cdot k}$.

Proof: (1)–(3) are easy.

(4) We will prove this by induction on k. For $k = 1$ we have $p_1 = 2 = 2^{1^1}$. For $k = 2$ we have $p_2 = 3 < 16 = 2^4 = 2^{2^2}$.

Now assume we already have $p_k\leq 2^{k^k}$ for some $k \geq 2$. Let p be the smallest prime factor of $p_1\cdot p_2\cdots p_k - 1$. Since p can not be in $p_1, \ldots, p_k$, we must have $p \geq p_{k+1}$. Hence

$$p_{k+1}\leq p < 2\cdot 3\cdots p_k \leq 2^{k^k}\cdot \;\cdots\; \cdot 2^{k^k} = 2^{k\cdot k^k} = 2^{k^{k+1}} \leq 2^{(k+1)^{k+1}}.$$

(5) $\#(m_1, \ldots, m_n) = 2^{m_1}\cdots p_k^{m_k}\leq (k^{k^k})^m\cdots (k^{k^k})^m = k^{k^k\cdot m\cdot k}$.

Remark: Using methods from number theory (and complex analysis) it is possible to show that p_k is approximately bounded by $k\cdot \log k$, so in particular also by k^2 for all large enough k, so the term 2^{k^k} is a vast overestimate. However, for our purposes **any** bounding term is sufficient.

4.10.2. Lemma. All the following formulas are equivalent to bounded formulas:

(a) $\mathsf{x}|\mathsf{y}$.

(b) $\mathit{prime}(\mathsf{p})$.

(c) $\mathit{nextprime}(\mathsf{p}, \mathsf{q})$.

(d) $\mathit{seq}(\mathsf{s})$.

(e) *products*(x).
(f) *nthprime*(p, k).
(g) *entry*(s, e, k).

Hence, the following functions are recursive:

(A) $f(k) = p_k$, the kth prime number (and $f(0) = 1$).
(B) $f(k, s) =$ the maximal e such that p_k^e divides s. We will call this function $expo(k, s)$.
(C) $f(k,s) = \begin{cases} \text{the unique } e \text{ with } \mathbb{N} \models entry(s,k,e) & \text{if this exists} \\ 0 & \text{otherwise.} \end{cases}$

Proof: (a) $\exists z{\le}y x{\cdot}z = y$.
(b) $\underline{1} < p \wedge \forall x{\le}p(x|p \to x = \underline{1} \vee x = p)$.
(c) $prime(x) \wedge prime(y) \wedge x < y \wedge\ \forall z{\le}y\big(prime(z) \to z{\le}x \vee z = y\big)$.
(d) $\forall p{\le}s\,\forall q{\le}p : \big(prime(p) \wedge prime(q) \wedge p < q \wedge q|c \to p|c\big) \wedge c > \underline{1}$.
(e) $2|s \wedge \neg 4|s \wedge seq(s) \wedge \forall p{\le}s\,\forall q{\le}s\,\forall e{\le}s\big(nextprime(q,p) \wedge p|x \to (q^e|x \leftrightarrow p^{Se}|x)\big)$.
(f) $prime(p) \wedge \exists w{\le}(p^k)^k\, products(w) \wedge p^k|w \wedge \neg p^{k+\underline{1}}|w$.
(g) $seq(c) \wedge (k > 0) \wedge (e > 0) \wedge \forall p{\le}c\big(prime(p) \wedge nthprime(p,k) \to p^e|c \wedge \neg p^{e+1}|c\big)$.
(A) $f(k) = p$ iff $\mathbb{N} \models nthprime(k, p)$.
(B) $f(k, s) = e$ iff $\mathbb{N} \models \exists p\, nthprime(p,k) \wedge p^s|e \wedge p^{s+1} \nmid e$, so f is recursive by 4.9.15.
(C) $f(s, k) = e$ iff $\mathbb{N} \models entry(s,k,e) \vee [e = 0 \wedge \forall e'{\le}s\, \neg entry(s,k,e')]$.

4.10.3. Lemma. The following formulas are equivalent to bounded formulas:

(a) *var*(x).
(b) *zero*(x).
(c) *yieldtm*(t_1, t_2, o, t).
(d) *obtained*(c, k, l, o, n).
(e) *atomicterm*(c, n).
(f) *buildTerm*(c, t).

Proof:

(a) $\exists y{\le}x\ x = \underline{2}{\cdot}y + \underline{1}$.
(b) clear.
(c) clear.
(d) $\exists e{\le}c\,\exists f{\le}c\,\exists g{\le}c : entry(c,k,e) \wedge entry(c,l,f) \wedge entry(c,n,g) \wedge yieldtm(e,f,o,g)$. (This works because of 4.10.1.)
(e) $\exists e{\le}c\, entry(c,n,e) \wedge \big(zero(e) \vee var(e)\big)$ again because of 4.10.1.
(f) $seq(c) \wedge \exists n{\le}t\, entry(c,n,t) \wedge \forall n{\le}t\,\forall e{\le}t :$
$entry(c,n,e) \to atomicterm(c,n) \vee$
$\vee \exists k, l{<}n\, \exists o{\le}\underline{12}\, obtained(c,k,l,o,n)$ (Why does

the number 12 show up here? The only meaningful values for o are (see 4.5.2 and 4.5.3) $\ulcorner + \urcorner = 8$, $\ulcorner \cdot \urcorner = 10$ and $\ulcorner \uparrow \urcorner = 12$.)

4.10.4. Lemma. *term*(t) is equivalent to a bounded formula.

Proof: We have to find an upper bound for the term-building sequence leading up to t. Note that

(1) If τ_1 is a subterm of τ, then $\ulcorner \tau_1 \urcorner \leq \ulcorner \tau \urcorner$.

(2) If τ is a term, then there is a term-building sequence $\tau_1, \ldots, \tau_k$, where all the τ_i are subterms of τ, and $k \leq \ulcorner \tau \urcorner$, $\tau_k = \tau$.

By (1), (2) and 4.10.1 we can estimate

$$2^{\ulcorner \tau_1 \urcorner} \cdot \cdots \cdot p_k^{\ulcorner \tau_k \urcorner} \leq (p_k^{\ulcorner \tau \urcorner})^{\ulcorner \tau \urcorner} \leq (2^{\ulcorner \tau \urcorner^{\ulcorner \tau \urcorner} \cdot \ulcorner \tau \urcorner})^{\ulcorner \tau \urcorner}$$

So $\mathbb{N} \models term(t) \leftrightarrow \exists c \leq (2^{t^t \cdot t})^t \, buildTerm(c, t)$

By the same method the reader will be able to show the following:

4.10.5. Theorem. The formulas *formula* and *derivation* are equivalent to bounded formulas.

4.10.6. Conclusion. There exists a closed Π^0_1 formula which is true (=valid in $\mathbb{N}$) but not provable in Peano Arithmetic.

Proof: We can show (cf. the theorem above) that *Diag*(x, y) is equivalent to a bounded (or at least a Σ^0_1) formula. So the formula

$$formula(x_0) \wedge \forall y (Diag(x, y) \rightarrow \neg theorem(y))$$

is equivalent to a Π^0_1 formula. Using arguments similar to those in in 4.7 we can conclude, letting $\sigma = \varphi(\underline{\ulcorner \varphi \urcorner})$, $\mathbb{N} \models \sigma$ but $\nvdash_{PA} \sigma$.

What about *theorem*? A natural approach would be to show that it is equivalent to a bounded formula by showing that from the Gödel code of a provable formula we can find a bound for the length of its proof.

However, this is not possible. We will see below that the set of Gödel codes of theorems cannot be represented by a bounded formula — in fact, it is not even recursive (though it is easily seen to be r.e.).

4.10.7. Theorem. The set of provable closed Σ^0_1 formulas is r.e. but not recursive. More precisely, the set of Gödel numbers of provable closed Σ^0_1 formulas (= closed Σ^0_1 formulas which are valid in $\mathbb{N}$) is r.e. but not recursive.

Proof: We leave it to the reader to show that there is a Σ_1^0 formula (in fact, even a bounded formula) *closed*(x) such that φ is closed iff $\mathbb{N} \models$ *closed*$(\ulcorner\varphi\urcorner)$. Similarly, we can find a bounded formula *Sigma01*(x) which selects exactly the Gödel codes of Σ_1^0 formulas. This shows that the set of closed bounded formulas can be expressed by the formula *formula*(x) $\wedge$ *closed*(x) $\wedge$ *Sigma01*(x), and that the set

$$A := \{\ulcorner\varphi\urcorner : \varphi \text{ a closed } \Sigma_1^0 \text{ formula}, \vdash_{PA} \varphi\}$$

is expressed by the formula *formula*(x)$\wedge$*closed*(x)$\wedge$*Sigma01*(x)$\wedge$*theorem*(x).

Now assume that the complement of A is also r.e. So there is a Σ_1^0 formula χ such that $n \notin A \leftrightarrow \chi(n)$, i.e., in particular, for any closed Σ_1^0 formula σ, we have

$$\nvdash_{PA} \sigma \quad \leftrightarrow \quad \mathbb{N} \models \chi(\ulcorner\sigma\urcorner).$$

Since the formula *Diag* is equivalent to a Σ_1^0 formula, the formula

$$\mathit{formula}(x_0) \wedge \exists y(\mathit{Diag}(x_0, y) \wedge \chi(y))$$

is equivalent to a Σ_1^0 formula, which we will call $\psi(x_0)$. Hence $\psi(\ulcorner\psi\urcorner)$ is a closed Σ_1^0 formula. Let us call this formula σ. As before, from $\mathbb{N} \models \mathit{formula}(\ulcorner\psi\urcorner)$ and $\mathbb{N} \models \mathit{Diag}(\ulcorner\psi\urcorner, y) \leftrightarrow y = \ulcorner\sigma\urcorner$ we can conclude

$$(*) \qquad \mathbb{N} \models \sigma \leftrightarrow \chi(\ulcorner\sigma\urcorner).$$

By assumption, $\mathbb{N} \models \chi(\ulcorner\sigma\urcorner)$ iff $\nvdash_{PA} \sigma$. But since σ is a Σ_1^0 formula, we know that $\vdash_{PA} \sigma$ iff $\mathbb{N} \models \sigma$. So

$$\mathbb{N} \models \chi(\ulcorner\sigma\urcorner) \text{ iff } \mathbb{N} \not\models \sigma,$$

which contradicts $(*)$.

4.10.8. Remark. Exactly the same method works for any other reasonable axiom system which describes a structure rich enough to encode our logical system. For example, the incompleteness theorem also applies to the axiom system ZFC, the axiom system commonly used in set theory.

Exercise

1. Prove theorem 4.10.5.

4.11. More on Recursive Sets and Functions

Recall that we called a function from $\mathbb{N}$ to $\mathbb{N}$ recursive if it (i.e., its graph) is a recursive relation on $\mathbb{N} \times \mathbb{N}$.

4.11.1. Lemma. A set $A \subseteq \mathbb{N}^2$ is recursive, if the set

$$\hat{A} := \{2^k \cdot 3^m : (k, m) \in A\}$$

is a recursive subset of $\mathbb{N}$. More generally, a set $A \subseteq \mathbb{N}^k$ $(k \geq 2)$ is recursive, if the set

$$\hat{A} := \{2^{n_1} \cdot 3^{n_2} \cdots p_k^{n_k} : (n_1, \ldots, n_k) \in A\}$$

is recursive. The same is true if we replace "recursive" by "r.e."

Proof: Exercise.

The following (easy) lemma will often be needed:

4.11.2. Lemma (recursive functions are closed under substitution). If $h_1, \ldots, h_k$ are recursive functions from $\mathbb{N}^m$ to $\mathbb{N}$, and g is a recursive function from $\mathbb{N}^k$ to $\mathbb{N}$, then the function $f : \mathbb{N}^m \to \mathbb{N}$, defined by

$$f(n_1, \ldots, n_m) = g(h_1(n_1, \ldots, n_m), \ldots, h_k(n_1, \ldots, n_m))$$

is recursive.

Proof: Exercise.

To understand the next two lemmas better, think of a relation A on $\mathbb{N} \times \mathbb{N}$ as a multi-valued function, i.e., a function which assigns to each number m not a single value $f(m)$ but a whole set of values $A_m = \{n : (m, n) \in A\}$. Then the next lemma tells us that if A was r.e., and the set A_m is nonempty for all n, then we can select a single value $f(m)$ from each A_m using a recursive function f.

4.11.3. Lemma (Selection Principle).

(a) Let $R \subseteq \mathbb{N} \times \mathbb{N}$ be a recursive relation with the property that for all n there exists an m such that $(n, m) \in R$. Then there is a recursive function f such that for all n, $(n, f(n)) \in R$. (We say that f "**uniformizes**" R or that f is a "selector" for R.) In fact, we can choose f to be

$$f(m) = \min\{n : (m, n) \in R\}.$$

(b) Let $R \subseteq \mathbb{N} \times \mathbb{N}$ be an r.e. relation with the property that for all n there exists an m such that $(n, m) \in R$. Then there is a recursive function f such that for all n, $(n, f(n)) \in R$. (But we do not claim that we can get this f by **minimization** as in (a). See exercise 2.)

Proof: Since R is recursive, there are Σ_1^0 formulas φ_1 and φ_2 expressing R and its complement:

$$(m, n) \in R \text{ iff } \mathbb{N} \models \varphi_1(m, n)$$
$$(m, n) \notin R \text{ iff } \mathbb{N} \models \varphi_2(m, n).$$

So the formula

$$\varphi(\mathtt{x}, \mathtt{y}) := \varphi_1(\mathtt{x}, \mathtt{y}) \wedge \forall \mathtt{z} < \mathtt{y}\, \varphi_2(\mathtt{x}, \mathtt{y})$$

is equivalent to a Σ_1^0 formula, and clearly it describes the graph of the function f.

(b) Let $A = \{(m, n) : \mathbb{N} \models \exists \mathtt{y}\, \psi(m, n, \mathtt{y})\}$, where ψ is a bounded formula. Our recursive function f will pick, for each m, the first component n of a "minimal" pair (n, k) such that $\mathbb{N} \models \psi(m, n, k)$. To this end, we first have to translate our ternary formula ψ into a binary formula ψ', letting $\mathtt{t}$ code the pair $(\mathtt{y}, \mathtt{z})$:

$$\psi'(\mathtt{x}, \mathtt{t}) := \exists \mathtt{y} {\leq} \mathtt{t}\, \exists \mathtt{z} {\leq} \mathtt{t}\, \left(\underline{2}^{\mathtt{y}} \cdot \underline{3}^{\mathtt{z}} = \mathtt{t} \wedge \psi(\mathtt{x}, \mathtt{y}, \mathtt{z})\right)$$

Note that ψ' is a bounded formula, and that $\mathbb{N} \models \psi'(m, 2^n \cdot 3^k) \leftrightarrow \psi(m, n, k)$. In particular, for each m there is a j such that $\mathbb{N} \models \psi'(m, j)$. By (a) we can find a recursive function g such that for each m, $\mathbb{N} \models \psi'(m, g(m))$. Looking at the formula ψ', we see that each $g(m)$ must be of the form $2^n \cdot 3^k$, where $\mathbb{N} \models \psi(m, n)$. Now let $f(m) = expo(2, g(m))$ (see 4.10.2(B))

4.11.4. Lemma (Primitive Recursion). If $a_0 \in \mathbb{N}$, and $g : \mathbb{N} \times \mathbb{N}$ is a recursive function, then the function $f : \mathbb{N} \to \mathbb{N}$, defined by

$$f(0) = a_0$$
$$f(n+1) = g(n, f(n))$$

is recursive. (We say that f is defined from g by primitive recursion.)

Proof: Essentially we will repeat the proof from 1.1.30. It is clear (by induction) that there is a unique function f satisfying the requirements. Why is f recursive? Remember that we can code any finite sequence of numbers into one number. So we will code the values $f(0), f(1), \ldots, f(k)$

into a single number m. For technical reasons (namely, to be able to use bounded quantification) we will require m to be greater than $k+2$. This will be achieved by a factor 2^{k+2}. So the function f up to the value $f(k)$ will be coded by the number

$$\#(k+2, f(0), \ldots, f(k)) = 2^{k+2} \cdot 3^{f(0)} \cdots p_{k+2}^{f(k)}.$$

First we claim that the set

$$A := \{\#(2, f(0)),\ \#(3, f(0), f(1)),\ \#(4, f(9), f(1), f(2)),\ \ldots\}$$

is recursive.

We will use the function *expo* from 4.10.2. We have that m is in A if m is of the form $p_1^{k+2} \cdot p_2^{e_0} \cdot p_3^{e_1} \cdots p_{k+2}^{e_k}$, $e_0 = a_0$, and for all $n < k$ we have $e_{n+1} = g(n, e_n)$. Note that $k+2$, p_1, …, p_k, e_1, …, e_k must all be $\leq m$. So, let $\varphi(\mathtt{m})$ be the following formula:

$$\begin{aligned} &\exists \mathtt{k} \leq \mathtt{m} \big[(\forall \mathtt{k}' \leq \mathtt{m}\, \mathtt{k}' > \mathtt{k} + \underline{2} \rightarrow expo(\mathtt{m}, \mathtt{k}') = 0) \land \\ &\quad \land\ expo(\mathtt{m}, \underline{1}) = \mathtt{k} + \underline{2} \land expo(\mathtt{m}, \underline{2}) = \underline{a_0} \\ &\quad \land\ \forall \mathtt{n} \leq \mathtt{k} \forall \mathtt{e} \leq \mathtt{m} \big(\mathtt{n} < \mathtt{k} \land \mathtt{n} > 1 \land expo(\mathtt{m}, \mathtt{n}) = \mathtt{e} \rightarrow \\ &\qquad\qquad \exists \mathtt{e}' \leq \mathtt{m}\, expo(\mathtt{m}, \mathtt{Sn}) = \mathtt{e}' \land \mathtt{e}' = g(\mathtt{n}, \mathtt{e}) \big) \big]. \end{aligned}$$

Strictly speaking, $expo(\mathtt{m}, \mathtt{Sn}) = \mathtt{e}'$ and $\mathtt{e}' = g(\mathtt{n}, \mathtt{e})$ are of course not really formulas, but we can use them as abbreviations for the Σ_1^0 or Π_1^0 formulas to which they are equivalent. Thus, φ is equivalent to both a Σ_1^0 and a Π_1^0 formula, so we have that $A = \{m : \mathbb{N} \models \varphi(m)\}$ is a recursive set. Let $\psi(\mathtt{x}, \mathtt{y})$ be the formula

$$\exists \mathtt{m}\, \varphi(\mathtt{m}) \land\ expo(\mathtt{m}, 1) = \mathtt{x} + \underline{2}\ \land\ expo(\mathtt{m}, \mathtt{x} + \underline{2}) = \mathtt{y}$$

then ψ defines the graph of f, so f is a recursive function.

4.11.5. Example. The "factorial" function $n!$ is recursive, where $n! = n \cdot (n-1) \cdots 1$ (and $0! = 1$).

Proof: We have $0! = 1$ and $(n+1)! = n! \cdot (n+1)$. So we can use the previous lemma with $a_0 = 1$ and $g(n, k) = (k+1) \cdot n$.

The previous lemma can be seen as a special case ($m = 0$) of the following lemma:

4.11.6. Lemma (recursive functions are closed under primitive recursion). If $h : \mathbb{N}^m \rightarrow \mathbb{N}$ is recursive, and $g : \mathbb{N}^{m+2} \rightarrow \mathbb{N}$ is recursive, then there is a unique recursive function f satisfying for all $n_1, \ldots, n_m, n$:

$$\begin{aligned} f(n_1, \ldots, n_m, 0) &= h(n_1, \ldots, n_m) \\ f(n_1, \ldots, n_m, n+1) &= g(n_1, \ldots, n_m, n, f(n_1, \ldots, n_m, n)). \end{aligned}$$

4.11.7. Example. The Fibonacci function *Fib* is defined by $Fib(0) = 0$, $Fib(1) = 1$, $Fib(n+2) = Fib(n+1) + Fib(n)$ for all $n \geq 0$. A few values of the Fibonacci function are given in the table below:

n	0	1	2	3	4	5	6	7	8	9	10	11
$Fib(n)$	0	1	1	2	3	5	8	13	21	34	55	89

The Fibonacci function is recursive.

Proof: We will first consider the auxiliary function g defined by $g(n) = 2^{f(n)}.3^{f(n+1)}$, which codes two consecutive values of the function *Fib*. We have $f(n) = expo(1, g(n))$ and $f(n+1) = expo(2, g(n))$. So

$$\begin{aligned} g(n+1) = 2^{f(n+1)}.3^{f(n+2)} &= 2^{f(n+1)}.3^{f(n)+f(n+1)} \\ &= 2^{expo(2,g(n))}.3^{expo(2,g(n))+expo(1,g(n))} = G(g(n)) \end{aligned}$$

where the function $G(k) = 2^{expo(2,k)}.3^{expo(2,k)+expo(1,k)}$ is recursive. Hence also g is recursive, and since $Fib(n) = expo(1, g(n))$, *Fib* is also recursive.

Often the following lemma is helpful when dealing with recursive sets and recursive functions:

4.11.8. Lemma. $A \subseteq \mathbb{N}^k$ is recursively enumerable iff A is empty or there is a recursive function with range $= A$.
A set $A \subseteq \mathbb{N}^k$ is recursive iff its "characteristic function," that is, the function $\chi_A : \mathbb{N}^k \to \{0, 1\}$, defined by

$$\chi_A(n_1, \ldots, n_k) = \begin{cases} 1 & \text{if } (n_1, \ldots n_k) \in A \\ 0 & \text{if } (n_1, \ldots n_k) \notin A \end{cases}$$

is recursive.

Proof: Exercise.

Exercises

1. Prove lemma 4.11.1.

2. Find an r.e. set $R \subseteq \mathbb{N} \times \mathbb{N}$ and a NONrecursive function f such that for every m, $f(m)$ is the minimal n satisfying $(m, n) \in R$. (Hint: Let $A \subseteq \mathbb{N}$ be r.e., but not recursive. Consider the characteristic function of A.)

3. Show that a set $A \subseteq \mathbb{N}$ is recursive iff A is finite or there is a strictly increasing recursive function f with range A.

4. Use lemma 4.11.2 to show that the function G used in the proof of 4.11.7 is indeed recursive.

5. Assume that $f : \mathbb{N}\times\mathbb{N}\times\mathbb{N} \to \mathbb{N}$ is a recursive function, and $R \subseteq \mathbb{N}\times\mathbb{N}\times\mathbb{N}$ is a recursive relation. We can define a function g as follows:

> To compute $g(n)$, start out with $i = 0$ and $k = 0$. If $R(n, i, k)$ holds, output k as the value of $g(n)$. Otherwise, replace k by $f(n, i, k)$, then replace i by $i + 1$, and check again if $R(n, i, k)$ holds. If it does, output k, otherwise again replace ...
> If this procedure never stops, $g(n)$ is undefined.

In other words, g is the function computed by the program

```
input(n)
i:=0
k:=0
until R(n,i,k) do
begin
  k:=f(n,i,k)
  i:= i+1
end
output k
```

Show that if $g(n)$ is defined for all n, then g is a recursive function.

(Hint: First show that the function H defined by $H(n,0) = 0$, $H(n, i+1) = f(n, H(n,i), i)$ is recursive. Now consider the function h defined by $h(n) = \min\{i : R(n, i, H(n,i)) \text{ holds}\}$.)

6. Nonrecursive functions.
(a) Show that there is a nonrecursive function.
(b) Assuming that $f : \mathbb{N} \to \mathbb{N}$ is nonrecursive, find a function $g : \mathbb{N} \times \mathbb{N} \to \mathbb{N}$ which is not recursive, such that each of the functions g_n, defined by $g_n(k) = g(n, k)$, IS recursive. (Hint: You can even get these functions to be constant.)

7. Let $k \geq 2$. Assume $g_1, \ldots, g_k$ are recursive functions from $\mathbb{N}^{k+1}$ into $\mathbb{N}$, and $a_1, \ldots, a_k$ are natural numbers. Show that there are recursive functions $f_1, \ldots, f_k$ from $\mathbb{N}$ into $\mathbb{N}$, such that

$$f_1(0) = a_1, \ldots, f_k(0) = a_k$$
$$f_1(n+1) = g_1(n, f_1(n), \ldots, f_k(n))$$
$$\vdots$$
$$f_k(n+1) = g_k(n, f_1(n), \ldots, f_k(n))$$

$f_1, \ldots, f_k$ are said to be defined by "simultaneous recursion" from $g_1, \ldots, g_k$.

(Hint: See 4.11.7.)

8. ("Course-of-values" induction) Assume that g is a recursive function. Then there is a unique function f satisfying

$$f(0) = g(0, 1)$$
$$f(n+1) = g(n, \#(f(0), \ldots, f(n)))$$

and moreover, f is recursive.

9. Prove 4.11.8.

10. We define the **projection functions** P_j^k for $1 \leq j \leq k$ by

$$P_j^k(n_1, \ldots, n_k) = n_j.$$

The successor function s is defined by $s(n) = n+1$, and the zero function Z is defined by $Z(n) = 0$. The projection functions, the zero function and the successor functions will be called **basic primitive recursive functions.**

The primitive recursive functions are the functions obtained from these basic primitive recursive functions by (repeated applications of) substitution (4.11.2) and primitive recursion (4.11.4 and 4.11.6).

Describe the natural inductive structure on the set of primitive recursive functions. (Hint: E.g. to take care of primitive recursion as in 4.11.6, there will be an binary operator *Primitive* which assigns to each pair (h, g) of functions a function $f = \mathit{Primitive}(h, g)$ such that

If for some $m > 0$, h is m-ary and g is $m+2$-ary, then f is the function defined by g and h as in 4.11.6,

otherwise $f = Z$. You also need a set of operators $\mathit{Substitute}_{k,m}$ to take care of 4.11.2.)

11. Show that the following functions are primitive recursive:

$$f(m) = 11492$$
$$f(m,n) = m + n$$
$$f(m,n) = m \cdot n$$
$$f(m,n) = m^n$$
$$f(m) = \begin{cases} 1 & \text{if } m > 0 \\ 0 & \text{if } m = 0 \end{cases}$$
$$f(m,n) = \begin{cases} m - n & \text{if } m \geq n \\ 0 & \text{otherwise.} \end{cases}$$

12. A set $A \subseteq \mathbb{N}^k$ is called primitive recursive iff its characteristic function is primitive recursive.
(1) Show that all sets defined by bounded formulas are primitive recursive. (Hint: First show that if f is a primitive recursive k+1-ary function, then the k-ary function Σf defined by

$$(\Sigma f)(n_1, \ldots, n_k, j) = f(n_1, \ldots, n_k, 0) + \cdots + f(n_1, \ldots, n_k, j)$$

is primitive recursive.)
(2) Show that the function *expo* is primitive recursive.

13. If f is a k+1-ary function such that for all $n_1, \ldots, n_k$ there is a j with $f(n_1, \ldots, n_k, j) = 0$, then define the k-ary function μf by

$$(\mu f)(n_1, \ldots, n_k) = \min\{j : f(n_1, \ldots, n_k, j) = 0\}$$

(1) Show that if f is recursive and μf is well-defined, then also μf is recursive.
(2) Show that every recursive function f is of the form $f(n_1, \ldots, n_k) = h(\,(\mu g)(n_1, \ldots, n_k)\,)$ where both h and g are primitive recursive. (Hint: You can even have the same g for all f, namely, $h(n) = expo(n, 1)$, see the proof of 4.11.3(b).)

14. Find a recursive function $F : \mathbb{N} \times \mathbb{N} \to \mathbb{N}$ such that for all primitive recursive functions $f : \mathbb{N} \to \mathbb{N}$ there is a number s such that for all n, $f(n) = F(s, n)$. (See exercise 10.)

Conclude that the function $F(n, n) + 1$ is recursive but not primitve recursive.

(Hint: Interpret s as a sequence of instructions for building f from the basic primitive recursive functions using substitution and primitive recursion. Now let $F(s, m) = n$ iff there is a sequence of finite partial functions which obey the instructions given by s and are all defined on a large enough set, such that the final function, applied to m, yields n.)

15. The Ackermann function $A : \mathbb{N} \times \mathbb{N} \to \mathbb{N}$ is defined by the following conditions:

$$A(0, n) = n + 1$$
$$A(c+1, 0) = A(c, 1)$$
$$A(c+1, n+1) = A(c, A(c+1, n)).$$

(1) Compute $A(1, n)$, $A(2, n)$, $A(3, n)$ for all n.
(2) Show that $A(c, n)$ is well-defined for all c, n.
(3) Show that A is recursive.
(4) Show that for each $c \in \mathbb{N}$ the function A_c, defined by $A_c(n) = A(c, n)$ is primitive recursive.

16. Let A be the Ackermann function (see previous exercise). First show that

(a) $A(c, n) > n$ for all c, n.
(b) If $n_1 \leq n_2$, then $A(c, n_1) \leq A(c, n_2)$.
(c) $A(c, n) > n$ for $c \geq 2$.

Now show by induction on the inductive structure from exercise 10 that for every primitive recursive function $f : \mathbb{N}^k \to \mathbb{N}$ there is a natural number c such that for all $n_1, \ldots, n_k$ we have $f(n_1, \ldots, n_k) < A(c, \max(n_1, \ldots, n_k))$. (Hint: If f is obtained by primitive recurison from g and h, and $c > 2$ works for g and h, then $c + 2$ will work for f.) Conclude that the function $a(n) = A(n, n)$ is recursive but grows faster than any primitive recursive function.

Index

Milton Keynes UK
Ingram Content Group UK Ltd.
UKHW040107071024
449327UK00019B/877

9 780367 401696